Nutrient Requirements of Laboratory Animals

Fourth Revised Edition, 1995

Subcommittee on Laboratory Animal Nutrition
Committee on Animal Nutrition
Board on Agriculture
National Research Council

NATIONAL ACADEMY PRESS
Washington, D.C.   1995

NATIONAL ACADEMY PRESS • 2101 Constitution Avenue, N.W. • Washington, D.C. 20418

NOTICE: The project that is the subject of this report was approved by the Governing Board of the National Research Council, whose members are drawn from the councils of the National Academy of Sciences, the National Academy of Engineering, and the Institute of Medicine. The members of the committee responsible for the report were chosen for their special competencies and with regard for appropriate balance.

This report has been reviewed by a group other than the authors according to procedures approved by a Report Review Committee consisting of members of the National Academy of Sciences, the National Academy of Engineering, and the Institute of Medicine.

This study was supported by the National Institutes of Health of the U.S. Department of Health and Human Services under cooperative agreement No. 5 R01 RR06161-03. Additional support was provided by Ziegler Brothers, Inc., and Harlan Sprague-Dawley Co.

The National Academy of Sciences is a private, nonprofit, self-perpetuating society of distinguished scholars engaged in scientific and engineering research, dedicated to the furtherance of science and technology and to their use for the general welfare. Upon the authority of the charter granted to it by the Congress in 1863, the Academy has a mandate that requires it to advise the federal government on scientific and technical matters. Dr. Bruce M. Alberts is president of the National Academy of Sciences.

The National Academy of Engineering was established in 1964, under the charter of the National Academy of Sciences, as a parallel organization of outstanding engineers. It is autonomous in its administration and in the selection of its members, sharing with the National Academy of Sciences the responsibility for advising the federal government. The National Academy of Engineering also sponsors engineering programs aimed at meeting national needs, encourages education and research, and recognizes the superior achievements of engineers. Dr. Robert M. White is president of the National Academy of Engineering.

The Institute of Medicine was established in 1970 by the National Academy of Sciences to secure the services of eminent members of appropriate professions in the examination of policy matters pertaining to the health of the public. The Institute acts under the responsibility given to the National Academy of Sciences by its congressional charter to be an adviser to the federal government and, upon its own initiative, to identify issues of medical care, research, and education. Dr. Kenneth I. Shine is president of the Institute of Medicine.

The National Research Council was organized by the National Academy of Sciences in 1916 to associate the broad community of science and technology with the Academy's purposes of furthering knowledge and advising the federal government. Functioning in accordance with general policies determined by the Academy, the Council has become the principal operating agency of both the National Academy of Sciences and the National Academy of Engineering in providing services to the government, the public, and the scientific and engineering communities. The Council is administered jointly by both Academies and the Institute of Medicine. Dr. Bruce M. Alberts and Dr. Robert M. White are chairman and vice-chairman, respectively, of the National Research Council.

Library of Congress Cataloging-in-Publication Data

Nutrient requirements of laboratory animals / Subcommittee on
   Laboratory Animal Nutrition, Committee on Animal Nutrition, Board on
   Agriculture, National Research Council. — 4th rev. ed.
       p.   cm. — (Nutrient requirements of domestic animals)
   Includes bibliographical references and index.
   ISBN 0-309-05126-6
   1. Laboratory animals—Feeding and feeds.   I. National Research
Council (U.S.). Subcommittee on Laboratory Animal Nutrition.
II. Series.
SF95.N32
[SF406.2]
636.08′542 s—dc20
[636′.93233]   94-43585
                                     CIP

Any opinions, findings, conclusions, or recommendations expressed in this publication are those of the author(s) and do not necessarily reflect the view of the organizations or agencies that provided support for this project.

Printed in the United States of America

iii

Preface

The first edition of *Nutrient Requirements of Laboratory Animals* was published in 1962. It summarized the nutrient requirements of the rat, mouse, guinea pig, hamster, monkey, and cat based on an evaluation of the literature. The second revised edition was published in 1972 and updated the information presented in the first edition. The third revised edition was published in 1978 and was expanded to include a chapter on general aspects of nutrition, and the species chapters incorporated information on expected growth and reproductive performance in addition to the nutrient requirements of the laboratory rat, mouse, guinea pig, hamster, gerbil, vole, and the nutrient requirements of fishes.

In this, the fourth revised edition, the subcommittee reviewed the literature and summarized the nutrient requirements of the rat, mouse, guinea pig, hamster, gerbil, and vole. The subcommittee structure was altered for this publication as members were assigned by nutrient across species rather than by nutrient within a species. This structure provided the subcommittee with nutrient expertise that could be applied to more than one species. To maintain a species expertise, one member of the subcommittee was designated as the species chair and integrated the information into the chapter. The species chair also developed a section on expected growth and reproduction of the various breeds within a species and reviewed the literature to assemble natural-ingredient and purified diets that should meet the needs of animals of the species used in long-term studies.

After its review of the literature, the subcommittee emphasized the need for experiments designed to determine nutrient requirements of laboratory animals. Work of that nature is of considerable value in compiling the information contained in a publication such as this. Thus, not all the requirements reported in this publication were derived from experiments specifically designed to estimate the requirement of a nutrient, and interpretation of published work was required to derive an estimate. Where appropriate, the subcommittee used information available for one species to estimate the requirements for another species. The text devoted to each nutrient includes a description of decisions made to obtain the requirement shown in the table.

The subcommittee thanks Mary Poos, Dennis Blackwell, and Janet Overton for their assistance during the development and preparation of this document.

N. J. Benevenga, *Chair*
Subcommittee on Laboratory Animal Nutrition

Contents

Tables and Figures

FIGURES

APPENDIX TABLES

Nutrient
Requirements
of Laboratory Animals

Overview

This fourth revised edition of *Nutrient Requirements of Laboratory Animals* integrates new information gained in the latest review of the world literature on nutrient requirements of laboratory animals. At our request many individuals provided the subcommittee with published materials from theses and other sources not recovered in a standard literature search.

The committee sought to make this publication a valuable reference to investigators whose expertise was other than nutrition. Examples of natural-ingredient and purified diets reported in the literature are provided. The subcommittee attempted to maintain clarity by identifying the chemical nature of the minerals and vitamins used in supplements to the diets. In addition, in the tables the subcommittee elected to use mass units of measure that can be used directly by diet manufacturers with minimal calculation. Molar and other units of measure can be found in the discussions included within each chapter. The references used in each chapter are listed as part of the chapter rather than compiled in a master reference list.

The intent of Chapter 1 is to provide a general narrative with critical references for those who are looking for a basic understanding of the formulation of diets for the expected feeding of laboratory animals used in biological research.

Chapter 2, which focuses on the laboratory rat, contains new information on amino acid, fatty acid, mineral, and vitamin requirements. To the extent possible, requirements for maintenance, growth, and reproduction are presented separately. A new section entitled "Potentially Beneficial Dietary Constituents" was included because these materials cannot be shown to be a dietary essential although their inclusion may be beneficial to animals maintained in laboratory settings. Examples of natural-ingredient diets and purified diets that can be used to support animals on long-term studies are also presented. These diets have been shown to be effective for the commonly used strains of rats and mice.

Chapter 3 focuses on the laboratory mouse and integrates new information to create an updated nutrient requirements table along with summary tables of sources used to estimate protein and amino acid requirements for various strains. As with the rat, those ingredients for which a requirement cannot be shown are included in the section on potentially beneficial dietary constituents.

The new information on the guinea pig is presented in Chapter 4, which also includes a new section on growth and reproduction and a new natural-ingredient diet and additional purified diets. In addition to updating the requirements reported previously, estimates are now available for the requirements for indispensable amino acids. A table summarizing the sources used to estimate protein and amino acid requirements is part of this chapter.

Chapter 5 summarizes the latest update on hamsters. New sections and summary tables on the origin of hamsters, their biological characteristics, growth, and reproductive development are included. New purified diets and a natural-ingredient diet that can be used for long-term studies are presented. A nutrient requirements table is not included. Few, if any, studies devoted to determining the nutrient requirements of the hamster could be found. The subcommittee concluded that the hamster, partly because of pregastric fermentation, is sufficiently different from the mouse and rat so that requirements identified for these species should not automatically be applied to the hamster.

Chapter 6 contains data on the gerbil. Information on the biology and origin of gerbils used in research is included. Little new information on nutrient requirements could be found.

Chapter 7 focuses on voles. New sections on the biological characteristics and husbandry are incorporated in this edition. The limited information available on the voles'

1

requirement for various nutrients is presented in the text.

New appendix tables are provided, detailing the amino acid and fatty acid composition of some ingredients commonly used in purified diets as well as molecular weights and international unit standards of various forms of vitamins. These tables may be of assistance in formulating purified diets.

General Considerations for Feeding and Diet Formulation

A laboratory animal's nutritional status influences its ability to reach its genetic potential for growth, reproduction, and longevity and to respond to pathogens and other environmental stresses. A nutritionally balanced diet is important both for the welfare of laboratory animals and to ensure that experimental results are not biased by unintended nutritional factors.

Laboratory animals require about 50 nutrients in appropriate dietary concentrations. Tables detailing the estimated minimum nutrient requirements of laboratory animals are presented in this report. It is important to recognize that the estimated requirements in these tables have been determined under specific restrictive conditions. Feed palatability and intake, nutrient absorption and utilization, and excretion can be affected by physicochemical characteristics of feeds such as physical form, sensory properties, naturally occurring refractory or antinutritive compounds, chemical contaminants, and conditions of storage. Many biological factors also affect nutrient requirements.

FACTORS AFFECTING NUTRIENT REQUIREMENTS

GENETICS

Genetic differences among species, breeds, strains, stocks, sexes, and individuals may affect nutrient requirements. For example, the lack of L-gulonolactone oxidase (a key enzyme required for the synthesis of ascorbic acid) in some species is apparently the consequence of genetic mutation (Chatterjee, 1978). L-gulonolactone oxidase activity differs among rodent species, among rat strains, and between sexes within rat strains (Jenness et al., 1980). A mutant rat has even been discovered that, like the guinea pig, lacks L-gulonolactone oxidase and has an obligatory dietary requirement for ascorbic acid (Mizushima et al., 1984; Horio et al., 1985). There is evidence that mouse strains may differ in requirements for riboflavin, pantothenic acid, and other nutrients (Fenton and Cowgill, 1947; Lee et al., 1953; Luecke and Fraker, 1979). Genetic differences in growth potential among species, strains, and sexes may influence the daily requirements for amino acids and other nutrients that are incorporated into tissues (Fenton, 1957; Goodrick, 1973).

STAGE OF LIFE

Nutrient requirements change during stages of the life cycle, especially in response to growth, pregnancy, or lactation. Synthesis of tissues or products requires amino acids, fatty acids, minerals, glucose, or other substrates as well as increased amounts of vitamins and associated cofactors. Research on farm animals demonstrates that rates of growth and of milk production affect nutrient requirements (National Research Council, 1984, 1985, 1988, 1989). The same is probably true for laboratory animals; however, few conclusive studies have been reported. As a result, for most nutrients, it is not currently possible to establish separate requirements for various stages of life for individual laboratory animal species.

ENVIRONMENTAL IMPACTS

Nutrient requirements are usually studied under controlled conditions with minimal diurnal or seasonal variation in temperature, light cycle, or other environmental conditions. Marked modification in these conditions may alter nutrient requirements. For example, exposure to temperatures below the lower threshold of the thermoneutral zone increases energy requirements as animals are obliged to expend energy to maintain a constant body temperature. The consequent increase in food intake may permit the feeding of diets of lower nutrient density without decreasing nutrient intakes. High temperature, disturbing stimuli, social conflict, or other environmental factors that reduce food intake may necessitate diets higher in nutrient concentrations to maintain adequate nutrient intakes.

Housing types can also affect the amounts of nutrients needed in diets. For example, laboratory rodents maintained in either galvanized cages or cages with solid bottoms may have a lower dietary requirement for zinc because of the availability of zinc from the feces and cage materials. Solubilized minerals in drinking water (such as copper from copper water lines) may affect the amounts of these minerals that must be supplied by the diet. If laboratory animals ingest bedding or other "nonfood" materials, these may provide an unintended source of some nutrients or toxins. In studies of the requirements of laboratory animals for constituents that might be needed at extremely low concentrations, even the air supply may be a significant source of contamination.

MICROBIOLOGICAL STATUS

Under normal rearing conditions, laboratory animals harbor populations of microorganisms in the digestive tract. These microorganisms generate various organic constituents as products or by-products of metabolism, including various water-soluble vitamins and amino acids. The extent to which these nutrients contribute to the nutrition of the host may be substantial but varies according to species, diet composition, and rearing conditions. In the rat and mouse, most of the microbial activity is in the colon, and many of the microbially produced nutrients are not available to the host unless feces are consumed, as is common for rats and other rodents (Stevens, 1988). Prevention of coprophagy may require an increase in the nutrient concentrations that must be supplied by the diet. The loss of some or all microbial symbionts in animals free of specific pathogens and germ-free animals, respectively, may also alter microbial nutrient synthesis and, thereby, influence dietary requirements. Adjustments in nutrient concentrations, the kinds of ingredients, and methods of preparation must be considered when formulating diets for laboratory animals reared in germ-free environments or environments free of specific pathogens (Wostmann, 1975).

RESEARCH CONDITIONS

Experimental procedures may produce stress or otherwise alter food intake. For example, surgical procedures or test substances in diets may lead to anorexia, necessitating the provision of more palatable diets or diets with elevated nutrient concentrations. Experimental protocols that require restriction of the amount of food offered alter the intakes of all nutrients unless dietary concentrations are altered to account for changes in food consumption.

NUTRIENT INTERACTIONS

Alterations in dietary energy density usually cause a change in feed intake. If high-energy diets are used, it may be necessary to increase nutrient concentrations in the diet to compensate for decreased food consumption. Other interactions occur between nutrients, such as competition for absorption sites among certain minerals that share common active transport systems. Thus in formulating diets containing unusual nutrient concentrations, the potential effects on other nutrients must be considered and adjustments made in nutrient concentrations, if appropriate.

FORMULATION OF DIET TYPES

Diet formulation is the process of selecting the kinds and amounts of ingredients (including vitamin and mineral supplements) to be used in the production of a diet containing planned concentrations of nutrients. Choice of ingredients will be influenced by the species to be fed and the experimental or production objectives. Target nutrient concentrations must take into account estimated nutrient requirements, possible nutrient losses during manufacturing and storage (National Research Council, 1973; Harris and Karmas, 1975), bioavailability of nutrients in the ingredients, and potential nutrient interactions.

Various types of diets are available for use with laboratory animals. Selection of the most appropriate type will depend on the amount of control required over nutrient composition, the need to add test substances, potential effects of feed microbes, diet acceptance by the animals, and cost. Wastage is also a problem with some types of diets, which may be a disadvantage if quantitative intake is to be measured.

The ideal diet for a particular animal colony will depend on production or experimental objectives. The diet must be sufficiently palatable to ensure adequate food consumption and must be nutritionally balanced so that the nutrients essential for the objectives are provided. It should also be free of substances or microorganisms that may be toxic or cause infection. Diets used in research also must be readily reproducible to ensure that the results can be verified by additional studies.

It is common to classify diets for laboratory animals according to the degree of refinement of the ingredients.

NATURAL-INGREDIENT DIETS

Diets formulated with agricultural products and by-products such as whole grains (e.g., ground corn, ground wheat), mill by-products (e.g., wheat bran, wheat middlings, corn gluten meal), high-protein meals (e.g., soybean meal, fishmeal), mined or processed mineral sources (e.g., ground limestone, bonemeal), and other livestock feed ingredients (e.g., dried molasses, alfalfa meal) are often called natural-ingredient diets. Commercial diets for laboratory animals are the most commonly used natural-ingredient diets, but special diets for research animals may also be of this type. This type of diet is relatively inexpensive to manufacture and, if appropriate attention is given to ingredient selection, is palatable for most laboratory animals. However, variation in the composition of the individual ingredients can produce changes in the nutrient concentrations of natural-ingredient diets (Knapka, 1983). Soil and weather conditions, use of fertilizers and other agricultural chemicals, harvesting and storage procedures, and manufacturing or milling methods can all influence the composition of individual ingredients, with the result that no two production batches of feed are identical. The potential for contamination with pesticide residues, heavy metals, or other agents that might compromise experimental data is another disadvantage (International Council for Laboratory Animal Science, 1987). Natural-ingredient diets are usually unsatisfactory for studies to determine micronutrient requirements, for toxicological studies that are sensitive to low concentrations of contaminants, or for immunological studies that may be influenced by antigens in diets.

Nutrient Concentrations

The formulation of natural-ingredient diets is complicated by the fact that each ingredient contains many if not most nutrients, so that an adjustment in the amount of any ingredient produces changes in the concentrations of most nutrients in the final product. Hence it is not possible to predetermine the concentration of each nutrient; rather diets are formulated to contain minimal concentrations of particular nutrients (such as crude protein, fiber, fat, calcium, and phosphorus), and other nutrients are added via vitamin and mineral premixes. In feeding domestic animals, cost considerations dictate that natural-ingredient diets be formulated using linear programming techniques that generate diet formulas that conform to set minimal and maximal nutrient concentrations, while minimizing ingredient costs. This has led to the marketing of variable formula products that differ in ingredient composition from batch to batch in response to changing ingredient prices. Such diets may be cost-effective for maintenance and rearing of laboratory animals, but they are too variable to be of use in nutritional, toxicological, or other types of experiments that may be affected by dietary constituents.

Fixed-Formula Diets

An alternative approach has been the development of fixed-formula diets in which the kinds and amounts of ingredients do not vary from batch to batch. These diets are often called *open-formula diets* when the formula is openly declared, as in specifications used to solicit competitive bids among manufacturers. A fixed-formula diet may contain multiple sources of protein, fat, and carbohydrate, thereby reducing the importance of variation in the composition of any particular ingredient from batch to batch (Knapka et al., 1974). A variety of ingredients also increases the probability that ultra-trace minerals of potential nutritional importance—such as chromium, nickel, and tin—will be provided at appropriate concentrations. Because it is difficult to demonstrate that these minerals are required, and because the amounts in most natural ingredients are apparently adequate, these minerals are not typically included in mineral premixes for natural-ingredient diets.

The steps in formulating a natural-ingredient diet are reviewed by Knapka (1985) and the International Council for Laboratory Animal Science (1987). It is important to recognize that the bioavailability of nutrients may be lower in natural-ingredient diets than in purified diets. Factors that may affect bioavailability include the chemical form of nutrients, constituents that may bind nutrients (such as phytate, tannins, and lignin), nutrient interactions, and effects of processing. Thus it is prudent to include nutrients at concentrations higher than the minimal requirements but within the safe range. Information on the vitamin and mineral tolerances of animals has been summarized in prior reports (National Research Council, 1980, 1987). Common practice is to use greater margins of safety for particularly labile vitamins and for trace minerals.

PURIFIED AND CHEMICALLY DEFINED DIETS

Purified Diets

Diets that are formulated with a more refined and restricted set of ingredients are designated *purified diets*. Only relatively pure and invariant ingredients should be used in these formulations. Examples of such ingredients are casein and soybean protein isolate (as sources of protein), sugar and starch (as sources of carbohydrate), vegetable oil and lard (as sources of fat and essential fatty acids), a chemically extracted form of cellulose (as a source of fiber), and chemically pure inorganic salts and vitamins. The nutrient concentrations in a purified diet are less variable and more easily controlled via formulation than in a natural-ingredient diet. However, even these ingredients may contain variable amounts of trace nutrients, and experimental diets intended to produce specific deficiencies may need to be even more restrictive as to ingredient specifications (International Council for Laboratory Animal Science, 1987). The potential for chemical contamination of these diets is also low. Purified diets are often used in studies of specific nutritional deficiencies and excesses. Unfortunately, they are not readily consumed by all species and are more expensive to produce than natural-ingredient diets.

Chemically Defined Diets

For studies in which strict control over nutrient concentrations and specific constituents is essential, diets have been made with the most elemental ingredients available, such as individual amino acids, specific sugars, chemically defined triglycerides, essential fatty acids, inorganic salts, and vitamins. Such diets are called *chemically defined diets*; they represent the highest degree of control over nutrient concentrations. Unfortunately, chemically defined diets are not readily consumed by most species of laboratory animals and are usually too expensive for general use. Although the nutrient concentrations in these diets are theoretically fixed at the time they are manufactured, the bioavailability of nutrients may be altered by oxidation or nutrient interactions during storage. Chemically defined diets that can be sterilized by filtration have been developed for use in germ-free and low-antigen studies (Pleasants, 1984; Pleasants et al., 1986).

Nutrient Concentrations

The ingredients used in purified and chemically defined diets have the advantage that each is essentially the source of a single nutrient or nutrient class, which greatly simplifies the task of formulation. Each ingredient must be carefully selected on the basis of purity, consistency of supply and composition, and physicochemical properties, but the decision of how much to use is primarily a function of the planned nutrient concentration. Attention must be paid to providing sources of all essential nutrients because inadvertent omission of trace and ultra-trace nutrients in purified and chemically defined diets is more likely than with natural-ingredient diets. Margins of safety above requirement concentrations should be modest and relate to potential losses caused by oxidative degradation or other reactions that may occur during and after manufacture. The ingredients and formulation of purified diets have been described by Navia (1977).

Impurities remain a major concern with purified diets (International Council for Laboratory Animal Science, 1987). Protein sources may supply variable but unknown amounts of vitamins, minerals, and essential fatty acids; starch may contain traces of lipid and essential fatty acids; and oils may contain fat-soluble vitamins. Thus it is necessary to select specific ingredients if strict control of a particular nutrient is required. Protein sources used to produce trace mineral deficiencies include Torula yeast for chromium and selenium; lactalbumin for cobalt; casein for copper, iron, and manganese; and dried egg white for zinc (International Council for Laboratory Animal Science, 1987). Casein contains phosphorus; soybean protein contains phytate, which binds minerals (Wise, 1982). Casein is often extracted to reduce vitamin content, but even "vitamin-free casein" may have significant residual amounts of vitamin B_6 (Quinn and Chan, 1979).

Chemically defined diets are formulated using chemically pure (analytical grade) nutrients such as amino acids, fatty acid esters, glucose, vitamins, and mineral salts. In selecting ingredients one must consider such factors as chemical stability and solubility (in liquid diets); obviously all essential nutrients must be added individually. The availability of the different chemical forms of nutrients is a primary concern in the formulation of chemically defined diets. For example, the *l*-isomeric forms of amino acids occur in natural food protein. However, the *d*-isomers of several of the essential amino acids will support growth in the rat. Of these, methionine alone appears to be as well utilized in either form (Wretlind and Rose, 1950). Details about the composition and use of chemically defined diets are provided by Pleasants and colleagues (Pleasants et al., 1970; Pleasants, 1984; Pleasants et al., 1986).

PHYSICAL FORM OF DIETS

Diets for laboratory animals can be provided in different physical forms. The most common form in use for laboratory animals is the pelleted diet, which is typically formed by adding water to the mixture of ground ingredients and then forcing it through a die. The size and shape of the holes in the die determine pellet shape and rotating blades control the length; the diet is then dried to firmness. Binders are sometimes used to improve pellet quality. Pelleted diets are easy to handle, store, and use; reduce dust in animal facilities; prevent animals from selecting choice ingredients; and tend to minimize wastage. It is not easy, however, to add test compounds or otherwise alter pelleted diets after manufacture.

Extruded diets are similar to pelleted diets except the meal is forced through a die under pressure and at high temperature after steam has been injected, so the product expands as it emerges from the die. Extruded diets are less dense than pelleted diets and are preferred by some animals (e.g., dogs, cats, and nonhuman primates). Extruded diets are not commonly used for laboratory rodents because of the increased wastage during feeding and higher production costs.

Diets in meal form are sometimes used because they permit incorporation of additives and test compounds after the diet has been manufactured. These diets are often inefficient, however, because large amounts may be wasted unless specially designed feeders are available. Also, meals cake under certain storage conditions. An additional problem is that dust generated from the feed may be hazardous if toxic compounds have been added. One solution to this problem is to add jelling agents and water to the meal to form a jelled mass that can be cut into cubes for feeding; however, the jelling agents may contain carbohydrate, amino acids, or minerals that must be accounted for in diet formulations. The gel diet requires refrigeration to retard microbial growth and must be fed daily or more frequently to maintain moisture content and thus food intake.

Crumbled diets are prepared by crushing pelleted or extruded diets and screening particles to the most appropriate size for a particular age or size of laboratory animal, including fish and birds. Crumbled diets offer a method of presenting small particles of diet that, theoretically, contain all dietary ingredients present in pelleted diets. Crumbled diets offer the convenience, without the problems, of diets in meal form; they are not frequently used for rodents, however.

Liquid diets have been developed to accommodate specific requirements such as filter sterilization. Liquid diets are often used in studies of the effects of alcohol on nutrient utilization and requirements. In some cases purified diets will take the form of a stable emulsion when blended with water (Navia, 1977). Neonatal animals are also fed liquid diets that are derived primarily from milk products. As with gel diets, care must be taken to store liquid diets properly to avoid microbial growth.

MANUFACTURE AND STORAGE PROCEDURES AND OTHER CONSIDERATIONS

The efficient manufacture of natural-ingredient diets requires a large capital investment for facilities, milling apparatus, and inventories of ingredients that are least expensive when purchased in bulk. Therefore, these laboratory animal diets are usually commercially manufactured. Laboratory animal diets should not be manufactured or stored in facilities used for farm feeds or any products containing additives such as rodenticides, insecticides, hormones, antibiotics, growth factors, or fumigants. Areas where ingredients and diets are stored and processed should be kept clean and enclosed to prevent entry of feral rodents, birds, and insects. Routine pest control is essential.

NATURAL-INGREDIENT DIETS

The initial step in manufacturing natural ingredient diets is to grind all ingredients to a similar particle size so they can be uniformly blended into a homogeneous mixture. Particle size depends on the pore size of the screen used in a hammer mill or other grinder. The optimal particle size of ground ingredients depends on the kind of ingredients involved and the planned physical form of the final product. Grinding may improve the digestibility of the ingredient by increasing the surface area that is exposed to digestive enzymes; however, grinding can also increase subsequent rates of destruction of nutrients by increasing exposure to

atmospheric oxygen and by releasing enzymes responsible for autocatalytic processes.

Ingredients used in large amounts are added directly, while those used in small amounts, such as vitamins and minerals, are added via premixes. Separate vitamin and mineral premixes should be used to minimize destruction of vitamins by oxidation reactions catalyzed by minerals. Premixes should be prepared with a carrier such that a sufficient amount is added (e.g., 1 percent of the diet) to avoid weighing errors and to ensure homogeneous distribution of these micronutrients. Errors such as omitting ingredients or adding incorrect amounts can be minimized by verifying on a check sheet each ingredient as it is added.

The length of time a particular combination of ingredients should be mixed for maximal homogeneity depends on a number of factors including particle size, particle density, mixer speed, and mixer size. Overmixing can occur, resulting in particle separation associated with differences in density, physical form, and susceptibility to static electrical charges that can develop in mixers (Pfast, 1976).

Ground, mixed feeds are often pelleted. Ingredient composition, amount of moisture and heat, die size, operating conditions, and other factors influence the size, hardness, and nutrient concentrations of pellets. Some loss of labile vitamins may occur during pelleting, especially if it is done at high temperatures; however, the heat of the pelleting process may also inactivate enzymes, reduce bacterial populations in the diet, and, in some cases, improve digestibility (Slinger, 1973; International Council for Laboratory Animal Science, 1987). Many species prefer pelleted products and, therefore, increase voluntary food intake. Pelleting also allows for a reduction in wastage.

PURIFIED AND CHEMICALLY DEFINED DIETS

Purified or chemically defined diets can be efficiently prepared in laboratories or diet kitchens with a minimum amount of special apparatus. All diets for laboratory animals should be prepared in facilities used only for this purpose and under strict rules to prevent contamination or errors in the kinds and amounts of ingredients used. Navia (1977) presents a detailed discussion regarding the preparation of purified diets. Purified diets may be pressed into tablets, pelleted, or fed as a powder, paste, or gel. Great care must be exercised to ensure homogeneous mixing when test substances are added to purified diets (International Council for Laboratory Animal Science, 1987).

ENVIRONMENTAL CONDITIONS OF STORAGE AREAS

Nutrient stability of feeds generally increases as temperature and humidity decrease. The shelf-life of any particular lot of feed depends on the environmental conditions of the storage area. Feed stored where temperature and humidity are high can deteriorate within several weeks. Natural-ingredient diets stored in air-conditioned areas should be used within 180 days of manufacture; diets containing vitamin C should be used within 90 days of manufacture (National Institutes of Health, 1985). Vitamins C and A are especially labile. Diets stored for long periods or under unusual environmental conditions should be assayed for nutrients prior to use. Diets formulated without antioxidants or with large amounts of highly perishable ingredients, such as fat, may require special handling or storage procedures. Sterilization of diets is essential for germ-free and specific-pathogen-free animals and is often advisable for conventionally reared animals. Autoclaving at temperatures greater than 100° C can be effective in achieving complete sterility so long as steam penetrates the entire load for a sufficient amount of time, but excessive exposure should be avoided as this exacerbates vitamin losses and affects protein quality (Zimmerman and Wostmann, 1963; Coates, 1984). Some autoclaves permit rapid heating to high temperatures under vacuum, with consequent reduction of exposure time and nutrient losses. Diets can be sterilized by ionizing radiation with less damage to nutrients than is caused by heat sterilization as long as diets are packed under vacuum or nitrogen and little moisture is present (Ley et al., 1969; Coates, 1984). It has been suggested that supplements of heat-labile vitamins be increased two-to fourfold in diets to be sterilized to compensate for potential losses during sterilization (International Council for Laboratory Animal Science, 1987).

High-lipid diets require several formulation and storage precautions. Unsaturated lipid in the diet is susceptible to oxidation, which reduces the amount of available essential fatty acids (EFA). Rancid characteristics of oxidized lipid may reduce diet acceptability. An antioxidant (butylated hydroxytoluene or ethoxyquin at 0.01 to 0.02 percent of oil) should be added to the oil (American Institute of Nutrition, 1980). As an additional precaution to reduce decomposition, diets should be stored at temperatures ≤4° C in a container that has been flushed with argon or nitrogen before sealing (Fullerton et al., 1982). When very highly unsaturated oils are fed (e.g., fish oils), the diet should be changed every 24 to 48 hours (Johnston and Fritsche, 1989). In addition, extra DL-α-tocopherol (e.g., 5 to 10 times the concentration in low-lipid diets) may need to be included in the diets to prevent in vivo peroxidation (Garrido et al., 1989; Johnston and Fritsche, 1989). (The fatty acid composition of several common dietary oils is shown in Appendix Table 1 to help researchers choose the dietary lipid source most appropriate for a particular experimental protocol.)

QUALITY ASSURANCE AND POTENTIAL CONTAMINANTS

Given the potential importance of diet quality to consistent experimental results, a routine program to assay nutrients should be implemented to verify the composition of diets fed to laboratory animals. Although accidental omission or inadvertent inclusion of ingredients is uncommon, when it does occur it can have disastrous consequences. Discrepancies between expected and actual nutrient concentrations in laboratory animal feeds can occur as a result of errors in formulation, losses of labile nutrients during manufacture and storage, and variation of the nutrient content of ingredients from average values presented in tables (e.g., National Research Council, 1976, 1982).

Assaying is particularly important when commercial diets of undeclared formula are used because nutrient concentrations may deviate from those published by the manufacturer. For example, because commercial corn starch can contain significant quantities of linoleic acid (Holman, 1968), diets designed to induce essential fatty acid deficiency are more effective when sucrose, rather than starch, is used. Batch-to-batch variation in nutrient composition may be substantial even in fixed-formula diets made from natural ingredients. For example, in 94 batches of a fixed-formula diet assayed, concentrations varied about sixfold for vitamin A, nearly fourfold for thiamin, and twofold for calcium (Rao and Knapka, 1987). However, some of this variation may have been the result of sampling or analytical error. Variation in purified diets, although of lesser magnitude, may be important if nutrients are provided at requirement concentrations.

Samples for assaying should be taken from multiple bags or containers of feed. Care must be exercised to obtain a representative well-mixed subsample, especially if any settling or segregation of diet particles has occurred. Nutrient analyses should be conducted by a reputable laboratory and in accordance with Association of Official Analytical Chemists methods of analysis (Association of Official Analytical Chemists, 1990). Analyses should at least include proximate constituents (moisture, crude protein, ether extract, ash, and crude or acid-detergent fiber) and any nutrients of particular interest. Some vitamins and other nutrients are difficult to assay because of low concentrations or interfering compounds or both.

Potential chemical and biological contaminants of feeds are a major source of concern for toxicological and immunological research but may impact on other types of experiments as well. The International Council for Laboratory Animal Science noted seven unwanted substances in laboratory animal diets (International Council for Laboratory Animal Science, 1987):

1. pesticides;
2. pests (especially insects and mites);
3. bacteria, bacterial toxins, and mycotoxins;
4. natural plant toxins;
5. breakdown products of nutrients;

6. nitrates, nitrites, and nitrosamines; and
7. heavy metals.

In addition, errors in formulation or manufacture can result in hazardous amounts of those nutrients, such as vitamins A and D, and copper, that can be toxic at concentrations not greatly in excess of requirements. The greater potential for contaminants and other unwanted substances in natural-ingredient diets may make these diets unsuitable for certain types of research. However, fixed-formula diets can omit ingredients that tend to be particularly variable (such as some fish and meat meals) and rigorous pretesting of raw ingredients for specific contaminants may eliminate most potential problems. For example, in the manufacture of a fixed-formula rodent diet, it was necessary to restrict the fish meal to batches that had been demonstrated to be low in nitrosamine concentrations (Rao and Knapka, 1987).

Recommended maximum acceptable concentrations of chemical contaminants have been published by various agencies (e.g., Food and Drug Administration, 1978; Environmental Protection Agency, 1979; International Council for Laboratory Animal Science, 1987). Based on observed contaminant amounts and potential toxic effects, Rao and Knapka (1987) provide a list of recommended limits for about 40 contaminants, including aflatoxins, nitrosamines, heavy metals, chlorinated hydrocarbons, organophosphates, polychlorinated biphenyls, nitrates and nitrites, preservatives, and estrogenic activity. They also proposed a scoring system for diets to be used in chemical toxicology studies that permits separation of tested diets into those acceptable for long-term use, those acceptable only for short-term (transitory) use, and those that should be rejected. Testing for contaminants should be routine in toxicological research and may be valuable on at least an occasional basis in other studies.

Good manufacturing technique, appropriate storage conditions, and feeders that prevent fecal and urinary contamination of diets will minimize, but not eliminate, bacterial and other biological agents in diets. Diet is a potential source of pathogens for laboratory animals (Williams et al., 1969). Clarke et al. (1977) described procedures for sampling and assaying feeds for various pathogenic organisms as well as standards regarding the number and kinds of organisms acceptable in diets. As mentioned previously, sterilization procedures are employed for diets fed to germfree and specific-pathogen-free animal colonies. Because microbial residues may be unacceptable in the low-antigen diets required for immunologic studies, the use of chemically defined diets may be necessary.

DIETARY RESTRICTION

Traditionally, maximal growth and reproduction have been used as criteria for the evaluation of laboratory animal diets. However, evidence from a number of studies indicate that restricting the caloric intake of laboratory animals may have beneficial effects on life span, the incidence and severity of degenerative diseases, and the onset and incidence of neoplasia (Weindruch and Walford, 1988; Snyder and Towne, 1989; Yu, 1990; Bucci, 1992). Based on these results, allowing animals to eat ad libitum to produce maximum growth and reproduction may not be consistent with objectives of long-term toxicological and aging studies.

It is important to achieve caloric restriction of test animals without producing unintended nutrient deficiencies. Elevation of nutrient concentrations in the diet may be necessary to ensure that the nutrient intake of animals whose eating is restricted is comparable to that of animals allowed to eat ad libitum. Unfortunately, relatively little information is available about the extent to which caloric restriction affects nutrient requirements.

REFERENCES

American Institute of Nutrition. 1980. Second report of the ad hoc committee on standards for nutritional studies. J. Nutr. 110:1726.

Association of Official Analytical Chemists. 1990. Official Methods of Analyses of the Association of Official Analytical Chemists, 15th Ed., K. Helrich, ed. Arlington, Va.: Association of Official Analytical Chemists.

Bucci, T. J. 1992. Dietary restriction: Why all the interest? An overview. Lab Anim. 21:29–34.

Chatterjee, I. B. 1978. Ascorbic acid metabolism. World Rev. Nutr. Dietet. 30:69–87.

Clarke, H. E., M. E. Coates, J. K. Eva, D. J. Ford, C. K. Milner, P. N. O'Donoghue, P. P. Scott, and R. J. Ward. 1977. Dietary standards for laboratory animals: Report of the Laboratory Animals Centre Diets Advisory Committee. Lab. Anim. 11:1–28.

Coates, M. E. 1984. Sterilization of diets. Pp. 85–90 in The germfree animal in biomedical research, M. E. Coates, and B. E. Gustafsson, eds. Laboratory Animals Handbooks 9. London: Laboratory Animals Ltd.

Environmental Protection Agency. 1979. Proposed health effects test standards for toxic substances control act. Test rules. Good laboratory practice standards for health effects. Federal Register, Part 2, 27334–27375; Part 4, 44054–44093.

Fenton, P. F. 1957. Hereditary factors in protein nutrition. Am. J. Clin. Nutr. 5:663–665.

Fenton, P. F., and G. R. Cowgill. 1947. The nutrition of the mouse. I. A difference in the riboflavin requirements of two highly inbred strains. J. Nutr. 34:273–283.

Food and Drug Administration. 1978. Nonclinical laboratory studies. Good laboratory practice regulations. Federal Register, Part 2, 59986–60025.

Fullerton, F. R., D. L. Greenman, and D. C. Kendall. 1982. Effects of storage conditions on nutritional qualities of semipurified (AIN-76) and natural ingredient (NIH-07) diets. J. Nutr. 112:567–573.

Garrido, A., F. Garrido, R. Guerro, and A. Valenzuela. 1989. Ingestion of high doses of fish oil increases the susceptibility of cellular membranes to the induction of oxidative stress. Lipids 24:833–835.

Goodrick, C. L. 1973. The effects of dietary protein upon growth of inbred and hybrid mice. Growth 37:355–367.

Harris, R. S., and E. Karmas, eds. 1975. Nutritional Evaluation of Food Processing, 2nd Ed. Westport, Conn.: AVI Publishing.

Holman, R. T. 1968. Essential fatty acid deficiency. Prog. Chem. Fats Other Lipids 9:275–348.

Horio, F., K. Ozaki, A. Yoshida, S. Makino, and Y. Hayashi. 1985. Requirement for ascorbic acid in a rat mutant unable to synthesize ascorbic acid. J. Nutr. 115:1630–1640.

International Council for Laboratory Animal Science. 1987. ICLAS Guidelines on the Selection and Formulation of Diets for Animals in Biomedical Research, M. E. Coates, ed. London: International Council for Laboratory Animal Science.

Jenness, R., E. C. Birney, and K. L. Ayaz. 1980. Variation of L-gulonolactone oxidase activity in placental mammals. Comp. Biochem. Physiol. B 67:195–204.

Johnston, P. V., and K. L. Fritsche. 1989. Nutritional methodology in dietary fat and cancer research. Pp. 9–25 in Carcinogenesis and Dietary Fat, S. Abraham, ed. Boston: Kluwer Academic.

Knapka, J. J. 1983. Nutrition. Pp. 51–67 in The Mouse in Biomedical Research. Vol. 3, H. L. Foster, J. D. Small, and J. G. Fox, eds. New York: Academic Press.

Knapka, J. J. 1985. Formulation of diets. Pp. 45–59 in Methods for Nutritional Assessment of Fats, J. Beare-Rogers, ed. Champaign, Ill.: American Oil Chemists Society.

Knapka, J. J., K. P. Smith, and R. J. Judge. 1974. Effect of open and closed formula rations on the performance of three strains of laboratory mice. Lab. Anim. Sci. 24:480–487.

Lee, Y. C. P., J. T. King, and M. B. Visscher. 1953. Strain difference in vitamin E and B_{12} and certain mineral trace-element requirements for reproduction in A and Z mice. Am. J. Physiol. 173:456–458.

Ley, F. J., J. Bleby, M. E. Coates, and S. J. Patterson. 1969. Sterilization of laboratory animal diets using gamma radiation. Lab. Anim. 3:221–254.

Luecke, R. W., and P. J. Fraker. 1979. The effect of varying zinc levels on growth and antibody mediated response in two strains of mice. J. Nutr. 109:1373–1376.

Mizushima, Y., T. Harauchi, T. Yoshizaki, and S. Makino. 1984. A rat mutant unable to synthesize vitamin C. Experientia 40:359–361.

National Institutes of Health. 1985. Guide for the Care and Use of Laboratory Animals, Publication No. 86–23. Bethesda, Md.: National Institutes of Health.

National Research Council. 1973. Effect of Processing on the Nutritional Value of Feeds. Washington, D.C.: National Academy Press.

National Research Council. 1976. Atlas of United States and Canadian Feeds. Washington, D.C.: National Academy Press.

National Research Council. 1980. Mineral Tolerance of Domestic Animals. Washington, D.C.: National Academy Press.

National Research Council. 1982. United States—Canadian Tables of Feed Composition. Washington, D.C.: National Academy Press.

National Research Council. 1984. Nutrient Requirements of Beef Cattle, Sixth Revised Edition. Washington, D.C.: National Academy Press.

National Research Council. 1985. Nutrient Requirements of Sheep, Sixth Revised Edition. Washington, D.C.: National Academy Press.

National Research Council. 1987. Vitamin Tolerance of Animals. Washington, D.C.: National Academy Press.

National Research Council. 1988. Nutrient Requirements of Swine, Ninth Revised Edition. Washington, D.C.: National Academy Press.

National Research Council. 1989. Nutrient Requirements of Dairy Cattle, Sixth Revised Edition. Washington, D.C.: National Academy Press.

Navia, J. M. 1977. Animal Models in Dental Research. Tuscaloosa, Ala.: University of Alabama Press.

Pfast, H. B., ed. 1976. Feed Manufacturing Technology. North Arlington, Va.: Feed Production Council, American Feed Manufacturing Association.

Pleasants, J. R. 1984. The germ-free animal fed chemically defined ultrafiltered diet. Pp. 91–109 in The Germ-Free Animal in Biomedical Research. Laboratory Animals Handbook 9, M. E. Coates, and B. E. Gustafsson, eds. London: Laboratory Animals Ltd.

Pleasants, J. R., M. H. Johnson, and B. S. Wostmann. 1986. Adequacy of chemically defined, water-soluble diet for germfree BALB/c mice through successive generations and litters. J. Nutr. 116:1949–1964.

Pleasants, J. R., B. S. Reddy, and B. S. Wostmann. 1970. Qualitative adequacy of chemically defined liquid diet for reproducing germ-free mice. J. Nutr. 100:498–508.

Quinn, M. R., and M. M. Chan. 1979. Effect of vitamin B_6 deficiency on glutamic acid decarboxylase activity in rat olfactory bulb and brain. J. Nutr. 109:1694–1702.

Rao, G. N., and J. J. Knapka. 1987. Contaminant and nutrient concentrations of natural ingredient rat and mouse diet used in chemical toxicology studies. Fund. Appl. Toxicol. 9:329–338.

Snyder, D. L., and B. Towne. 1989. The effect of dietary restriction on serum hormone and blood chemistry changes in aging Lobund-Wistar rats. Prog. Clin. Biol. Res. 287:135–146.

Slinger, S. J. 1973. Effect on pelleting and crumbling methods on the nutritional value of feeds. Pp. 48–66 in Effect of Processing on the Nutritional Value of Feeds: Proceedings. Washington, D.C.: National Academy of Sciences.

Stevens, C. E. 1988. Comparative physiology of the vertebrate digestive system. New York: Cambridge University Press.

Weindruch, R., and R. L. Walford. 1988. The Retardation of Aging and Disease by Dietary Restriction. Springfield, Ill.: Charles C Thomas.

Williams, L. P., J. B. Vaughn, A. Scott, and V. Blanton. 1969. A ten-month study of salmonella contamination in animal protein meals. J. Am. Vet. Med. Assoc. 155:167–174.

Wise, A. 1982. Interaction of diet and toxicity—The future role of purified diet in toxicological research. Arch. Toxicol. 50:287–299.

Wostmann, B. S. 1975. Nutrition and metabolism of the germ-free animal. World Rev. Nutr. Dietet. 22:40–92.

Wretlind, K. A. J., and W. C. Rose. 1950. Methionine requirement for growth and utilization of its optical isomers. J. Biol. Chem. 187:697–703.

Yu, B. P. 1990. Food restriction: Past and present status. Rev. Biol. Res. Aging 4:349–371.

Zimmerman, D. R., and B. S. Wostmann. 1963. Vitamin stability in diets sterilized for germfree animals. J. Nutr. 79:318–322.

2 Nutrient Requirements of the Laboratory Rat

The laboratory rat (*Rattus norvegicus*) has long been favored as an experimental model for nutritional research because of its moderate size, profligate reproduction, adaptability to diverse diets, and tractable nature. It is now the species of choice for many experimental objectives because of the large body of available data and the development of strains with specific characteristics that facilitate the study of disease and other processes.

ORIGIN OF THE LABORATORY RAT

The laboratory rat is a domesticated Norway rat, which in nature is one of the most widespread and abundant of the more than 70 species of the genus *Rattus* (family Muridae). The albinos of the Norway rat were first domesticated in Europe in the early 19th century and came into use as experimental animals shortly thereafter (Lindsey, 1979). The Norway rat is not indigenous to Europe; however, it is believed to have originated in Asia and to have taken advantage of human movement in expanding its range worldwide (Nowak, 1991). The origin and historical development of the major strains of laboratory rats have been reviewed by Lindsey (1979).

In the wild, the Norway rat exhibits both territorial and colonial behavior and typically occupies underground burrows (Calhoun, 1963). Females produce 1 to 12 litters per year, and those in a colony nurture their young collectively. The Norway rat is omnivorous, eating a wide variety of seeds, grains, and other plant matter as well as invertebrates and small vertebrates (Nowak, 1991). Other than the fact that it lacks a gallbladder, the rat's digestive tract resembles that of other omnivorous rodents in that the stomach contains both nonglandular and glandular regions, the small intestine is of moderate length, and the cecum is relatively well developed (Bivin et al., 1979; Vorontsov, 1979).

GROWTH AND REPRODUCTIVE PERFORMANCE

Growth and reproductive performance are two key indicators of dietary adequacy. It is important that investigators monitor performance of experimental animals in relation to expected patterns of weight gain and reproduction. Given the large number of strains and different genotypes, it is not possible to describe a single growth pattern or reproductive performance applicable to all laboratory rats. Poiley (1972) summarized body weight gains from birth to about 24 weeks of age for 18 inbred strains and 3 outbred strains of rats. Examples of his mean values for 5 major inbred strains (Brown Norway, Fischer 344, Long-Evans, Osborne-Mendel, and Sprague-Dawley) are illustrated in Figure 2-1. Males gain weight more quickly and become larger than females of the same strain, but there are considerable differences in growth rates and adult body mass among the strains. In view of ongoing genetic selection of the strains of rats, and improvements in diets, the mean growth rates shown in Figure 2-1 may not represent desired performance at present. However, Reeves et al. (1993a) found similar growth rates in Sprague-Dawley rats fed a commercial rat diet for 16 weeks. Investigators should obtain expected or desired growth curves for their experimental rats from the supplier or breeding colony from which animals originate.

Reproductive characteristics such as age at the time reproduction begins, fertility, litter size, growth rates of suckling young, and preweaning mortality also vary among strains. C. T. Hansen (Veterinary Resources Program, National Center for Research Resources, National Institutes of Health, personal communication, 1993) recently summarized selected reproductive parameters of rat strains maintained in breeding colonies at the National Institutes of Health, including 37 monogamously mated inbred strains, 42 harem-mated congenic strains, 5 harem-mated mutant

11

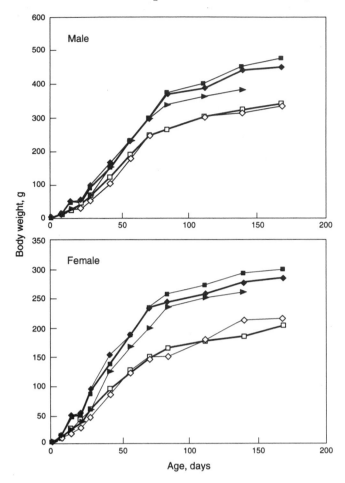

FIGURE 2-1   Mean body weight of male and female rats of five inbred strains: □, Brown Norway; ◇, Fischer 344; ▶, Long-Evans; ■, Osborne-Mendel; and ♦, Sprague-Dawley. SOURCE: Data adapted from Poiley (1972).

stocks, and 3 monogamously mated outbred stocks. Differences in the percentage of sterile matings, litter size, and preweaning mortality were evident among major inbred strains such as Brown Norway, Fischer 344, Osborne-Mendel, Sprague-Dawley, and Wistar (Table 2-1). Outbred strains tended to have higher reproductive performance (Table 2-1). Thus the expected reproductive performance of rats in experimental studies may vary according to strain and system of breeding.

ESTIMATION OF NUTRIENT REQUIREMENTS

Although the nutrient requirements of the laboratory rat are better known than those of other laboratory animals, there can be considerable disparity in estimated requirements as a consequence of the criteria used (Baker, 1986). For example, the amounts of nutrients required to sustain maximum growth of young rats may be different from the

TABLE 2-1   Some Reproductive Characteristics of Representative Strains of Inbred and Outbred Rat Colonies Maintained at the National Institutes of Health

Strain	Sterile Matings (%)	Mean Litter Size	Preweaning Mortality (%)
Inbred strain			
Brown Norway	30	5.9	1
Fischer 344	14	7.6	10
Osborne-Mendel	38	7.6	9
Sprague-Dawley	16	4.4	6
Wistar	31	7.9	8
Outbred strain			
Osborne-Mendel	6	7.4	6
Sprague-Dawley	2	7.1	2

NOTE: Both strains are from monogamously mated stocks.
SOURCE: Data summarized and provided by C. T. Hansen (Veterinary Resources Program, National Center for Research Resources, National Institutes of Health, personal communication, 1993).

amounts needed to maintain tissue concentrations or to maximize functional measures such as enzyme activities. Moreover, nutrient requirements are not static; they change according to developmental state, reproductive activity, and age. There is also evidence of differences in requirements between males and females as well as among various inbred and outbred strains. The nutrient requirements listed in this chapter represent average values, but they may not suffice in all circumstances. There is a need for further research that will identify the sources of variation in nutrient requirements.

Recommended nutrient concentrations in this report have not been increased to allow a margin of safety for variation in dietary ingredients or for differences among rats. The data on which requirements are based were reported from many different laboratories that used various colony management practices. One may assume that the recommendations are adequate for rats in most laboratory conditions, but particular experimental protocols such as maintenance of germ-free colonies or testing of experimental drugs (see Chapter 1) may alter the requirements for one or more nutrients. In some cases sufficient data were available to differentiate the nutrient requirements for adult maintenance from those for growing, pregnant, or lactating rats; hence, estimates of requirements are provided for maintenance, growth, and reproduction (Table 2-2). If data available were insufficient to determine requirements, adequate concentrations are reported on the basis of long-term feeding. If cited papers provided nutrient intakes per day but did not specify dietary concentrations, the values have been converted to dietary content by assuming a dietary intake of 15 g/rat/day for growing rats or adult rats at maintenance, 15 to 20 g/rat/day during pregnancy, and 30 to 40 g/rat/day during lactation.

TABLE 2-2 Estimated Nutrient Requirements for Maintenance, Growth, and Reproduction of Rats

Nutrient	Unit	Amount, per kg diet		
		Mainte-nance	Growth	Repro-duction (Female)
Fat	g	50.0	50.0	50.0
Linoleic acid (n-6)	g	[a]	6.0[a]	3.0[a]
Linolenic acid (n-3)	g	R	R	R
Protein	g	50.0[b]	150.0[b]	150.0
Amino Acids[c]				
Arginine	g	ND	4.3	4.3
Aromatic AAs[d]	g	1.9	10.2	10.2
Histidine	g	0.8	2.8	2.8
Isoleucine	g	3.1	6.2	6.2
Leucine	g	1.8	10.7	10.7
Lysine	g	1.1	9.2	9.2
Methionine + cystine[e]	g	2.3	9.8	9.8
Threonine	g	1.8	6.2	6.2
Tryptophan	g	0.5	2.0	2.0
Valine	g	2.3	7.4	7.4
Other (including nonessentials)	g	[f]	66.0	66.0
Minerals				
Calcium	g	[g]	5.0	6.3
Chloride[h]	g	[g]	0.5	0.5
Magnesium	g	[g]	0.5	0.6
Phosphorus	g	[g]	3.0	3.7
Potassium[h]	g	[g]	3.6	3.6
Sodium	g	[g]	0.5	0.5
Copper	mg	[g]	5.0	8.0
Iron	mg	[g]	35.0	75.0
Manganese	mg	[g]	10.0	10.0
Zinc[i]	mg	[g]	12.0	25.0
Iodine	μg	[g]	150.0	150.0
Molybdenum	μg	[g]	150.0	150.0
Selenium	μg	[g]	150.0	400.00
Vitamins				
A (retinol)[j]	mg	[g]	0.7	0.7
D (cholecalciferol)[k]	mg	[g]	0.025	0.025
E (RRR-α-tocopherol)[l]	mg	[g]	18.0	18.0
K (phylloquinone)	mg	[g]	1.0	1.0
Biotin (d-biotin)	mg	[g]	0.2	0.2
Choline (free base)	mg	[g]	750.0	750.0
Folic acid	mg	[g]	1.0	1.0
Niacin (nicotinic acid)	mg	[g]	15.0	15.0
Pantothenate (Ca-d-pantothenate)	mg	[g]	10.0	10.0
Riboflavin	mg	[g]	3.0	4.0
Thiamin (thiamin-HCl)[m]	mg	[g]	4.0	4.0
B₆ (pyridoxine)[n]	mg	[g]	6.0	6.0
B₁₂	μg	[g]	50.0	50.0

NOTE: Nutrient requirements are expressed on an as-fed basis for diets containing 10% moisture and 3.8–4.1 kcal ME/g (16–17 kJ ME/g) and should be adjusted for diets of differing moisture and energy concentrations. Unless otherwise specified, the listed nutrient concentrations represent minimal requirements and do not include a margin of safety. Higher concentrations for many nutrients may be warranted in natural-ingredient diets. R, required but no concentration determined;

TABLE 2-2 (*Continued*)

other long-chain n-3 polyunsaturated fatty acids may substitute for linolenic acid. ND, not determined.

[a]Females require only 2 g/kg linoleate for growth. Separate requirements for maintenance have not been determined for linoleate. Requirements presented for growth will meet maintenance requirements.

[b]Estimates based on highly digestible protein of balanced amino acid composition (e.g., lactalbumin).

[c]Asparagine, glutamic acid, and proline may be required for very rapid growth (see text).

[d]Phenylalanine plus tyrosine. Tyrosine may supply up to 50 percent of aromatic acid requirement.

[e]Cystine may supply up to 50% of the methionine plus cystine requirement on a weight basis.

[f]41.3 g/kg diet as a mixture of glycine, L-alanine, and L-serine.

[g]Separate requirements for maintenance have not been determined for minerals and vitamins. Requirements presented for growth will meet maintenance requirements.

[h]Estimate represents adequate amount, rather than true requirement.

[i]Higher concentration is required when ingredients that contain phytate (such as soybean meal) are included in the diet.

[j]Equivalent to 2,300 IU/g. Requirement may also be met by 1.3 mg β-carotene/kg diet. Higher vitamin A concentration is needed under conditions of stress (e.g., surgical recovery).

[k]Equivalent to 1,000 IU/kg.

[l]Equivalent to 27 IU/kg. Higher concentration may be required if high-fat diets are fed.

[m]Higher concentration may be required with low-protein, high-carbohydrate diets.

[n]Estimate represents adequate amount, rather than true requirements.

EXAMPLES OF DIETS FOR RATS

The type of diet and its nutrient composition will vary according to experimental objectives (see Chapter 1). Most practical diets include nutrient concentrations that exceed requirements as a margin of safety. Examples of natural-ingredient diets based on detailed formula specifications and used successfully to maintain rat colonies at the National Institutes of Health and at other facilities are provided in Table 2-3. The ingredient specifications, however, have not been updated for some years and are not entirely in agreement with recommendations in this report.

As indicated in Chapter 1, natural-ingredient diets do not offer the same control over nutrient concentrations or potential contaminants as do purified diets. An example of a purified diet that has often been used in rat studies is given in Table 2-4. However, as there has been concern about the standardization and concentrations of some constituents, as well as the clinical observation of nephrocalcinosis in females when this diet is used (Reeves, 1989; Reeves et al., 1993a), a change in formulation is warranted. A committee of the American Institute of Nutrition (AIN) has reformulated and tested new purified diets for the

TABLE 2-3 Examples of Natural-Ingredient Diets Used for Rat and Mouse Breeding Colonies at the National Institutes of Health

Ingredient	Conventional (NIH-07)	Autoclavable (NIH-31)
Basic Diet, g/kg diet		
Dried skim milk	50.0	
Fish meal (60% protein)	100.0	90.0
Soybean meal (48% protein)	120.0	50.0
Alfalfa meal, dehydrated (17% protein)	40.0	20.0
Corn gluten meal (60% protein)	30.0	20.0
Ground #2 yellow shelled corn	245.0	210.0
Ground hard winter wheat	230.0	355.0
Ground whole oats		100.0
Wheat middlings	100.0	100.0
Brewer's dried yeast	20.0	10.0
Dry molasses	15.0	
Soybean oil	25.0	15.0
Salt	5.0	5.0
Dicalcium phosphate	12.5	15.0
Ground limestone	5.0	5.0
Mineral premix	1.2	2.5
Vitamin premix	1.3	2.5
Mineral Premix, mg/kg diet		
Cobalt (as cobalt carbonate)	0.44	0.44
Copper (as copper sulfate)	4.40	4.40
Iron (as iron sulfate)	132.30	66.20
Manganese (as manganous oxide)	66.20	110.00
Zinc (as zinc oxide)	17.60	11.00
Iodine (as calcium iodate)	1.54	1.65
Vitamin Premix, per kg diet		
Stabilized vitamin A palmitate or stearate	6,060.00 IU	24,300.00 IU
Vitamin D_3 (D-activated animal sterol)	5,070.00 IU	4,190.00 IU
Vitamin K (menadione activity)	3.09 mg	22.10 mg
All-*rac*-α-tocopheryl acetate	22.10 mg	16.50 mg
Choline chloride	617.00 mg	772.00 mg
Folic acid	2.43 mg	1.10 mg
Niacin	33.10 mg	22.10 mg
Ca-*d*-pantothenate	19.80 mg	27.60 mg
Pyridoxine-HCl	1.87 mg	2.21 mg
Riboflavin supplement	3.75 mg	5.51 mg
Thiamin mononitrate	11.0 mg	71.7 mg
d-Biotin	0.15 mg	0.13 mg
Vitamin B_{12} supplement	0.004 mg	0.015 mg

NOTE: Amounts listed for mineral and vitamin premixes represent the mass or IU of the specific mineral element or vitamin rather than the added compounds.

SOURCES: Knapka (1983), Knapka et al. (1974), and National Institutes of Health (1982).

TABLE 2-4 Example of a Commonly Used Purified Diet (AIN-76A) for Rats

Ingredient	Amount, g/kg diet
Basic Diet	
Sucrose	500.0
Casein (≥85% protein)	200.0
Cornstarch	150.0
Corn oil (including 0.01-0.02% antioxidant[a])	50.0
Fiber source (cellulose-type)	50.0
Mineral mix (listed below)	35.0
Vitamin mix (listed below)	10.0
DL-Methionine	3.0
Choline bitartrate	2.0
Mineral Premix	
Calcium phosphate, dibasic ($CaHPO_4$)	500.00
Potassium citrate, monohydrate ($K_3C_6H_5O_7 \cdot H_2O$)	220.00
Sodium chloride	74.00
Potassium sulfate	52.00
Magnesium oxide	24.00
Ferric citrate (16–17% Fe)	6.00
Manganous carbonate (43–48% Mn)	3.50
Zinc carbonate (70% ZnO)	1.60
Chromium potassium sulfate [$CrK(SO_4)_2 \cdot 12H_2O$]	0.55
Cupric carbonate (53–55% Cu)	0.30
Potassium iodate (KIO_3)	0.01
Sodium selenite ($Na_2SeO_3 \cdot 5H_2O$)	0.01
Sucrose, finely powdered	118.03
Vitamin premix	
Nicotinic acid or nicotinamide	3.000
Calcium *d*-pantothenate	1.600
Pyridoxine-HCl	0.700
Thiamin-HCl	0.600
Riboflavin	0.600
Folic acid	0.200
d-Biotin	0.020
Cyanocobalamin (vitamin B_{12})	0.001
Retinyl palmitate or acetate (vitamin A)	+[b]
α-Tocopheryl acetate (vitamin E)	+[c]
Cholecalciferol (vitamin D_3)	0.0025
Menaquinone (vitamin K)	0.005
Sucrose, finely powdered	To make <1,000 g

SOURCE: American Institute of Nutrition (1977, 1980).

[a]Betahydroxytoluene or Santoquin.

[b]As stabilized powder to provide 400,000 IU vitamin A activity (120,000 retinol equivalents).

[c]As stabilized powder to provide 5,000 IU vitamin E activity.

growth (AIN-93G) and maintenance (AIN-93M) of rats and mice (Reeves et al., 1993a,b; Reeves et al., 1994). These diets are included in Table 2-5.

ENERGY

Purified diets containing 5 to 10 percent fat have gross energy (GE) values of about 4.0 to 4.5 Mcal/kg (17 to 19 MJ/kg). The digestible energy (DE) of most purified diets ranges from 90 to 95 percent of GE (Hartsook et al., 1973; McCraken, 1975; Deb et al., 1976). The metabolizable energy (ME) varies from 90 to 95 percent of DE (Hartsook et al., 1973; Pullar and Webster, 1974; McCraken, 1975; Deb et al., 1976). These values may be somewhat lower when diets formulated from natural ingredients are used

TABLE 2-5   Examples of Recently Tested Purified Diets for Rapid Growth of Young Rats and Mice or for Maintenance of Adult Rats and Mice

Ingredient	Growth (AIN-93G)	Maintenance (AIN-93M)
Basic Diet, g/kg diet		
Cornstarch	397.486	465.692
Casein (≥85%)	200.000	140.000
Dextrinized cornstarch[a]	132.000	155.000
Sucrose	100.000	100.000
Soybean oil (no additives)	70.000	40.000
Fiber source (cellulose)[b]	50.000	50.000
Mineral mix (listed below)	35.000	35.000
Vitamin mix (listed below)	10.000	10.000
L-Cystine	3.000	1.800
Choline bitartrate (41.1% choline)	2.500	2.500
Tert-butylhydroquinone (TBHQ)	0.014	0.008
Mineral Premix, g/kg mix[c]		
Calcium carbonate, anhydrous (40.0% Ca)	357.00	357.00
Potassium phosphate, monobasic (22.8% P, 28.7% K)	196.00	250.00
Potassium sulfate (44.9% K, 18.4% S)	46.60	46.60
Potassium citrate tri-K, monohydrate, (36.2% K)	70.78	28.00
Sodium chloride (39.3% Na, 60.7% Cl)	74.00	74.00
Magnesium oxide (60.3% Mg)	24.00	24.00
Ferric citrate (16.5% Fe)	6.06	6.06
Zinc carbonate (52.1% Zn)	1.65	1.65
Manganous carbonate (47.8% Mn)	0.63	0.63
Cupric carbonate (57.5% Cu)	0.30	0.30
Potassium iodate (59.3% I)	0.01	0.01
Sodium selenate anhydrous (41.8% Se)	0.01025	0.01025
Ammonium paramolybdate 4 hydrate (54.3% Mo)	0.00795	0.00795
Sodium meta-silicate 9 hydrate (9.88% Si)	1.4500	1.4500
Chromium potassium sulfate 12 hydrate (10.4% Cr)	0.2750	0.2750
Lithium chloride (16.4% Li)	0.0174	0.0174
Boric acid (17.5% B)	0.0815	0.0815
Sodium fluoride (45.2% F)	0.0635	0.0635
Nickel carbonate (45% Ni)	0.0318	0.0318
Ammonium vanadate (43.6% V)	0.0066	0.0066
Powdered sucrose	221.0260	209.8060
Vitamin Premix, g/kg mix[d]		
Nicotinic acid	3.000	3.000
Calcium pantothenate	1.600	1.600
Pyridoxine-HCl	0.700	0.700
Thiamin-HCl	0.600	0.600
Riboflavin	0.600	0.600
Folic acid	0.200	0.200
d-Biotin	0.020	0.020
Vitamin B_{12} (cyanocobalamin 0.1% in mannitol)	2.500	2.500
All-rac-α-tocopheryl acetate (500 IU/g)[e]	15.000	15.000
Retinyl palmitate (500,000 IU/g)[e]	0.800	0.800
Vitamin D_3 (cholecalciferol 400,000 IU/g) Vitamin K_1 (phylloquinone)	0.075	0.075
Powdered sucrose	974.655	974.655

[a]Ninety percent tetrasaccharides or higher oligosaccharides such as Dyetrose (Dyets Inc., Bethlehem, PA) Lo-Dex (American Maize Co., Hammond, NJ) or equivalent.

TABLE 2-5   (Continued)

[b]Solka-Floc 200 FCC (FS&D Corp., St. Louis, MO) or equivalent.
[c]Mineral mix for growth referred to as AIN-93G-MX and for maintenance as AIN-93M-MX.
[d]Vitamin mixes referred to as AIN-93-VX
[e]Dry gelatin-matrix form.
SOURCE: Adapted from Reeves et al. (1993b).

(Yang et al., 1969; Peterson and Baumgardt, 1971a). The addition of cellulose to natural-ingredient diets decreases energy digestibility (Yang et al., 1969; Peterson and Baumgardt, 1971a) even though 15 to 60 percent of the cellulose is digested (Conrad et al., 1958; Yang et al., 1969; Peterson and Baumgardt, 1971b). Some of the decrease in energy digestibility is caused by the low digestibility of cellulose and some is caused by increased fecal nitrogen losses (Meyer, 1956).

In general the rat will consume food to meet its energy requirement (Brody, 1945; Mayer et al., 1954; Sibbald et al., 1956, 1957; Yoshida et al., 1958; Peterson and Baumgardt, 1971b; Kleiber, 1975). Yoshida et al. (1958) reported that the daily caloric intake remained constant when the diet contained from 0 to 30 percent fat. A proportionate increase in consumption of diet occurs when the diet is diluted with inert materials. A maximum concentration of 40 percent dilution of the diet could be made for weanling female rats before caloric intake was reduced, whereas 50 percent dilution could be made for mature females (Peterson and Baumgardt, 1971b). The energy requirement of the lactating female is high, and dilution of the diet with only 10 percent of inert material results in a significant decrease in DE intake (Peterson and Baumgardt, 1971b). Inadequate dietary protein may decrease energy intake (Menaker and Navia, 1973).

Temperature, age, and activity influence the energy requirement of the rat. The lower critical temperature of the fasting rat is 30° C (Swift and Forbes, 1939; Brody, 1945; Kleiber, 1975). The lower critical temperature is that environmental temperature below which heat production must be increased to maintain body temperature. The basal metabolic rate of the rat can be estimated from the general formula

$$H_{kcal} = 72 \ BW_{kg}^{0.75}$$

where H_{kcal} is the heat production in kcal per day, BW is the body weight in kilograms, and 72 is the average heat production (kcal) per $kg^{0.75}$ of 26 groups of rats studied (Kleiber, 1975). This formula is valid only for estimating the basal heat production of mature animals. If kilojoules (kJ) are used as the unit of energy, the formula is $H_j = 301 \ BW_{kg}^{0.75}$.

MAINTENANCE

The maintenance energy requirement can be generally defined as that portion of the total energy requirement that is separate from the needs for growth, pregnancy, and lactation. Animals fed at maintenance are in energy equilibrium. Advantages and disadvantages of the methods by which the maintenance energy requirement may be estimated have been reviewed by van Es (1972). The maintenance energy requirement is usually expressed as energy required per unit of body weight in kilograms to the 0.75 power ($BW_{kg}^{0.75}$), and is based on the work of Brody (1945) and Kleiber (1975).

An estimate of the maintenance requirement for the adult rat (300 g) can be made by increasing the value for basal heat production (300 $BW_{kg}^{0.75}$; Kleiber et al., 1956) by approximately 20 percent (Morrison, 1968) to cover the expected requirement for activity in a laboratory setting. This value, which is in net energy units, can be converted to ME units by dividing by 0.75, based on an efficiency of 75 percent in conversion of ME to net energy (McCraken, 1975). The daily maintenance energy requirement in ME units for adult rats, based on these assumptions, is 114 kcal ME/$BW_{kg}^{0.75}$ (477 kJ ME/$BW_{kg}^{0.75}$). This value is remarkably similar to direct estimates of the maintenance energy requirement in ME units at 100 kcal (418 kJ; Pullar and Webster, 1974), 106 kcal (444 kJ; McCraken, 1975), 130 kcal (544 kJ; Deb et al., 1976), and 91 kcal (381 kJ; Ahrens, 1967). However, the requirement for fat rats (e.g., obese Zucker) is approximately 15 percent lower than the requirement for normal rats (Pullar and Webster, 1974; Deb et al., 1976). As this difference in maintenance energy requirement between the adult normal and adult obese Zucker is not determined by body weight, the practice of estimating maintenance energy requirement solely from body weight must be used with caution (Pullar and Webster, 1974; Webster et al., 1980).

It is well established that resting heat production per $BW_{kg}^{0.75}$ is greater in working and producing animals than in nonworking animals (Brody, 1945). A portion of this difference is attributable to differences in amount of intake. Walker and Garrett (1971) demonstrated that a decrease in food intake of rats results in a decrease in energy expenditure and in the maintenance energy requirement.

Previous plane of nutrition also influences maintenance energy requirement. Rats fed at a high plane of nutrition (39.6 g/$BW_{kg}^{0.75}$/day) for a 3-week period had a 38 percent greater heat production when compared to rats on a low plane (28.8 g/$BW_{kg}^{0.75}$/day) of nutrition (Koong et al., 1985). Experiments with sheep indicate that changes in organ mass associated with amount of intake and sequence of feeding may be largely responsible for altered heat production (Koong et al., 1985). Thus an accurate prediction of the maintenance energy requirement of the rat requires consideration of the amount of intake, previous nutritional history, physiological state, and other factors including the composition of the body (see Baldwin and Bywater, 1984, for a review). Nonetheless, data suggest that maintenance energy requirements of the rat will be met in most cases by consumption of 112 kcal/$BW_{kg}^{0.75}$/day (470 kJ/$BW_{kg}^{0.75}$/day).

GROWTH

It is difficult to estimate the energy requirement for growth because of variation in the composition of weight gain (Meyer, 1958; Schemmel et al., 1972; Hartsook et al., 1973; McCraken, 1975; Deb et al., 1976) and in the energetic efficiency of net protein and fat synthesis. Kielanowski (1965) used a multiple regression model that partitioned total ME intake (MEI) into a component proportional to body weight, representing maintenance energy requirement, a component proportional to energy gain as fat, and a component proportional to energy gain as protein or lean body mass. Separation of the efficiencies of energy gain in fat and lean is problematic, however, leading to inconsistent and sometimes biologically impossible results (see Chapter 3, which deals with the mouse). Pullar and Webster (1974) attempted to circumvent some of these difficulties by using obese and lean Zucker rats, which partition energy differently between gain in fat and in lean at all stages of growth. The energetic efficiencies of fat and net protein synthesis were estimated at 65 and 43 percent, respectively. These estimates assumed that the maintenance energy requirement remained constant. In a subsequent experiment that did not require assumptions about maintenance energy requirements, Pullar and Webster (1977) determined the energetic efficiencies of fat and net protein deposition as 73.5 and 44.4 percent, respectively. Kielanowski (1976) concluded from a review of previous work that the energetic efficiency of net protein deposition in growing rats is approximately 43 percent.

The requirement of the rat for maintenance and growth can be met by diets with a wide range of energy densities. Peterson and Baumgardt (1971b) reported that weanling and mature rats consumed 225 and 150 kcal DE/$BW_{kg}^{0.75}$ (940 and 630 kJ/$BW_{kg}^{0.75}$), respectively, when the energy density of the diet varied from 2.5 to 5.0 kcal DE/g (10.5 to 20.9 kJ/g). When the energy density in the diet fell below 2.9 kcal/g (12.1 kJ/g), the weanling rat could not meet its energy requirement. The mature rat could meet its energy requirement until DE density fell to values below 2.5 kcal DE/g (10.5 kJ/g). These values are equivalent to diluting the diet with 40 and 50 percent inert material for weanling and mature rats, respectively. During the 4-week growth period after weaning at 21 days postpartum, the average daily energy requirement is at least 227 kcal/$BW_{kg}^{0.75}$ (950 kJ/$BW_{kg}^{0.75}$) and may be greater. A diet containing at least 3.6 kcal ME/g (15.0 kJ ME/g) will meet the energy requirement for maintenance and growth if rats are allowed free access to food and the diet is not deficient in other nutrients.

GESTATION AND LACTATION

The energy requirement for gestation appears to be 10 to 30 percent greater than that of the mature but nonreproductive female rat (Morrison, 1956; Menaker and Navia, 1973). Food intake of rats fed diets adequate in protein increased 10 to 20 percent (Menaker and Navia, 1973) or 20 to 30 percent during the first days of gestation and up to 140 percent by days 16 to 18 of gestation (Morrison, 1956). Total heat production in pregnant rats increased approximately 10 percent above that of nonpregnant female rats (Brody et al., 1938; Kleiber and Cole, 1945; Morrison, 1956; Champigny, 1963). Approximately one-third of the 100 to 201 kcal (420 to 840 kJ) stored during gestation is deposited in fetal tissues (Morrison, 1956). Restriction of the diet during gestation decreases the size and viability of the young and may induce resorption (Perisse and Salmon-Legagneur, 1960; Berg, 1965). Protein appears to be more critical than energy for satisfactory reproduction (Hsueh et al., 1967; Menaker and Navia, 1973).

The daily ME requirement of the rat is about 143 kcal/$BW_{kg}^{0.75}$ (600 kJ/$BW_{kg}^{0.75}$) in early gestation and may increase to 265 kcal/$BW_{kg}^{0.75}$ (1,110 kJ/$BW_{kg}^{0.75}$) during the later stages of gestation. Lactating rats consume from two to four times more energy than nonlactating female rats, and the magnitude of increase depends on the number of offspring being nursed (Nelson and Evans, 1961; Peterson and Baumgardt, 1971b; Menaker and Navia, 1973; Grigor et al., 1984). Some of the increase in measured intake late in the lactation period may be caused by the consumption of diet by the litter, but this is not significant until about 15 to 17 days postpartum. In spite of the large increase in feed consumption, however, rats are generally in negative energy balance at peak lactation. Losses of both body fat and protein can occur. In general it appears that lipid is stored in the maternal body during gestation and then mobilized to support the lactation process (Sampson and Janson, 1984). Naismith et al. (1982) concluded that hormonal factors were responsible for the storage of lipid during gestation and the mobilization of body fat during lactation. They suggested that body fat supplies a major portion of the energy required to support lactation. Sainz et al. (1986) fed lactating rats diets containing 12, 24, and 36 percent protein and 4.50, 4.37, and 4.04 kcal ME/g (18.8, 18.3, and 16.9 kJ/g). Although fed ad libitum, the lactating rats lost body fat from days 7 to 14 of lactation regardless of diet. It is difficult to establish an energy "requirement" for the lactating rat based on body energy change. At peak lactation, rats rearing 8 pups produced about 41 g milk/day, representing a milk energy output of about 239 kcal/$BW_{kg}^{0.75}$/day (1,000 kJ/$BW_{kg}^{0.75}$/day) (Kametaka et al., 1974; Oftedal, 1984). It is likely that during peak lactation, the dam's daily ME requirement to support lactation will be at least 311 kcal/$BW_{kg}^{0.75}$ (1,300 kJ/$BW_{kg}^{0.75}$), but the ME requirement will vary with litter size.

LIPIDS

Lipid is an important component of the rat diet because it provides essential fatty acids (EFA) and a concentrated energy source, aids in the absorption of fat-soluble vitamins, and enhances diet acceptability.

ESSENTIAL FATTY ACIDS

n-6 Fatty Acids

The rat requires fatty acids from the n-6 family as a component of membranes, for optimal membrane-bound enzyme function, and to serve as a precursor for prostaglandin formation (Mead, 1984; Dupont, 1990; Clandinin et al., 1991). Linoleic acid [18:2(n-6)] cannot be synthesized endogenously but can be elongated and desaturated to form arachidonic acid [20:4(n-6)]; thus, the requirement for n-6 fatty acids can be met through dietary linoleic acid. Early estimates of the requirement for n-6 fatty acids were made by using growth and skin condition as response criteria. The requirement for growing male rats was estimated to be 50 to 100 mg/day and that of growing female rats to be 10 to 20 mg/day (Greenberg et al., 1950). That the requirement differs between sexes is in agreement with early observations that male rats are more susceptible than female rats to the development of EFA deficiency signs (Loeb and Burr, 1947). A series of later experiments indicated that other lipid components of the diet [e.g., oleic acid (Lowry and Tinsley, 1966) and cholesterol (Holman and Peiffer, 1960)] increased the need for n-6 fatty acids. Also, Holman (1960) demonstrated that more linoleic acid is needed in high-lipid diets versus low-lipid diets in order to meet the EFA requirement of the rat. Therefore it has become common practice to express the requirement as a percent of dietary ME rather than as a percent of diet weight. The Δ^6 desaturase enzyme involved in the synthesis of long-chain polyunsaturated fatty acids uses the fatty acid substrates in the following order: n-3 > n-6 > n-9 (Johnston, 1985).

Holman (1960) found that the ratio of 20:3(n-9) to 20:4(n-6) (also called the triene:tetraene ratio) was relatively constant in tissues until clinical signs of EFA deficiency began to develop. Because this biochemical measure was both objective and relatively stable (the precise ratio varied slightly among tissues), Holman (1960) initiated the use of the triene:tetraene ratio to indicate inadequate EFA status. Diets high in n-3 fatty acids also lower the triene concentration (Morhauer and Holman, 1963; Rahm and Holman, 1964) because of the competition for desaturation mentioned above. Therefore the triene:tetraene ratio is a valid indicator of EFA status only when n-6 fatty acids are the principal unsaturated dietary fatty acids. Pudelkewicz et al. (1986) used this technique (in conjunction with dermatitis and growth) to estimate the linoleic acid requirement of growing female and male rats to be 0.5 percent and 1.3 percent of dietary ME, respectively.

The requirement for pregnancy is met by diets adequate for growth, while that for lactation is somewhat higher. Deuel et al. (1954) conducted two experiments to estimate these requirements. In the first experiment, 0, 10, 40, 100, 200, 400, and 1,000 mg cottonseed oil (approximately 50 percent linoleic acid) were fed daily. The number of litters born and the average number of pups per litter increased and then plateaued at 10 and 100 mg dietary cottonseed oil (e.g., 5 and 50 mg linoleate), respectively. When pure linoleate was fed as the fat source in a second experiment in which rats were provided with 0, 2.5, 5.0, 10, 20, 40, and 80 mg linoleate daily, the maximum litters born and average number of pups per litter increased and then plateaued at 2.5 and 20 mg linoleate, respectively. This higher amount of linoleic acid intake will be achieved if the pregnant rat consumes 18 g/day and the diet contains 0.25 percent of dietary ME as linoleic acid.

The requirement for lactation also can be estimated from these experiments. Indicators of successful lactation (the maximum average weight per pup in a litter at 21 days and minimum percent mortality between 3 and 21 days) reached a plateau between 100 to 200 mg cottonseed oil (50 to 100 mg linoleate) in the first experiment. In the second experiment, these criteria for successful lactation reached a maximum at 80 mg of linoleate. Because greater amounts of linoleate were not tested, it is not possible to determine whether the optimal requirement for lactation is more than 80 mg/day. In addition, the authors did not indicate the range of experimental error; therefore, a precise requirement cannot be established. Assuming a requirement of 100 mg linoleate/day, a lactating rat consuming 35 g diet/day will consume sufficient linoleate for lactation (100 mg/day) if the diet contains 0.68 percent of dietary ME as linoleate (approximately 0.30 percent of the diet).

Bourre et al. (1990) proposed that the n-6 fatty acid requirement be established as the percent of dietary linoleate that results in a constant concentration of tissue arachidonic acid. They found that constant concentrations of arachidonic acid were achieved at 150 mg (nerve tissue), 300 mg (testicle), 800 mg (kidney), and 1,200 mg (liver, lung, heart) linoleate/100 g diet. The minimal requirement then was expressed as 1,200 mg linoleic acid/100 g diet (or approximately 2.5 percent of dietary ME) because some tissues require a higher concentration of arachidonic acid to reach a plateau. Because the authors did not relate the arachidonic acid concentration plateau to any functional phenomena, the validity of this method of estimating the n-6 fatty acid requirement remains to be established.

n-3 Fatty Acids

The essentiality of the n-3 fatty acids has been equivocal until recently. It was initially demonstrated that the n-3 fatty acid, α-linolenic acid [18:3(n-3)], could substitute, in part, for the requirement of n-6 fatty acids (Greenberg et al., 1950). Tinoco et al. (1971) failed to show any change in growth of rats raised for three generations on diets lacking in n-3 fatty acids in comparison to litter mates fed diets containing 1.25 percent linoleate and 0.25 percent linolenate. However, the sequestering of n-3 fatty acids in specific tissues (retina, cerebral cortex, testis, sperm; Tinoco, 1982) and the tenacity with which they retain these fatty acids, despite the variation in dietary concentration (Crawford et al., 1976), led many researchers to speculate that n-3 fatty acids were required for some function in the body.

Bernsohn and Spitz (1974) fed rats lipid-free diets for 4 months and measured slightly decreased amounts (38 percent of control values) of monoamine oxidase and 5'-mononucleotidase in cerebral cortex that responded to dietary α-linolenic but not to linoleic acid. Retinal function may be negatively impacted in offspring of rats fed a low linolenate oil for two to three generations (Okuyama et al., 1987). Lamptey and Walker (1976) found reduced exploratory behavior in second generation rats fed safflower oil. A large percent of safflower oil is linoleic acid, and a very small percent is linolenic acid. This was confirmed in studies by Enslen et al. (1991) who showed a reduction in exploratory behavior in 16- to 18-week-old rats from dams fed safflower oil 6 weeks prior to mating and throughout gestation and lactation. These researchers found that when rats were switched to α-linolenic acid at weaning, they did not recover exploratory behavior, further suggesting a specific requirement for n-3 fatty acids during development.

Yamamoto et al. (1987, 1988) found a reduction in brightness discrimination learning in offspring from rats fed safflower oil through two generations in comparison to rats similarly fed perulla oil, a rich source of α-linolenic acid.

Bourre et al. (1989) measured impairment of nerve terminal Na^+,K^+-ATPase activity, brain 5'-nucleotidase, and 2',3'-cyclic nucleotide-3'-phosphodiesterase in offspring from rats fed 1.8 percent sunflower oil in comparison to those from dams fed 1.9 percent soybean oil through two generations. They estimated the requirement for n-3 fatty acids as the least amount of α-linolenic acid in the diet that resulted in a higher concentration of brain n-3 fatty acids. Using this methodology, they estimated a requirement of 2 g/kg food (or 0.4 percent of the total dietary ME). These experiments indicate that n-3 fatty acids are required; further study is needed to determine the amount below which functional impairment occurs.

Although a requirement has not been defined, it may be advisable to include a source of n-3 fatty acids when dietary oils such as sunflower or safflower are fed through two or more generations. Homeostatic mechanisms appear to sequester n-3 fatty acids to protect the rat from n-3 deficiency during short-term dietary deprivation.

SIGNS OF EFA DEFICIENCY

A deficiency of EFA results in a plethora of gross clinical signs, anatomical changes, and physiological changes as discussed by Holman (1968, 1970). Classical overt signs include diminished growth, dermatitis, caudal necrosis, fatty liver, impaired reproduction, increased triene:tetraene ratio in the tissue and blood, and increased permeability of skin, with impaired water balance. There are many other less noticeable but equally severe changes that have been reported, including kidney lesions and a decrease in urine volume (Sinclair, 1952), lipid-containing macrophages in the lung (Bernick and Alfin-Slater, 1963), increased metabolic rate (Wesson and Burr, 1931), decreased capillary resistance (Kramar and Levine, 1953), and aberrant ventricular conduction (Caster and Ahn, 1963).

The EFA content of the rat's diet prior to the feeding of EFA-deficient diets affects reserve stores of EFA (Guggenheim and Jurgens, 1944). Weanling rats rapidly exhibit signs of EFA deficiency after consuming a lipid-free diet for 9 to 12 weeks, while mature rats may require an extensive period of starvation, then refeeding of a lipid-free diet in order to develop EFA deficiency signs (Barki et al., 1947). Dermal signs resulting from EFA deficiency were reported after feeding adult rats a lipid-free diet for 35 weeks (Aaes-Jorgensen et al., 1958). Pups from dams fed EFA-deficient diets exhibit the most severe signs of EFA deficiency and usually die within 3 days to 3 weeks after birth, depending on the duration of the EFA feeding to the dam (Guggenheim and Jurgens, 1944; Kummerow et al., 1952). Other dietary factors [e.g., cholesterol (Holman and Peiffer, 1960) and 18:2(n-6) trans,trans fatty acid (Hill et al., 1979; Kinsella et al., 1979)] accelerate the development of EFA deficiency in rats fed EFA-deficient diets. Males may develop signs of EFA deficiency more quickly than females because males have a greater EFA requirement than do females (Morhauer and Holman, 1963; Pudelkewicz et al., 1968). Prevention of coprophagy will accelerate the development of EFA deficiency in rats fed lipid-free diets (Barnes et al., 1959a).

The relative capacities of n-6 and n-3 fatty acids to alleviate some of the deficiency signs are shown in Table 2-6. The classical signs of EFA deficiency appear to be more amenable to amelioration by n-6 fatty acids. It was hypothesized that the sole function of linoleic acid was as a precursor to arachidonic acid [20:4(n-6)] (Rahm and Holman, 1964; Yamanaka et al., 1980). Linoleic acid concentration is much lower than that of arachidonic acid in membrane lipids (Sprecher, 1991). However, linoleate-rich O-acyl sphingolipids have been identified in the epidermis of pigs and humans (Gray et al., 1978). The structures of pig and human epidermal acyl ceramide and acyl glucosyl ceramide were confirmed by Wertz et al. (1986). Hansen and Hensen (1985) fed EFA-deficient rats oleic [18:1(n-9)], linoleic [18:2(n-6)], columbinic [18:3(n-6)], α-linolenic [18:3(n-3)], and arachidonic [20:4(n-6)] acid esters and measured epidermal sphingolipids and trans-epidermal water loss. Only n-6 fatty acid esters restored the water barrier; however, among n-6 fatty acids, only linoleate was esterified in substantial amounts in the sphingolipids. The authors suggested that columbinate and arachidonate result in linoleate mobilization from other tissues for incorporation into the epidermal sphingolipids. The relationship between linoleate-containing epidermal sphingolipids and trans-epidermal water loss awaits further study.

DIGESTIBILITY OF LIPIDS

Most commercial sources of dietary lipid consist exclusively of triglycerides and contain a high percentage of 18-carbon fatty acids. During digestion, lipase activity releases fatty acids from the 1 and 3 positions of the triglyceride. Free fatty acids and 2-monoacyl glycerol are absorbed.

Digestibility differs among lipid sources. Crockett and Deuel (1947) demonstrated that lipid digestibility was reduced when the melting point of the lipid was greater than 50° C. Fatty acid composition also may affect digestibility. Mattson (1959) showed that digestibility was reduced with increasing content of simple triglycerides (same fatty acid at each position) composed of 18-carbon saturated fatty acids. Saturated fatty acids and monounsaturated trans-fatty acids of 18-carbon chain length or longer are poorly

TABLE 2-6   Relative Ability of n-6 and n-3 Fatty Acids to Alleviate Several Signs of EFA Deficiency in Rats

Sign	Ability to Alleviate Sign
Diminished growth	n-6 better able than n-3
Impaired reproduction	n-6 will alleviate; n-3 ineffective
Dermatitis	n-6 will alleviate; n-3 ineffective
Increased triene:tetraene ratio in tissues and blood	n-6 equal to n-3
Capillary fragility	n-6 equal to n-3

absorbed as free fatty acids but easily absorbed in the 2-monoacyl glycerol form (Linscher and Vergroeson, 1988). Increasing the number of double bonds in the fatty acid improved absorption; increasing the fatty acid chain length decreased absorption (Chen et al., 1985, 1987a). Very long chain n-3 fatty acids were poorly hydrolyzed in in vitro experiments (Brockerhoff et al., 1966; Bottino et al., 1967) but were well absorbed in unesterified forms (Chen et al., 1985).

The rate of triglyceride digestion and subsequent absorption of fatty acids depends on the nature of the fatty acid (chain length and number and position of double bonds) and its molar frequency and position in the triglyceride (Apgar et al., 1987; Nelson and Ackman, 1988). Tables of fatty acid composition of dietary fats that include their positional specificity are available (Small, 1991).

Utilization of dietary lipid may be affected by other dietary components (Vahouny, 1982; Carey et al., 1983). Estimates of utilization by using the lipid source in the absence of other dietary ingredients may not correctly define the utilization of lipid in a complex diet. Nelson and Ackman (1988) reviewed the literature on the use of ethyl esters of lipid to study absorption and concluded that absorption and transport may not be identical to naturally occurring (triglyceride) lipid sources. The digestibility of many dietary lipids has been determined and certain of these experimental results are summarized in Table 2-7.

DIETARY LIPID CONCENTRATION

Better growth, reproduction, and lactation performance result when rats are fed diets in which lipid content is increased from 5 to 40 percent (Deuel et al., 1947). These observations and those of Forbes et al. (1946a,b) led Deuel (1950, 1955) to conclude that 30 percent lipid is the optimal dietary concentration. These response criteria alone are no longer considered sufficient to establish the optimal amount of dietary lipid.

Maximum growth also is associated with a decrease in longevity (French et al., 1953). Rats fed diets containing 10 or 20 percent corn oil had higher growth rates of a transplantable mammary tumor than those fed diets containing 2 or 5 percent corn oil (Kollmorgen et al., 1983). Rolls and Rowe (1982) demonstrated diminished growth and survival of pups suckled by dams fed high lipid diets. Another study showed that rats consuming diets containing 3 or 20 percent lipid had superior reproduction (total numbers of offspring; percent of young weaned) compared to rats consuming diets containing 36 and 50 percent lipid (Richardson et al., 1964). It appears, then, that maximum growth should not be the only predictor of optimal dietary lipid content.

Data from the following experiments serve as justification for maintaining the previously recommended amount of 5 percent dietary lipid for both males and females during rapid growth and for adult females during reproduction and lactation (National Research Council, 1978). Swift and Black (1949) showed that the greatest improvement in energy retention occurred when dietary lipid content was increased from 2 to 5 percent; additional increments in energy retention were smaller when lipid content was above 5 percent. Deuel et al. (1947) reported that the greatest reduction in number of days required to reach puberty occurred when the percentage of lipid in the diet was increased from 0 to 5 percent. Relatively small changes occurred when lipid was greater than 5 percent of the diet. Burns et al. (1951) demonstrated that 5 percent lipid was satisfactory for absorption of carotene and vitamin A. Loosli et al. (1944) reported only slight improvement in weight gain of rat pups when lactating females were fed diets that contained more than 5 percent lipid. Furthermore, many lipids provide sufficient EFA when included in the diet at these concentrations. Reeves et al. (1993b: pp. 1941–1942) noted the following:

Bourre et al. (1989, 1990) used the method of dietary titration of 18:2(n-6) and 18:3(n-3) to determine linoleic and linolenic acid requirements, respectively. They used tissue saturation of 20:4(n-6) and 22:6(n-3) to make the assessments and concluded that 12 g of linoleic acid and 2 g of α-linolenic acid per kilogram of diet were the minimal requirements for the rat. This amounts to approximately 3 percent soybean oil in the diet. However, to reach the plateau for maximal concentrations of these fatty acids in many tissues of growing rats, an amount of fat equivalent to 5–6 percent soybean oil was required.

Lee et al. (1989) suggest that a n-6:n-3 ratio of five and a polyunsaturate:saturate (P:S) ratio of two are the points of greatest influence on tissue lipids and eicosanoid production. Bourre et al. (1989) suggested that the optimal n-6:n-3 ratio is between one and six. Soybean oil is a source of dietary fat that may meet these criteria. The oil contains about 14 percent saturated fatty acids, 23 percent monounsaturated fatty acids, 51 percent linoleic acid, and 7 percent linolenic acid. This gives a n-6:n-3 ratio of seven, and a P:S ratio of approximately four. The fatty acid composition of commercial lipid sources must be monitored because of the widespread practice of hydrogenation and the emergence of new cultivars with different fatty acid compositions.

CARBOHYDRATES

Although no definite carbohydrate requirement has been established, rats perform best with glucose or glucose precursors (such as other sugars, glycerol, glucogenic amino acids) in their diets. Diets containing 90 percent of dietary ME from fatty acids and 10 percent from protein were unable to support growth of young male rats. The substitution of neutral fats (soybean oil) for fatty acids or the addition of glycerol equivalent to that in the triglyceride allowed growth but not at rates equivalent to that achieved

TABLE 2-7 Digestibility of Some Selected Dietary Fats

Test Fat	Fat in Diet	Fat Digestibility, %	Test Animal and Method
		Experiment 1 (Hoagland and Snider, 1943)	
Butterfat	5%	87.9*	Male albino weanling rats
	15%	90.2*	
Mutton, tallow	5%	74.2†	
	15%	85.0*,≠	
Cocoa butter	5%	63.3§	
	15%	81.6‡	
Soybean oil	5%	98.7¶	
	15%	98.3¶	
Corn oil	5%	97.5‡	
	15%	98.3¶	
Coconut oil	5%	98.9¶	
	15%	96.5¶	
		Experiment 2 (Crockett and Devel, 1947)	
Margarine (mp = 34° C)	15%	97.0¶	Adult female rats; calculations similar to Experiment 1
Bland lard (mp = 43° C)	15%	94.3¶	
Hydrogenated lard (mp = 55° C)	15%	63.2*	
		Experiment 3 (Apgar et al., 1987)	
Corn oil	5%	92.9*	Male weanling Sprague-Dawley rats; calculations similar to Experiment 1
	10%	96.7¶	
	20%	96.3¶	
Cocoa butter	5%	58.8†	
	10%	60.3†	
	20%	71.7‡	
		Experiment 4 (Chen et al., 1987b)	
Corn oil	170 mg	100.0¶	Adult male Wistar rats, 250–350 g; lipid emulsion infused via duodenal catheters; lipid collected as lymph from thoracic duct catheters; method measures absorption; corn oil set as control = 100%
Palm kernal oil	170 mg	82.3¶	
Cocoa butter	130 mg	63.0*	
		Experiment 5 (Chen et al., 1989)	
Corn oil	170 mg	100.0¶	Same as Experiment 4
Menhaden oil	170 mg	56.6*	
Max EPA® (eicosapentanoic acid-containing fish oil concentrate)	170 mg	47.1*	

NOTE: Different symbols following measurements in the "Fat Digestibility" column indicate that digestibility mean differs significantly ($P < 0.5$).

with a 78 percent starch diet (Konijn et al., 1970; Carmel et al., 1975). When carbohydrate-free diets containing 80 percent of dietary ME from fatty acids and 20 percent from protein were fed, rats were capable of weight gain, but growth increased when the diet was supplemented with glucose or neutral fats (Goldberg, 1971; Akrabawi and Salji, 1973). Rats fed low protein (10 percent of dietary ME), carbohydrate-free diets were hypoglycemic and demonstrated abnormal glucose tolerance curves (Konijn et al., 1970; Carmel et al., 1975); rats fed higher protein (18 percent of dietary ME), carbohydrate-free diets had normal blood glucose concentrations but still demonstrated slightly abnormal glucose tolerance curves (Goldberg, 1971). When neutral fats replaced fatty acids in carbohydrate-free diets (20 percent of dietary ME from protein), growth did not improve when rats were allowed to eat ad libitum but was greater with diets containing the neutral fats when rats were meal-fed once daily (Akrabawi and Salji, 1973).

Over wide ranges of dietary fat:carbohydrate ratios (0.2 to 1.4, ME basis), the heat increment was found to be constant at 47.5 percent of ME, indicating that carbohydrate and lipid are used with equal efficiency (Hartsook et al., 1973).

A large number of carbohydrates can be used by the rat. Those most commonly used in rat diets include glucose, fructose, sucrose, starch, dextrins, and maltose. (See "Fiber" section for a discussion of fiber sources.) These carbohydrate sources support similar rates of growth; however, in diets adequate in other respects, fructose (and sucrose as a source of fructose) can lead to several abnormalities when compared to glucose or glucose-based polymers. Because the initial metabolic steps in fructose utilization are mediated by fructokinase and aldolase B, fructose metabolism bypasses the control of glycolysis at phosphofructokinase and, thus, increases the flux through glycolysis. Feeding of fructose or sucrose leads to increases in liver weight, liver lipid, liver glycogen, and activities of liver lipogenic enzymes: glucose-6-phosphate dehydrogenase, malic enzyme, ATP citrate lyase, and fatty acid synthetase (Worcester et al., 1979; Narayan and McMullen, 1980; Michaelis et al., 1981; Cha and Randall, 1982; Herzberg and Rogerson, 1988a,b). Hypertriglyceridemia associated with fructose feeding has been attributed to both increased hepatic synthesis (Herzberg and Rogerson, 1988b) and decreased peripheral clearance (Hirano et al., 1988) of triglyceride. Increases in kidney weight and nephrocalcinosis also were observed when diets containing 55 percent sucrose (Kang et al., 1979) or 63 percent fructose (Koh et al., 1989) were fed. Starch was more easily metabolized than sucrose by rats fed low-protein diets (12.5 percent casein; Khan and Munira, 1978) or protein-free diets (Yokogoshi et al., 1980). Essential fatty acid deficiencies may be exacerbated by high sucrose diets (Trugnan et al., 1985).

Poor performance and cataract formation occurred in rats fed lactose or galactose (Day and Pigman, 1957); diarrhea was also observed in weanling rats fed α- or β-lactose (Baker et al., 1967). Xylose is toxic to rats; lens opacity and diarrhea were observed in rats fed diets containing 15 percent or more xylose (Booth et al., 1953). Sorbose, a slowly absorbed sugar, decreases feed intake and growth rate when added to rat diets but appears to supply the rat with a significant amount of energy, presumably, in part, as end products of hindgut fermentation (Furuse et al., 1989). Mannose (up to 8 percent of the diet) improved growth of rats fed a carbohydrate-free diet, suggesting that it can be metabolized, at least in low concentrations (Keymer et al., 1983). Leucrose [D-glucosyl-α(1–5)D-fructopyranose, a bond isomer of sucrose] appears to be metabolized as well as sucrose (Ziesenitz et al., 1989). Of the sugar alcohols, lactitol and xylitol decrease feed intake and growth when added to diets at 16 percent of dry matter, although rats appear to adapt, at least in part, to these two sugar alcohols within 2 weeks (Grenby and Phillips, 1989). Sorbitol can be metabolized by rat liver (Ertel et al., 1983).

A series of experiments defined the rats' need for carbohydrate for successful reproduction. In all these experiments a low-protein diet was required in order to demonstrate the need for carbohydrate. Rats fed carbohydrate-free diets [ME = 4.25 kcal/g (17.8 kJ/g), 12 percent of dietary ME from protein] were unable to maintain pregnancy. Although 78 percent of embryos were normal following 6 days of gestation for rats fed carbohydrate-free diets (compared to 91 percent for controls), only 25 percent (control = 89 percent) were classified as normal following 8 days and 0.6 percent (control = 90 percent) following 10 days of the carbohydrate-free diet. By day 12 of gestation, all embryos from rats fed carbohydrate-free diets had been resorbed (Taylor et al., 1983). A carbohydrate-free diet [ME = 4.11 kcal/g (17.2 kJ/g), 10 percent of dietary ME from protein] fed to gestating rats had to be supplemented with 4 percent carbohydrate (as glucose or an equivalent amount of glycerol) to maintain pregnancy to term, 6 to 8 percent glucose to produce normal maternal weight gain and normal fetal weight, and 12 percent glucose to produce fetal liver glycogen concentrations one-half as large as controls fed a 62 percent carbohydrate diet (Koski et al., 1986). Survival was poor for pups from dams fed low-glucose diets (9.5 percent protein) from day 9 of gestation through day 7 of lactation. From dams fed 6 percent or less glucose, no pups survived 7 days postpartum. From those dams fed 8 or 12 percent glucose, pup survival at 7 days was 6 and 30 percent, respectively. Control pups whose dams were fed 62 percent glucose diets had 93 percent survival (Koski and Hill, 1986). Poor rat pup survival caused by feeding dams a low (4 percent)-glucose diet (10 percent of calories from protein) could be markedly improved by feeding dams a high-carbohydrate diet for the final 2 days of gestation and through lactation (Koski and Hill, 1990). Lactation is not supported by carbohydrate-free diets. Milk production can occur for rats fed 6 percent glucose diets, but the milk contains low concentrations of carbohydrate and lipid, which is associated with retarded postnatal growth of pups (Koski et al., 1990). In general, fructose appears to be an adequate source of carbohydrate in diets fed to pregnant rats. However, when low (4 percent)-carbohydrate diets are fed during lactation, neither fructose- nor glycerol-supplemented diets will support deposition of as much fetal liver glycogen as 4 percent glucose diets (Fergusson and Koski, 1990). Essential fatty acid deficiencies may be more likely to occur in gestating rats fed sucrose-based (61.5 percent) diets than in those fed glucose-based diets (Cardot et al., 1987).

PROTEIN AND AMINO ACIDS

In establishing the protein requirements at different stages of life, three factors must be considered: (1) energy concentration in the diet, (2) amino acid composition of the protein (see Appendix Table 2), and (3) bioavailability of the amino acids.

PROTEIN AND GROWTH

Protein requirements are most accurately expressed as a protein:energy ratio to take into account the large differences in energy concentration that may occur among diets. In earlier studies that used egg protein as a highly digestible and balanced source of amino acids, the minimal amount of protein required for maximum weight gain in young rats was 25 to 31 mg/kcal GE (6.0 to 7.4 mg/kJ GE) (Hamilton, 1939; Barnes et al., 1946; Hoagland et al., 1948; Mitchell and Beadles, 1952). Similar results were obtained with pure amino acid mixtures (Rose et al., 1948) and with casein supplemented with sulfur amino acids (Breuer et al., 1963; Hartsook and Mitchell, 1956); as expected, greater amounts of protein were required when unsupplemented casein was used (Yoshida et al., 1957). These data indicate that a dietary protein concentration of 10 to 15 percent is required for maximum growth when a low-fiber diet containing a balanced amino acid pattern, 5 percent fat and 4 kcal ME/g (17 kJ ME/g), is fed. The 1978 edition of this report concluded that the protein requirement for maximum growth of the rat is 12 percent when highly digestible protein of balanced amino acid pattern is used.

Computation of the percentage of dietary protein required for maximum growth when the diet contains a mixture of proteins requires that both the content and bioavailability of the amino acids in the different proteins be considered. Historically, most methods have assumed that protein quality is constant over a range of dietary protein concentrations. For example, in the slope-ratio procedure, test proteins are fed at several concentrations and the value of a protein is determined by linear regression (Hegsted and Chang, 1965). However, the value of protein as used for maintenance differs from the value of protein as used for growth, and the difference is not linear (Phillips, 1981; Finke et al., 1987a,b, 1989; Mercer et al., 1989; Schulz, 1991). Finke et al. (1987a,b, 1989) used a four-parameter logistical model to describe the effect of utilization of protein from a variety of sources on growth rates and nitrogen gain of young rats (Sprague-Dawley strain). Although 1.11 times more casein than lactalbumin was required to achieve 95 percent of the maximum nitrogen gain, 1.43 times more casein than lactalbumin was required to achieve maintenance or zero nitrogen gain. In another comparison, twice as much soybean protein as lactalbumin was required to support 95 percent of the maximum nitrogen gain, but only 1.54 times as much soybean protein was required to meet maintenance needs. Thus, use of nonlinear models to describe an animal's growth response to dietary protein or protein mixtures indicates that the relative value of protein is not constant and that the value of a protein for maintenance may not predict its value for growth. This may be an expression of the different amino acid patterns required for maintenance versus growth. Finally, nonlinear response models, in which marginal efficiency (response per unit input) changes with response level, seem to be more accurate than linear (constant marginal efficiency, i.e., broken stick) models in predicting the relative capacity of proteins to support maximum gain or nitrogen retention.

In determining the relative value of proteins to support growth using nonlinear models, it is important to include test diets that produce a maximal response or response plateau so that the "diminishing-returns" portion of the response curve can be defined. The weight gain response per unit of protein added (diminishing-returns) is expected to vary with type of dietary protein. By applying a saturation kinetics model, Mercer et al. (1989) used data from Peters and Harper (1985) to demonstrate that 19 percent unsupplemented casein (about 17 percent crude protein) in the diet was necessary to give 95 percent of the maximum growth response and that about 26 percent unsupplemented casein (23 percent crude protein) was needed to produce 100 percent of the maximum growth response. Given that 1.11 times as much casein as lactalbumin is required to support maximum gain, the requirement for 95 percent maximum growth response of rats fed lactalbumin is about 15 percent crude protein. This concentration is adopted as the protein requirement for rats fed a diet containing a balanced protein source and 4 kcal ME/g (17 kJ ME/g). Additional studies with nonlinear-response models and rapidly growing rat strains are needed to refine this requirement. In practice, natural-ingredient diets that contain 18 to 25 percent crude protein have supported high rates of postweaning growth.

PROTEIN AND MAINTENANCE

Although protein requirement declines with age after weaning, the problem has not been studied extensively (Forbes and Rao, 1959; Hartsook and Mitchell, 1956). Hartsook and Mitchell (1956) estimated from carcass analyses that the requirement declined from about 28 percent of the diet (14 mg net protein[1]/kJ GE) at 30 days of age to 10 percent (about 5 mg net protein/kJ GE) at 50 days of age. The higher value agrees with that calculated from analysis of rat milk (Luckey et al., 1954). Using carcass nitrogen as the dependent variable, Sheehan et al. (1981) found that a dietary protein concentration averaging 4 percent was required for 12-month-old Sprague-Dawley female rats. Baldwin and Griminger (1985) were able to maintain nitrogen balance in 12- and 24-month-old male rats with an amino acid mixture simulating casein provided in the diet at 4.5 percent to 6.0 percent. Dibak et al. (1986) found that minimal concentrations of casein and wheat

---

[1] "Net protein" is protein retained in the body for use in maintenance and production.

gluten of 4.86 percent and 7.12 percent were required for positive nitrogen balance of 6-month-old male rats. Therefore the maintenance requirement is about 5 percent protein when the source is of high quality. In natural-ingredient diets a concentration of 7 percent crude protein is suggested by Bricker and Mitchell (1947).

AMINO ACIDS AND GROWTH

As with estimation of the protein requirement, it is necessary to consider the energy concentration of the diets when estimating the amount of each amino acid needed to support growth (Wretlind and Rose, 1950; Rosenberg and Culik, 1955). The sample amino acid patterns given in Table 2-8 are intended for use in a diet that contains 5 percent fat. Extrapolation of the requirements to diets of different caloric densities can probably be safely made by maintaining a constant amino acid:energy ratio and allowing for variations in amino acid digestibility (Kornberg and Endicott, 1946; Guthneck et al., 1953; Schweigert and Guthneck, 1953, 1954; Lushbough et al., 1957; Rogers and Harper, 1965).

Amino acid requirements are related to dietary protein concentration (Grau, 1948; Almquist, 1949; Brinegar et al., 1950; Becker et al., 1957; Bressani and Mertz, 1958). In general, the requirement for an amino acid, expressed as a percent of the diet, tends to increase as dietary protein

concentration increases but may remain constant or decrease slightly when expressed as percent of protein (Forbes et al., 1955; Bressani and Mertz, 1958).

As with protein quality assessment, a nonlinear model best describes the growth response of rats fed varying amounts of amino acids (Yoshida and Ashida, 1969; Heger and Frydrych, 1985; Gahl et al., 1991). A nonlinear model most accurately describes the diminishing-returns portion of the response curve. Heger and Frydrych (1985) and Gahl et al. (1991) used different nonlinear models to assess the maintenance and maximum response of young rats to dietary concentrations of individual amino acids. Gahl et al. (1991) added incrementally a mixture of amino acids to the diet to obtain a growth response rather than adding a test amino acid to a diet devoid of that test amino acid to obtain a growth response. The test amino acid was incorporated into the amino acid mixture at a concentration 35 percent below that of the other amino acids. Addition of incremental amounts of the mixture to the diet was used to obtain a growth response to the test amino acid. This approach ensured that the limiting amino acid remained first limiting. Figure 2-2 shows the response curves for nitrogen gain as a function of lysine and sulfur amino acid intake. These curves were generated from data reported by Gahl et al. (1991) and Benevenga et al. (1994). Similar curves for each indispensable amino acid were used to generate the requirement estimated to support growth (Table 2-9). Estimates were also made for the amount of amino acid required for nitrogen gain based on carcass nitrogen gain. The estimated amino acid requirements based on nitrogen gain were 1.1 to 1.7 times those required for weight gain. Because weight gain per se does not reflect a change in body composition, nitrogen gain may be a more dependable response criterion. The substitution value of tyrosine for phenylalanine and cystine for methionine could not be estimated from the results used to generate the requirements shown in Table 2-9. The replacement of phenylalanine by tyrosine was determined by Stockland et al. (1971) by comparing the growth of rats fed diets containing phenylalanine alone as part of an amino acid mix and with five phenylalanine:tyrosine ratios. They found the requirement for phenylalanine alone was 0.70 percent of the diet, while that for phenylalanine plus tyrosine was 0.69 percent of the diet. Tyrosine without phenylalanine would not support growth, and at least 0.38 percent L-phenylalanine had to be in the diet for tyrosine to be of benefit. Tyrosine could provide 45 percent of the aromatic amino acid requirement. Estimates of the replacement value of cystine for methionine have been made (Sowers et al., 1972; Stockland et al., 1973). The requirement for methionine alone was 0.49 percent of the diet and cystine could replace 48 to 58 percent of methionine. The diet had to have at least 0.17 percent methionine for cystine to be of benefit. Rat growth was between 4 and 5.5 g/day in these studies. The estimates

TABLE 2-8 Examples of Amino Acid Patterns Used in Studies with Purified Diets Containing 5 Percent Fat

Amino Acid	Amount	
	g/kg diet	mg/g nitrogen
Arginine	4.3	254
Histidine	2.8	163
Isoleucine	6.2	367
Leucine	10.7	626
Lysine	9.2	540
Methionine	6.5	381
Cystine	3.3	190
Phenylalanine	6.8	397
Tyrosine	3.4	198
Threonine	6.2	366
Tryptophan	2.0	115
Valine	7.4	435
Alanine	4.0	235
Aspartic acid	4.0	235
Glutamic acid	40.0	2,351
Glycine	6.0	353
Proline	4.0	235
Serine	4.0	235
Asparagine	4.0	235

NOTE: This pattern was demonstrated by Gahl et al. (1991) to support weight gain of 6 g/day for 65 g rats over a 21-day period.

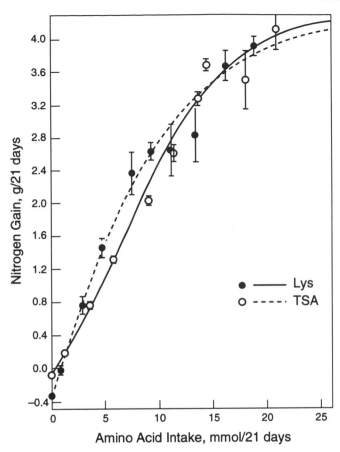

FIGURE 2-2 Nitrogen gain response curves generated using the parameter estimates for the logistic equation as described by Gahl et al. (1991) for rats fed diets limiting in one indispensable amino acid. Each data point represents the mean (SEM) for four rats.

TABLE 2-9   Comparison of National Research Council Estimates of Indispensable Amino Acid Requirements of Rats for Growth

| | Amount, g/kg diet | | |
L-Amino Acid	NRC, 1972	NRC, 1978	This Report
Arginine	6.0	6.0	4.3
Histidine	3.0	3.0	2.8
Isoleucine	5.5	5.0	6.2
Leucine	7.5	7.5	10.7
Lysine	9.0	7.0	9.2
Methionine[a]	6.0	6.0	9.8
Phenylalanine[b]	8.0	8.0	10.2
Threonine	5.0	5.0	6.2
Tryptophan	1.5	1.5	2.0
Valine	6.0	6.0	7.4

NOTE: Comparison is based on diet with 90 percent dry matter.

[a]One-half of L-methionine may be replaced by L-cystine (see NRC, 1978).

[b]One-half of L-phenylalanine may be replaced by L-tyrosine (see NRC, 1978).

means, respectively: arginine, 4.3, 2.7; histidine, 2.8, 2.3; isoleucine, 6.2, 5.3; leucine, 10.7, 7.9; lysine, 9.2, 7.4; total sulfur amino acids, 9.7, 7.5; total aromatic amino acids, 10.1, 7.2; threonine, 6.2, 4.5; tryptophan, 2.0, 1.6; valine, 7.4, 7.5 (M. Gahl and N. J. Benevenga, University of Wisconsin, personal communication, 1994).

The difference between the total nitrogen requirement and the essential amino acid nitrogen requirement should be made up with mixtures of nonessential amino acids. Stucki and Harper (1962) reported that amino acid diets that contained both essential and nonessential amino acids supported greater growth in rats than diets that contained only essential amino acids. Ratios of essential amino acid nitrogen to nonessential amino acid nitrogen of 0.5 to 4.0 were satisfactory in diets that contained 9.4 to 15.0 percent protein. Arginine, asparagine, glutamic acid, and proline must be included in the nonessential amino acid mixture to support maximum growth (Breuer et al., 1964; Hepburn and Bradley, 1964; Ranhotra and Johnson, 1965; Rogers and Harper, 1965; Adkins et al., 1966; Breuer et al., 1966; Newburg et al., 1975). The responses to these amino acids are presumed to reflect the inability of the rat to synthesize the quantities required for rapid growth. However, as pointed out by Breuer et al. (1964) and Crosby and Cline (1973), rats appear to adapt to diets devoid of certain of the nonessential amino acids and resume nearly maximum growth.

It is evident that specific requirements for the nonessential amino acids cannot be given because of the metabolic relationships among them. Therefore, the values given in Table 2-8 represent a pattern that has been used successfully in studies with purified diets. The value of 40.0 g/kg for glutamic acid is based on the data of Hepburn and

of these replacement values may not be applicable to rats with high growth potential such as those used by Gahl et al. (1991).

The estimates for indispensible amino acids reported in earlier editions of this publication are, on average, 23 percent lower than the current estimates (see Table 2-9 for comparisons). The reason for this difference is the methods used to estimate the requirement. Estimates reported earlier (National Research Council, 1972, 1978) were based on significant differences between means by use of a multiple-range test, which produced values similar to results obtained with broken-stick models. Estimates made in this way will be lower than those made by the four-parameter logistical model and, as observed in the guinea pig (Chapter 5), may underestimate the requirement for indispensible amino acids by about 20 percent. A reanalysis of the original data of Benevenga et al. (1994) using these two analytical approaches gave the following requirement estimates, as grams per kilogram of diet, for 95 percent of Rmax (maximum rate) for growth versus multiple-range tests of group

Bradley (1964) and Breuer et al. (1964); that for asparagine is 4.0 g/kg, as found by Breuer et al. (1966) to be required for maximum growth. Proline at 4 g/kg is the concentration used by Adkins et al. (1966). To raise the dietary crude protein limit to that planned, a mixture of alanine, glycine, and serine can be used.

Amino acid imbalances and antagonisms can result in increased requirements for individual amino acids, an area reviewed by Harper et al. (1970) and Benevenga and Steele (1984). The effects of imbalances and antagonisms on the requirement for maximum growth may be small or nonexistent if dietary protein concentration is adequate, but the effect in diets that contain suboptimal concentrations of protein may be considerable. The immediate response of an imbalance is decreased food intake (Harper et al., 1970).

AMINO ACIDS AND MAINTENANCE

The determination of amino acid requirements for adult rats is difficult because of the flat dose-response curves that occur for many amino acids (Smith and Johnson, 1967; Said and Hegsted, 1970). The indispensable amino acid requirements for maintenance of adult rats are based on reports by Benditt et al. (1950), Smith and Johnson (1967), and Said and Hegsted (1970). The data for each amino acid from these reports were averaged and are expressed on the basis of metabolic body size as follows (mg/$BW_{kg}^{0.75}$): histidine, 23.5; isoleucine, 90.4; leucine, 53.1; lysine, 32.2; methionine, 67.2; phenylalanine, 54.5; threonine, 53.1; tryptophan, 15.6; and valine, 67.1. Assuming a basal energy requirement of 117 kcal/$BW_{kg}^{0.75}$ (490 kJ/$BW_{kg}^{0.75}$) for a 300-g rat, these data have been incorporated into Table 2-2 as g/kg of diet.

GESTATION AND LACTATION

Nelson and Evans (1953) reported that 5 percent protein as unsupplemented casein was the minimum amount needed to support reproduction, while optimal performance occurred at 15 to 20 percent. Later, Nelson and Evans (1958) reported that 18 percent casein supported maximum growth in suckling young but that 24 percent was required to provide for weight gain in the dam during lactation. Supplementary cystine was added to both diets. Sucrose was used as the source of carbohydrate, a factor that may have influenced food intake, and thus protein utilization, at the lower protein concentrations in their studies (Harper and Elvehjem, 1957; Harper and Spivey, 1958). Gander and Schultze (1955) reported that 15 to 16 percent protein derived from a combination of casein, methionine, and mixed cereals supported reproduction and lactation in rats. More recently, Turner et al. (1987) compared the protein requirements for growth and reproduction in Sprague-Dawley female rats. Protein intakes of 8.6 percent (as whole-egg powder) delayed puberty but met the needs for subsequent reproductive function except pup weight. Concentrations of 15.6 percent (the highest used in their study) were needed for maximum growth from weaning to breeding. Subsequent response surface analysis revealed that a concentration of 21 percent would have been needed for maximum responses with a minimum concentration of between 9 to 11 percent whole-egg powder. It seems that the net protein requirement for gestation and lactation as a percentage of the diet does not differ significantly from that for growth of weanling rats (see Table 2-2).

The amino acid requirements for gestation and lactation have not been studied in depth. A concentration of 0.11 percent tryptophan in diets that contained 1 or 2 percent nitrogen (6.25 to 12.5 percent crude protein) was found to be adequate to support normal pregnancy in rats (Lojkin, 1967). Nelson and Evans (1958) reported that the sulfur amino acid requirement for lactation was 1 percent of the diet, one-half of which could come from cystine. Newburg and Fillios (1979) reported an apparent requirement for dietary asparagine in pregnant rats because its omission from the diet may have been associated with impaired neurological development of pups. Data are inadequate at this time to conclude that the concentration of amino acids in the diet required to support gestation and lactation exceed that required for growth in young rats.

SIGNS OF PROTEIN DEFICIENCY

Protein deficiency in young rats results in reduced growth, anemia, hypoproteinemia, depletion of body protein, muscular wasting, emaciation, and, if sufficiently severe, death. In adults a loss of weight and body nitrogen occurs (Cannon, 1948), and chronic deficiency may lead to edema (Alexander and Sauberlich, 1957). Estrus becomes irregular and may cease, fetal resorption occurs, and newborns are weak or dead. A lack of protein for pregnant and lactating rats may result in offspring that are stunted in growth (Hsueh et al., 1967) and have reduced concentrations of DNA and RNA in various tissues (Zeman and Stanbrough, 1969; Ahmad and Rahman, 1975). Low-protein diets also result in reduced food intake (Black et al., 1950). The reproductive capacity of the male is impaired by consumption of diets with inadequate concentrations of protein (Goettsch, 1949).

Removal of a single indispensable amino acid results in an immediate reduction in feed consumption, a situation that can return to normal within a day after replacement of that amino acid. A lack of an indispensable amino acid in the diet tends to be reflected in the concentration of the amino acid in the blood plasma (Longnecker and Hause, 1959; Kumta and Harper, 1962). Lack of specific amino acids has been reported to manifest as specific signs:

- lack of tryptophan—cataract formation, corneal vascularization, and alopecia (Cannon, 1948; Meister, 1957);
- lack of lysine—dental caries, impaired bone calcification, blackened teeth, hunched stance, and ataxia (Harris et al., 1943; Cannon, 1948; Kligler and Krehl, 1952; Bavetta and McClure, 1957; Likins et al., 1957; Meister, 1957);
- lack of methionine—fatty liver (Follis, 1958);
- lack of arginine—increased excretion of urinary urea, citrate, and orotate (Milner et al., 1974) and increased plasma and liver glutamate and glutamine (Gross et al., 1991).

The accumulation of a porphyrin-like pigment on the nose and paws has been observed in rats deficient in tryptophan, methionine, and histidine (Cole and Robson, 1951; Forbes and Vaughan, 1954), but this condition is also observed in other deficiency states.

MINERALS

Recommended mineral intakes have often been based on estimates of the amounts of minerals that promote maximum growth in short-term studies with little consideration of potential toxicological problems and of nutrient interactions. However, high safety margins added to recommended intakes of one mineral may affect requirements for another. For example, ingestion of extra calcium has been found to decrease absorption of iron and zinc (Greger, 1982, 1989), and excess intake of manganese may decrease iron utilization as these two elements are antagonistic (Davis et al., 1990).

Because of concerns about the consequences of feeding purified diets to rats for more than 6 months in studies on aging, hypertension, and cancer, a separate discussion of the role of dietary minerals in the development of nephrocalcinosis follows the discussions of the individual minerals.

MACROMINERALS

Six mineral elements occur in living tissues in substantial amounts and are commonly called "macrominerals" to distinguish them from mineral elements present in lesser quantities and designated as "trace elements." Although this distinction arose historically because of difficulties in the accurate analysis of the latter (Underwood and Mertz, 1987), it is still of some practical use because the trace elements are typically added to diets in premixes rather than via the formulated proportions of the primary ingredients.

Calcium and Phosphorus

The dietary requirements for calcium and phosphorus are closely linked and depend on the availability of each mineral from the dietary source. In the 1978 edition of *Nutrient Requirements of Laboratory Animals*, the recommendation for the minimal concentration of calcium and phosphorus to maximize bone calcification during growth was 5 and 4 g/kg, respectively. This gives a Ca:P molar ratio of 0.96. However, Bernhart et al. (1969) had shown previously that adequate mineralization could be accomplished with smaller amounts of dietary calcium and phosphorus. They held the dietary molar ratio of Ca:P to 0.91 and noted that Sprague-Dawley rats attained maximum weight gain when fed diets containing as little as 1.6 g Ca/kg, but the rats required at least 3.5 g Ca/kg to attain maximum body accumulation of calcium. In the same experiment, the amount of dietary phosphorus required to maximize body weight was ≥1.5 g/kg, and to maximize body phosphorus concentration it was ≥3.5 g/kg. Several groups of investigators have observed normal growth and tissue concentrations of calcium and phosphorus in bones and other tissues of rats (RIVm:TOX and Sprague-Dawley strains) fed from 2.85 to 3.2 g P/kg diet (Kaup et al., 1991b; Shah and Belonje, 1991). Ritskes-Hoitinga et al. (1993) fed rats 2.0 and 4.0 g P/kg diet with 5.2 g Ca/kg diet and 0.6 g Mg/kg diet for three successive generations. They found that 2.0 g P/kg diet sustained reproduction but delayed bone mineralization in offspring. Kaup et al. (1991b) noted that growth and bone calcium concentrations in male Sprague-Dawley rats fed 2.8 g P/kg diet increased as dietary calcium concentrations increased from 2.1 to 3.4 g Ca/kg diet but were similar for rats fed 3.6 and 4.4 Ca/kg diet.

These studies suggest that dietary concentrations of calcium and phosphorus at 3.5 and 3.0 g/kg, respectively, with a molar ratio 0.9, would be sufficient. However, other studies have shown that a larger Ca:P ratio is required to prevent specific abnormalities in Sprague-Dawley rats. Draper et al. (1972), for example, showed that a molar ratio of 1.5 was better than 0.8 for the prevention of osteoporosis.

Variations in calcium and phosphorus intake have been associated with soft tissue calcification, especially nephrocalcinosis, in rats. However, a variety of other dietary factors can influence the development of nephrocalcinosis.

The recommendations for calcium and phosphorus intakes reflect the somewhat conflicting needs to maximize growth and maximize bone calcium and phosphorus concentrations without inducing nephrocalcinosis. A Ca:P molar ratio of 1.3 is needed to prevent nephrocalcinosis in female rats. However, a dietary phosphorus concentration of more than 2.0 g/kg diet is needed to prevent inadequate bone mineralization in successive generations. Therefore, the recommended concentrations of dietary calcium and phosphorus under normal conditions for growth and maintenance of nonlactating rats are 5.0 and 3.0 g/kg, respectively.

Considering the demands placed on the dams during lactation, it might be prudent to increase the amount of dietary calcium and phosphorus during this period. It has been estimated that a lactating dam will produce 70 mL of milk per day (Brommage, 1989). This amounts to approximately 200 mg calcium and 140 mg phosphorus transferred to milk in a 24-hour period. Brommage (1989) showed that this demand for calcium and phosphorus was compensated for by increases in food intake and dramatic increases in intestinal absorption of these minerals. Ritskes-Hoitinga et al. (1993) showed that 2 g P/kg diet, compared to 4 g P/kg, sustained reproductive performance but delayed growth and bone mineralization in successive generations of female rats. To help relieve some of the stress of lactation, it is recommended that the dietary calcium and phosphorus be increased by 25 percent (6.3 g Ca, 3.7 g P/kg diet) during this period. This could be especially helpful for those females used in continuous-breeding programs. It should be mentioned, however, that when lactating and nonlactating rats were given a choice of diets containing various concentrations of calcium and phosphorus, the lactating rats chose a diet that contained a Ca:P molar ratio of 2, whereas the nonlactating rats chose a diet that contained a ratio of 1.5 (Brommage and DeLuca, 1984).

Factors Affecting Calcium and Phosphorus Requirements   Certain dietary factors can affect the biological availability of calcium and phosphorus and thus affect requirements for them in the diet. When these factors are present in the diet, appropriate adjustments to the dietary concentrations of calcium or phosphorus should be made. Low concentrations of vitamin D in the diet will reduce the absorption of calcium. Kaup et al. (1990) showed reduced phosphorus absorption in rats fed 10 g Ca/kg diet when compared to those fed only 2.5 g/kg. High dietary phosphorus will in turn reduce the apparent absorption of calcium (Schoenmakers et al., 1989) but may depend on the concentration of dietary magnesium (Bunce et al., 1965). Increasing the amount of fat in the diet from 5 to 20 percent reduced phosphorus absorption in older rats but not young rats (Kaup et al., 1990). High-fat diets also have been shown to decrease the absorption of calcium in mature rats but not young rats (Kane et al., 1949; Kaup et al., 1990). Calcium absorption is decreased in rats fed diets containing sources of oxalate (Weaver et al., 1987; Peterson et al., 1992), and phosphorus availability is reduced in diets containing phytate (inositol hexaphosphate) (Taylor, 1980; Moore et al., 1984). Some protein sources, including soybean protein isolates and other plant products, contain phytate. Phosphorus availability from these products should be considered when they are used in an animal's diet.

Other factors enhance calcium or phosphorus absorption. Bergstra et al. (1993) showed that the absorption of phosphorus was stimulated in rats fed diets high in fructose. Dietary disaccharides such as lactose and sucrose stimulate calcium absorption in rat intestine (Armbrecht and Wasserman, 1976). However, dietary lactose was better than sucrose in improving bone growth and development in intact vitamin D-deficient rats (Miller et al., 1988). Buchowski and Miller (1991) showed that 20 percent lactose added to diets containing a variety of calcium sources such as calcium carbonate, milk, and cheese significantly increased the amount of calcium in the tibia compared to rats fed diets without lactose. The enhancement was evident in 21-day-old rats fed the diets for 8 days but not in older rats.

Protein sources contain varying amounts of phosphorus, and the bioavailability of this phosphorus may not be the same in all sources. It is advisable, therefore, to analyze the source before using it in the diet and to be aware of whether the phosphorus is biologically available. Although few data are available in rats (Moore et al., 1984), data obtained for common feedstuffs in other species may be useful (National Research Council, 1988, 1994).

Signs of Calcium and Phosphorus Deficiency   Boelter and Greenberg (1941, 1943) fed 0.1 g Ca/kg diet to young rats for 8 weeks. The rats exhibited growth retardation, decreased food consumption, increased basal metabolic rate, reduced activity and sensitivity, osteoporosis, rear leg paralysis, and internal hemorrhage. Males failed to mate; females did not lactate properly. Day and McCollum (1939) fed 0.17 g P/kg diet to young rats. The animals survived up to 9 weeks and exhibited lethargy, pain, and cessation of bone growth with massive losses of calcium in urine.

When Sprague-Dawley rats were fed diets with moderate restrictions in calcium (2.1 to 3.4 g Ca/kg diet) for 28 days, the rats grew normally but had reduced bone weight and bone calcium concentrations (Kaup et al., 1991b). Improved calcium absorption and reduced urinary calcium losses allowed the rats to compensate for these marginally low calcium intakes, but these mechanisms would not be sufficient to prevent growth retardation if lesser amounts of calcium had been fed.

Chloride

Voris and Thacker (1942) obtained a reduction of 25 percent in growth in a 10-week, paired-feeding comparison of rats fed 0.2 g Cl/kg diet as compared to 2.9 g Cl/kg diet. Picciano (1970) found no increase in weight gain of young rats fed 2 g Cl/kg compared to rats fed 0.5 g Cl/kg. Miller (1926) reported that 5 mg/day (0.17 to 0.25 g Cl/kg diet) was acceptable for reproduction and lactation. On the basis of these limited studies, the estimated requirement is 0.5 g

Cl/kg diet, but future work may indicate this can be reduced.

Signs of Chloride Deficiency   The rat tenaciously conserves its supply of tissue chloride by reducing drastically the urinary excretion within hours of consuming a diet deficient in the element. As a result, the signs of deficiency develop slowly. Rats fed 0.12 g Cl/kg diet exhibited poor growth, reduced efficiency of feed utilization, reduced blood chloride, reduced urinary chloride excretion, and increased blood CO_2 content (Greenberg and Cuthbertson, 1942). Rats fed 0.5 g Cl/kg diet for more than 70 days had reduced growth, reduced tissue chloride concentrations, and extensive kidney damage; but some rats survived for more than a year (Cuthbertson and Greenberg, 1945).

Signs of Chloride Toxicity   Rats are also relatively insensitive to excess dietary chloride as judged by growth and tissue composition. Dahl (salt sensitive) and Sprague-Dawley rats fed excess chloride (15.6 to 26.6 g Cl/kg diet) as sodium or potassium chloride grew normally with unchanged chloride concentrations in kidneys and muscle (Whitescarver et al., 1986; Kaup et al., 1991a,c). However, the Sprague-Dawley rats fed 15.6 g Cl/kg had elevated blood pressure and enlarged kidneys (Kaup et al., 1991a,c). Dahl (salt sensitive) rats fed 4.86 g Cl/kg as NaCl also had elevated blood pressures, while rats fed a basal diet or a diet supplemented with $NaHCO_3$ did not (Kotchen et al., 1983).

Magnesium

Magnesium is required for numerous physiological functions in the rat. The amount required in the diet for adequate nutrition of the rat depends on numerous factors that affect availability of magnesium—the most important being the amount of dietary calcium, phosphorus, and vitamin D present. McAleese and Forbes (1961) found that a diet containing 0.1 g Mg/kg supported normal growth in weanling Sprague-Dawley rats. However, a diet containing 0.35 to 0.425 g/kg was required to maintain normal plasma magnesium concentrations of approximately 20 mg/L. More recently, Brink et al. (1991) found no significant difference in plasma magnesium between rats fed 0.4 and 0.6 g/kg diet; however, bone magnesium concentration was slightly (5 percent) but significantly higher in rats fed 0.6 g Mg/kg diet. On the other hand, rats fed only 0.2 g Mg/kg diet had significantly lower plasma and bone concentrations of magnesium than those fed 0.6 g/kg; 20 and 10 percent differences, respectively. Previously, Martindale and Heaton (1964) reported that 0.4 g Mg/kg was not sufficient to maintain normal bone and serum magnesium in adult Sprague-Dawley rats.

Other studies have shown questionable extremes for magnesium requirement. For example, Smith and Field (1963) estimated that a diet containing 0.05 g Mg/kg would replace endogenous losses in hooded Lister rats, but Clark and Belanger (1967) observed that a diet containing 2.5 g Mg/kg was needed for normal bone histology in Holtzman rats. These extremes may be the result of dietary factors that affect magnesium bioavailability. Brink et al. (1991) showed that magnesium absorption in rats fed diets with soybean protein was significantly less than in rats fed similar diets but with casein. Apparently this effect was caused by the presence of phytate in the soybean protein because when phytate was added to a casein diet, there was a similar reduction in absorption of magnesium. On the other hand, when lactose was added to the casein diet, magnesium absorption was enhanced.

Dietary calcium and phosphorus also affect magnesium requirement. O'Dell et al. (1960) showed that high-phosphorus diets enhanced the signs of magnesium deficiency in rats. Bunce et al. (1965) found that magnesium absorption was reduced by high dietary phosphorus. High dietary calcium also will depress magnesium absorption (O'Dell et al., 1960; Hardwick et al., 1987). Brink et al. (1992) have shown, however, that the effects of calcium and phosphorus on magnesium absorption is probably caused by the two minerals complexing magnesium in the gut lumen and rendering it insoluble. They showed that calcium was ineffective when phosphorus was not present and vice versa. [For an excellent review of factors affecting magnesium absorption, see Hardwick et al. (1991).]

Based on the review of these studies, and keeping the dietary calcium concentration at 5 g/kg and phosphorus at 3 g/kg, the dietary requirement for magnesium for growing and mature, nonpregnant rats is set at 0.5 g/kg diet. However, if diets contain factors, such as phytate, that might reduce the absorption of magnesium, a slightly higher dietary concentration might be required. In addition, because of the demands of lactation, an increase in dietary magnesium during this period is recommended. Wang et al. (1971) found that Sprague-Dawley rats were able to sustain gestation and lactation whether fed 0.8 g Mg/kg or 1.9 g Mg/kg diet.

Signs of Magnesium Deficiency   Signs of magnesium deficiency in growing Sprague-Dawley rats include vasodilation, hyperirritability, cardiac arrhythmias, spasticity, and fatal clonic convulsions. Vasodilation occurred after about 1 week and often disappeared and reappeared spontaneously. Convulsions occurred between 21 and 30 days (Kunkel and Pearson, 1948; Ko et al., 1962). In Sprague-Dawley rats renal calcification was common and was detected within 2 days after initiating a markedly deficient diet (Reeves and Forbes, 1972).

Tufts and Greenberg (1938) reported that lactating females fed a deficient diet bred successfully but did not suckle their young. Hurley et al. (1976a,b) reported that magnesium-deficient Sprague-Dawley dams resorbed their fetuses or bore malformed pups; they suggested that the malformations resulted from a concomitant zinc deficiency.

Potassium

In two studies, rats grew normally when fed 1.7 to 1.8 g K/kg diet (Kornberg and Endicott, 1946; Grunert et al., 1950). Two other studies indicated that rats may require 5 to 6 g K/kg diet during lactation (Heppel and Schmidt, 1949; Nelson and Evans, 1961); however, diets designed by the American Institute of Nutrition (1977) that contain 3.6 g K/kg support adequate growth and reproduction. Studies to determine potassium requirements for the rat are limited. Until more extensive research has been done, the minimal requirement is estimated to be 3.6 g K/kg diet. The dietary concentration of potassium may be increased to 5 g/kg during lactation, but such an increase does not seem necessary based on available data.

The potassium requirement of different strains of rats may vary. Sato et al. (1991) showed that increasing the dietary potassium to 42 g/kg diet protected spontaneously hypertensive rats against elevation of blood pressure induced by the ingestion of 8 percent sodium chloride. However, Sprague-Dawley rats developed elevated blood pressure when fed 15.5 g Cl/kg diet either as sodium chloride or potassium chloride (Kaup et al., 1991a,b).

Signs of Potassium Deficiency   Insufficient potassium markedly reduces appetite and growth. Animals become lethargic and comatose and may die within 3 weeks. They have an untidy appearance, cyanosis, short fur-like hair, diarrhea, distended abdomens with ascites, and are frequently hydrothoracic. Rats fed diets containing 1 g K/kg diet had a symmetrical loss of hair along the back and a 50 percent reduction in hair per follicular group (Robbins et al., 1965). Pathological lesions are widespread with potassium depletion (Schrader et al., 1937; Kornberg and Endicott, 1946; Newberne, 1964). The initial noninflammatory degeneration of myocardial fibers is followed by necrosis and cellular infiltration. Renal lesions include cast formation in proximal convoluted tubules, sloughing of tubular epithelium in the medulla, and accumulation of hyalin droplets in the epithelium of the collecting tubules.

Signs of Potassium Toxicity   Pearson (1948) fed weanling rats 5 percent potassium in the diet as potassium bicarbonate, and after 3 weeks observed reduced growth rate. More than 60 percent mortality was observed when dietary magnesium was low and only 17 percent when magnesium was adequate. Drescher et al. (1958) showed electrocardiograph changes when potassium intake exceeded 10 mg/kg

body weight. Potassium toxicity can cause hypertrophy of the adrenal zona glomerulosa, sodium depletion, and increased density of mitochondrial cristae in the kidney tubules (Hartroft and Sowa, 1964; Sealey et al., 1970; Pfaller et al., 1974).

Sodium

Grunert et al. (1950) estimated the sodium requirement of the rat to be 0.50 g Na/kg and to be independent of potassium intake. Forbes (1966) found that 0.48 g Na/kg diet was inadequate for maximum weight gain of weanling Sprague-Dawley rats over a 28-day period but that 2.20 g Na/kg gave maximum gains. Intermediate concentrations were not tested.

Pregnant Sprague-Dawley females fed low-sodium diets (0.3 g Na/kg diet) ate less food and showed languor and debility, particularly during the last week of pregnancy; however, reproduction was not seriously impaired (Kirksey and Pike, 1962). Ganguli et al. (1969a,b) suggested that the sodium requirement for gestation and lactation was 0.50 g Na/kg diet. The estimated requirement for growth, maintenance, gestation, and lactation is 0.5 g Na/kg diet.

Signs of Sodium Deficiency   The classic sodium-deficiency syndrome was described by Orent-Keiles et al. (1937). Rats fed a diet that contained 20 mg Na/kg diet exhibited growth retardation, corneal lesions, and soft bones. Males became infertile after 2 to 3 months, and sexual maturity was delayed in females. Death ensued in 4 to 6 months. At a concentration of 70 mg Na/kg diet, Kahlenberg et al. (1937) noted reduced appetite, poor growth, increased heat production, and reduced stores of energy, fat, and protein.

Signs of Sodium Toxicity   Rats are relatively insensitive to excess sodium as indicated by growth and tissue composition. Sprague-Dawley rats fed 10.1 to 11.2 g Na/kg diet as chloride, sulfate, carbonate, and bicarbonate salts grew normally and had concentrations of sodium in bones and kidneys similar to rats fed a basal diet but had higher concentrations of sodium in bone than rats fed the potassium forms of these salts (Kaup et al., 1991a). However, ingestion of excess sodium as NaCl is associated with elevated blood pressure in several different rat strains, including Sprague-Dawley, Dahl (salt sensitive), Dahl (salt resistant treated with deoxycorticosterone acetate), and spontaneously hypertensive rats (SHR) (Kaup et al., 1991a,c; Tobian, 1991).

TRACE MINERALS

Of the many trace mineral elements found in foods, only seven—copper, iodine, iron, manganese, molybdenum, selenium, and zinc—have been unequivocally demonstrated to be required by rats. Although there is some evidence that other mineral elements (such as chromium, lithium, nickel, sulfur, and vanadium) may be required, further research is needed to establish requirement amounts. These other elements are treated in the discussion "Potentially Beneficial Dietary Constituents."

Copper

There is a general consensus that weanling and adult rats housed individually in wire-bottom stainless steel cages have a dietary copper requirement on the order of 5 to 6 mg/kg diet, the value used in the AIN-76A and AIN-93 diets. Johnson et al. (1993) fed weanling male Sprague-Dawley rats casein-starch-based purified diets containing copper concentrations ranging from 0.2 to 5.4 mg/kg for 5 weeks. They found that numerous functional measures of copper status (including platelet cytochrome c oxidase activity, serum ceruloplasmin activity, plasma copper, copper, and zinc-superoxide dismutase activity) were depressed in rats fed diets with copper concentrations at or below 3 mg/kg.

Studies by Klevay and Saari (1993) showed similar results. When weanling male rats were fed dietary copper ranging from 0.2 to 5.2 mg/kg diet for 5 weeks, the concentrations of copper in liver and heart and serum ceruloplasmin activity were depressed in rats fed copper concentrations at less than 4 mg/kg diet. However, Failla et al. (1988) could find no significant differences in copper status of Lewis rats fed an egg white-starch-based purified diet containing 2.1 or 7.0 mg Cu/kg diet. When similar rats were fed diets with sucrose instead of starch, indicators of copper status were significantly reduced at dietary concentrations up to 2.9 mg Cu/kg when compared to diets containing 7.1 mg Cu/kg. These studies suggest that the minimal dietary requirement for copper, to maintain adequate copper status in young growing rats of different strains and different dietary conditions, is more than 4 mg/kg diet.

Spoerl and Kirchgessner (1975a,b) reported that increasing dietary copper from 5 to 8 mg/kg during pregnancy and lactation resulted in an improvement in serum copper, serum ceruloplasmin activity, and liver copper concentrations in dams and their offspring. Cerklewski (1979) reported higher copper concentrations in milk on day 14 postpartum and in pup liver on day 21 postpartum when dams were fed diets containing 9 versus 6 mg Cu/kg diet during pregnancy and lactation. The functional significance of the higher tissue copper concentrations was not investigated.

Based on these data, a dietary copper concentration of 5 mg/kg diet is recommended for growth and maintenance for a variety of rat strains and under different dietary conditions. A dietary concentration of 8 mg Cu/kg diet is recommended for pregnancy and lactation. It should be noted that the requirement for growth and maintenance has not changed from that reported in the 1978 edition of this volume. However, the requirement recommended for pregnancy and lactation has increased from 5 to 8 mg Cu/ kg diet. High dietary concentrations of zinc, cadmium, and ascorbic acid may increase the dietary requirement for copper (Davis and Mertz, 1987).

Signs of Copper Deficiency Copper deficiency develops rapidly in young rats and slowly in adult rats fed diets containing less than 1 mg Cu/kg. The copper deficiency develops more rapidly when fructose or sucrose is fed than when starch is fed, and males show a greater sensitivity to copper deficiency than females with respect to both the severity of signs and the duration of time until the onset of signs of deficiency (Fields et al., 1986; Koh, 1990; C.G. Lewis et al., 1990). Copper deficiency during early development can result in significant abnormalities in the cardiovascular, nervous, skeletal, reproductive, immune, and hematopoietic systems (Keen et al., 1982; Davis and Mertz, 1987). Weanling and adult rats fed copper-deficient diets can develop alterations in platelet function, impairments in both the acquired and innate arms of the immune system, altered exocrine pancreatic morphology and function, anemia (if dietary iron is marginal), alterations in thromboxane and prostaglandin synthesis, and impaired cardiovascular function (Cohen et al., 1985; Koller et al., 1987; Kramer et al., 1988; Dubick et al., 1989; Johnson and Dufault, 1989; Babu and Failla, 1990a,b; Allen et al., 1991; Medeiros et al., 1992; Saari, 1992). Weanling rats fed diets containing less than 2.7 mg Cu/kg show impaired neutrophil function within 5 weeks (Babu and Failla, 1990a).

Signs of Copper Toxicity Rats are particularly tolerant of high concentrations of dietary copper. Boyden et al. (1938) observed no adverse effects after feeding rats diets containing 500 mg Cu/kg, rats fed diets containing 2,000 mg/ kg showed marked weight loss, and diets in excess of 4,000 mg/kg resulted in severe anorexia and starvation. Evidence of liver and kidney pathology in rats fed diets containing more than 1,000 mg/kg has been reported (Haywood, 1979; Czarnecki et al., 1984; Haywood, 1985).

Iodine

Iodine is regarded as essential for the rat. Utilization of dietary iodine is very high, and absorption occurs all along the gastrointestinal tract. It is concentrated by many tissues (Gross, 1962) but functions primarily as an integral part of the thyroid hormones. The few studies done to determine the iodine requirement generally agree that it is between 100 and 200 μg/kg diet (Levine et al., 1933; Remington and Remington, 1938; Halverson et al., 1945; Parker et al.,

1951). Parker et al. (1951) found that 100 to 230 µg I/kg diet was satisfactory for reproduction; natural-ingredient diets that contained 330 µg/kg supported reproduction (Kellerman, 1934). The 1978 edition of this report set the iodine requirement at 150 µg/kg diet. Many commercially available natural-ingredient diets contain this amount and support adequate growth and reproductive performance in rats. There have been no recent studies to suggest that this concentration should be changed.

Signs of Iodine Deficiency   The most obvious sign of iodine deficiency in the rat is enlargement of the thyroid glands with the formation of trabecular-type nodules (Taylor and Poulson, 1956). Iodine-deficient rats also had more coarse and less dense hair than controls. Iodine deficiency results in impaired reproduction (Feldmann, 1960). One biochemical sign of iodine deficiency is a decrease in serum concentrations of the thyroid hormone thyroxine (T_4) (Abrams and Larsen, 1973); another is a dramatic increase in the concentrations of serum thyrotropin (Pazos-Moura et al., 1991). An increase in the activity of liver type I iodothyronine deiodinase also occurs in iodine-deficient rats. Arthur et al. (1991) showed that the activity of this enzyme in rats fed less than 100 µg I/kg diet was 60 percent higher than in controls fed 1,000 µg/kg.

Signs of Iodine Toxicity   The rat has a relatively high tolerance for dietary iodine. Adult female rats fed 500 to 2,000 mg/kg diet during pregnancy had increased neonatal mortality (Ammerman et al., 1964). However, when fed concentrations approaching 500 mg/kg diet, female rats had decreased milk production. The fertility of male rats fed as much as 2,500 mg I/kg diet for 200 days from birth was not affected.

Iron

For weanling and adult rats the iron requirement for growth and maintenance of maximum hemoglobin concentration is on the order of 35 mg/kg (McCall et al., 1962a; Ahlström and Jantti, 1969). For reproduction McCall et al. (1962b) reported that a diet containing 240 mg Fe/kg was adequate for maximum weight gain and hemoglobin concentrations. Ahlström and Jantti (1969) reported that considerably less iron was needed for reproduction: 28 mg Fe/kg for maximum hemoglobin concentration and 58 mg/kg for maximum iron stores in the offspring. Lin and Kirksey (1976) reported that growing, pregnant rats required between 10 and 50 mg Fe/kg diet for maximum hemoglobin concentration and between 50 and 250 mg/kg for maximum fetal liver iron stores. Shepard et al. (1980) reported that diets containing less than 10 mg/kg resulted in loss of embryos and fetuses. Kochanowski and Sherman (1983) have argued that although 35 mg Fe/kg diet is adequate

to maximize maternal weight gain and hemoglobin concentrations during pregnancy and lactation, iron concentrations between 75 and 250 mg/kg diet are needed for optimal iron status at the end of lactation and between 150 and 250 mg Fe/kg are needed to maximize iron stores in the pups. Given the above, the recommended dietary iron concentration during pregnancy and lactation is 75 mg/kg diet.

Signs of Iron Deficiency   In addition to anemia, iron-deficient rats can be characterized by multiple abnormalities including hyperlipidemia and low tissue carnitine concentrations (Bartholmey and Sherman, 1986), growth failure, elevated resting metabolic rate (Tobin and Beard, 1990; Borel et al., 1991), reduced exercise capacity (Willis et al., 1990), low milk folate concentrations (O'Connor et al., 1990), a compromised immune system including impaired phagocytosis and natural killer cell activity, and reduced antibody production (Hallquist et al., 1992; Kochanowski and Sherman, 1984, 1985).

Signs of Iron Toxicity   The general term used to describe iron toxicity in animals is "iron overload." Chronic consumption of large amounts of dietary iron results in an accumulation of large quantities of iron in various cells and tissues, especially the liver. Wu et al. (1990) fed young (4 month old) and old (20 month old) rats 25 g elemental Fe/kg diet as finely powdered carbonyl iron and found detrimental effects on growth and maintenance of body weight. Within 2 weeks of feeding the high-iron diet, young rats had stopped gaining weight and old rats had lost 12 percent of their body weight. Weight of the old rats continued to decrease for up to 10 weeks of feeding. Large increases in the concentrations of iron in liver and spleen were seen in both age groups. Reductions in concentrations of serum, liver, and heart copper were also observed in the iron-overloaded rats. Britton et al. (1991) observed similar results in rats fed 30 g Fe/kg diet. These large amounts of tissue iron result in lipid peroxidation and cellular damage (Houglum et al., 1990; Wu et al., 1990).

Manganese

There is a paucity of studies that address manganese requirements in rats. Holtkamp and Hill (1950) reported that the optimal manganese intake for growth is between 2 and 5 mg/kg diet; when dietary manganese concentration was increased to 40 mg/kg, the average weight gain was less than in the group fed 5 mg/kg diet. In contrast, Anderson and Parker (1955) reported a faster growth rate in weanling rats fed 50 mg Mn/kg compared to rats fed 5 mg/kg. The manganese requirement for reproduction has not been firmly established. Diets containing 1 mg Mn/kg are inadequate for normal reproduction; litters from dams fed this concentration of manganese are characterized by ataxia (due to inner ear defects), skeletal defects, and a high incidence of early postnatal death (Hurley and Keen, 1987). Litters from dams fed diets containing 3 mg Mn/kg have normal survival and growth rates; however, depending on the strain, they can still be characterized by a high incidence of ataxia (Baly et al., 1986; Hurley and Keen, 1987). An increased incidence of ataxia was not observed in litters from Sprague-Dawley dams fed diets containing 5 mg Mn/kg (C. L. Keen, University of California, Davis, personal communication, 1992). The above data suggest that a dietary concentration of 5 mg Mn/kg is probably adequate for normal growth and development. However, because there is a significant difference in how different strains respond to dietary manganese intake (Hurley and Bell, 1974; Kawano et al., 1987), the requirement is set at 10 mg/kg. It should be noted that this is lower than the recommendation of 50 mg/kg (National Research Council, 1978); however, given the lack of data supporting the need for such a high concentration of manganese in the diet, coupled with the possible negative effects of excess manganese on iron metabolism (Davis et al., 1990), reduction in the manganese requirement is warranted.

High concentrations of dietary iron, calcium, phosphorus, and copper have been reported to increase the requirement for dietary manganese (Hurley and Keen, 1987; Johnson and Korynta, 1992).

Signs of Manganese Deficiency   Diets containing less than 1 mg Mn/kg can result in reduced food consumption, poor growth, bone abnormalities, and early mortality. Reproduction can be impaired and is characterized by testicular degeneration in the male and by a delay in the opening of the vaginal orifice and defective ovulation in the female. If reproduction occurs, litters are characterized by ataxia, skeletal defects, marked abnormalities in glucose and lipid metabolism, and a high frequency of early postnatal death (Hurley and Keen, 1987). Manganese deficiency in weanling and adult rats can result in significant alterations in pancreatic exocrine and endocrine functions (Baly et al., 1985; Chang et al., 1990), impaired glucose transport and metabolism in adipose cells (Baly et al., 1990), an increase in tissue lipid peroxidation (Zidenberg-Cherr et al., 1983), abnormal lipoprotein metabolism (Davis et al., 1990; Kawano et al., 1987), decreased hepatic arginase activity (Brock et al., 1994), and marked inhibitions of osteoblast and osteoclast activities resulting in severe bone disease (Strause et al., 1987).

Signs of Manganese Toxicity   The postnatal growth of rats is unaffected by dietary manganese intakes as high as 1,000 to 2,000 mg/kg diet, provided dietary iron is adequate. If dietary iron is low (<20 mg/kg), dietary manganese concentrations in excess of 1,000 mg/kg can result in reduced weight gain and iron deficiency (Rehnberg et al., 1982). Diets in excess of 3,500 mg Mn/kg can result in severe growth retardation and mortality. Reproductive dysfunction resulting from long-term intake of excess manganese (>1,050 mg/kg) has been reported for both males and females (Laskey et al., 1982). Although the concentrations of dietary manganese needed for overt toxicity are quite high, weanling rats given water containing 55 µg Mn/mL for 3 weeks were reported to have reduced rates of brain RNA and protein synthesis (Magour et al., 1983). The mechanisms underlying the cellular toxicity of manganese have not been clearly identified but may involve manganese-initiated oxidative damage, disturbances in carbohydrate metabolism, and altered intracellular iron metabolism (Keen and Hurley, 1989).

Molybdenum

Molybdenum metabolism was reviewed by Mills and Davis (1989). Similar criteria used to establish selenium as an essential nutrient also can be used to establish the essentiality of molybdenum. Molybdenum is a cofactor for three known enzymes in the rat—xanthine oxidase/dehydrogenase (XDH), aldehyde oxidase (AO), and sulfite oxidase (SOX) (Rajagopalan, 1988). These enzymes catalyze redox reactions. When rats are fed diets with very low concentrations of molybdenum, activities of these enzymes in various tissues are depressed. However, it has not been demonstrated that this is detrimental to the animal. On the other hand, if rats are fed tungsten, an antagonist to molybdenum, the activities of molybdenum-dependent enzymes are scarcely measurable and signs of deficiency then become apparent. Genetic deficiencies of sulfite oxidase in humans have been shown to result in numerous pathologies (Mudd et al., 1967; Johnson et al., 1980; Abumrad et al., 1981).

Early studies used the effect of low dietary molybdenum on liver and intestinal XDH activities to establish the requirement for molybdenum. Studies by Higgings et al. (1956) concluded that 20 µg Mo/kg diet was sufficient to maintain normal growth and reproduction. Xanthine oxidase activity was impaired, however. Titration experiments showed that about 100 µg Mo/kg diet was required to maximize intestinal XDH activity. More recent studies showed that not all criteria affected by dietary molybdenum are maximized by the same concentrations of molybdenum. Wang et al. (1992) showed that 25 µg/kg diet satisfied growth requirements in female rats and that 50 µg/kg diet was the minimal concentration required to maintain maximum liver XDH and SOX activity as well as spleen and kidney molybdenum concentrations. It took 100 µg Mo/kg diet to maximize XDH activity in the intestine and 200 µg Mo/kg diet to maximize liver and brain concentrations. However, it may not be valid to use tissue concentration of a trace element as a criterion to establish requirements. In many instances, the element accumulates in the tissue above a certain concentration but does not have a physiological function. Based on these studies, the requirement for molybdenum is estimated to be about 150 μg/kg diet. There is a strong interaction among molybdenum, copper, and sulfur; thus the dietary requirement of molybdenum might depend on the amount of copper and sulfur in the diet.

Signs of Molybdenum Deficiency Outward signs of deficiency are difficult to produce when rats are fed purified diets with only molybdenum absent. Even when an antagonist of molybdenum, tungsten, was fed at a ratio of 2,000:1 (tungsten:molybdenum) there were no detrimental effects on weight gain, but the animals had very low activity concentrations of molybdenum-dependent enzymes in liver.

Signs of Molybdenum Toxicity The occurrence of signs of molybdenum toxicity when rats and other species are fed high concentrations depends on the amount of copper and sulfate in the diet. Gray and Daniel (1964) showed that as little as 10 mg Mo/kg diet caused weight loss in copper-deprived rats. This condition could be ameliorated with the addition of 3 mg Cu/kg diet. Miller et al. (1956) found that the reduction in growth rate of rats fed 100 mg Mo/kg diet could be prevented by sulfate supplementation. Molybdenum toxicity also causes elevated liver copper, decreased serum ceruloplasmin, and increased tissue concentrations of molybdenum. Most of these signs can be reversed by supplementing the diet with copper and/or sulfate.

Selenium

Selenium is found in living organisms as an integral part of selenoproteins (Sunde, 1990) in the form of selenomethionine or selenocystine. There are also selenium binding proteins. A number of important selenoenzymes have been discovered in mammalian systems—glutathione peroxidase (GSH-Px; Rotruck et al., 1973), selenoprotein P (^{75}Se-P; Burk and Gregory, 1982), phospholipid hydroperoxide GSH-Px (Ursini et al., 1985), and hepatic type I iodothyronine 5′-deiodinase (ITD-I; Berry et al., 1991). GSH-Px is found in most tissues and cells and catabolizes hydrogen peroxide and other free and membrane associated hydro- and phospholipid peroxides. ITD-I is found in liver, kidney, and thyroid and catalyzes the generation of 3,5,3′-triiodothyronine (T_3), the metabolically active thyroid hormone, from T_4, the main circulating thyroid hormone. Other functions for selenium are probable but have not been adequately defined (Beckett et al., 1989; Kim et al., 1991). Since the 1978 edition of *Nutrient Requirements of Laboratory Animals* (National Research Council, 1978),

there have been many excellent reviews on selenium metabolism (Burk, 1983; National Research Council, 1983; Combs and Combs, 1984; Levander and Burk, 1990; Sunde, 1990).

Dietary sources of selenium can be of two forms—inorganic, represented by selenite or selenate, and organic, represented by selenomethionine or selenocystine. Because of the complex metabolic fate of these different forms, and because of the influence of other possible antioxidants in the diet, it is difficult to establish an exact dietary requirement for selenium. Selenium is more readily transported across the intestinal cells as selenate than as selenite. Selenium from selenomethionine is more readily transported than either selenate or selenite (Vendeland et al., 1992).

Various criteria have been used to assess the requirement for selenium. Schwarz and Foltz (1957) showed that 40 μg Se/kg diet was required to prevent nutritional liver necrosis in rats that were also deficient in vitamin E. Hafeman et al. (1974) found that 50 μg Se/kg diet permitted maximum growth in rats, but 100 μg/kg was required to maintain maximum tissue activity of GSH-Px. More recent studies (Arthur et al., 1990) showed that less than 5 μg Se/kg diet permitted growth not different from that found when 100 μg Se/kg diet was fed. However, liver and plasma GSH-Px activities decreased significantly after only 2 weeks on low-selenium diets. The vitamin E content of these diets, supplied as α-tocopheryl acetate, was 200 mg/kg.

Yang et al. (1989) used the concentration of selenoprotein P (SeP) (Yang et al., 1987) in plasma and GSH-Px activities in plasma and liver to assess selenium nutriture in rats. Weanling rats were fed diets supplemented with selenium as sodium selenate ranging from 10 to 2,000 μg/kg diet. The SeP concentrations in plasma reached a plateau between 100 and 500 μg Se/kg of diet but they continued to rise when 2,000 μg Se/kg was fed. GSH-Px activities in plasma and liver were not maximized until dietary selenium had reached 500 μg/kg diet.

Whanger and Butler (1988) fed rats selenium, from 20 to 4,000 μg/kg, as sodium selenite. GSH-Px activities for numerous tissues, except red blood cells, were maximized at 200 μg Se/kg diet. Pence (1991) fed rats diets with selenium concentrations of 20, 120, and 520 μg/kg; Pence found that selenium-dependent GSH-Px as well as total GSH-Px activities in liver and colon of rats fed 120 μg/kg were only about 50 percent of the activities in those fed 520 μg/kg. L'Abbé et al. (1991) found that liver GSH-Px activity was about 20 percent higher in rats fed 1,000 μg Se/kg for 25 weeks than in those fed 100 μg/kg, but the difference was not significant. This may suggest, however, that 100 μg/kg diet is close to the minimum needed for maximum activity of GSH-Px and that more than 100 μg/kg would be optimal. In this regard Eckhert et al. (1993)

found evidence that the microvasculature of rats may have a unique requirement for selenium. They fed male rats diets high in sucrose to induce an elevation in blood triglycerides and cholesterol—a feeding regimen used by Lockwood and Eckhert (1992) to cause insult to the microvascular system. To this diet 100 or 200 μg Se/kg was added. The results showed that dietary selenium concentration had no effect on GSH-Px activity in the erythrocytes; however, there was a marked effect on the microvasculature of the retinae. In two different experiments there were twofold increases in the number of acellular segments and in the number of vessels over the optic nerve head in rats fed 100 μg Se/kg diet compared to those fed 200 μg Se/kg diet. In addition, the inner retinal pericyte:endothelial cell ratio of the vessels was increased in rats fed the higher concentration of selenium. These authors interpreted these results to suggest that the higher concentration of dietary selenium protected the retinal microvasculature, particularly the pericyte cells, from sucrose-induced metabolic insult. These data suggest that a minimal dietary requirement for selenium is more than 100 μg/kg.

It has been suggested that GSH-Px activity might be the best criterion for establishing a dietary requirement for selenium. However, recent investigations by Sunde et al. (1992) and Evenson et al. (1992) showed that the concentrations of GSH-Px-mRNA and ^{75}Se incorporation into selenoproteins may also be used. They fed selenium (as selenite) in concentrations ranging from 7 to 200 μg/kg diet in a titration scheme to determine the amount of dietary selenium required to maximize GSH-Px activity, GSH-Px mRNA concentration, and ^{75}Se incorporation into liver selenoproteins of growing rats. In all three cases, 100 μg/kg was the minimal amount required. Vadhanavikit and Ganther (1993) used liver and thyroidal 5′-deiodinase (type I) activities as well as GSH-Px activities to determine selenium requirements for the rat. They fed rats diets containing 10, 50, 100, and 500 μg Se/kg for 20 weeks. Liver GSH-Px activity was significantly decreased in rats fed 10, 50, and 100 μg Se/kg than in those fed 500 μg Se/kg; however, liver 5′-deiodinase activity was significantly decreased (90 percent) only in rats fed 10 μg Se/kg. GSH-Px activity in the thyroid was decreased in rats fed 10 μg Se/kg but not in those fed other concentrations. Thyroidal 5′-deiodinase activity was not significantly affected even at the lowest concentration of dietary selenium.

Therefore, considering all the criteria mentioned for the establishment of the selenium requirement, it is suggested that 150 μg Se/kg diet is the minimal requirement for the growing rat. This concentration may also be used for maintenance.

Other investigators have provided evidence which suggests that the minimal selenium requirement in the form of selenite for pregnant and lactating rats might be higher than 150 μg/kg diet (Smith and Picciano, 1986, 1987). In one experiment, Smith and Picciano (1986) raised dams through pregnancy and lactation on four concentrations of dietary selenium: 25, 50, 100, and 200 μg/kg. At 15 days of gestation, erythrocyte selenium concentrations and GSH-Px activities of the dams were not different among the three highest concentrations of dietary selenium. At day 18 of lactation, erythrocyte GSH-Px activity was significantly higher in dams fed 200 μg Se/kg compared to those fed 100 μg Se/kg diet. At this period, selenium concentration and GSH-Px activity in the liver of nonpregnant controls was not different among the three highest concentrations of selenium. In the lactating females, however, liver selenium concentration and GSH-Px activity were different among all groups. The highest selenium concentrations and GSH-Px activities were in those lactating rats fed 200 μg Se/kg; however, this value was not as high as that from nonpregnant controls fed the same amount. GSH-Px activity in the liver of 18-day-old pups was 1.7 times greater when dams were fed 200 μg Se/kg diet than when they were fed 100 μg/kg diet.

Smith and Picciano (1987) also found that the form of dietary selenium could influence selenium bioavailability and dietary requirement. A concentration of 250 μg Se/kg diet as selenomethionine resulted in maximum GSH-Px activity in the tissues of both dams and pups. However, when selenium was supplied as sodium selenite, 500 μg Se/kg was necessary. Studies by Lane et al. (1991) showed that the livers of 14-day-old pups fed 150 μg Se/kg diet as selenomethionine had twice as much GSH-Px activity as livers from similar rats fed selenite. GSH-Px activity in the livers of dams was not different between the two sources. Whanger and Butler (1988) and Vendeland et al. (1992) also showed that selenium from selenomethionine was more available to rats than that from sodium selenite. Based on these criteria, it appears that when selenium is supplied as selenite, the minimal requirement during pregnancy and lactation is at least 400 μg Se/kg diet. If the dietary source of selenium is selenate or selenomethionine, the requirement could be less.

Signs of Selenium Deficiency   Outward signs of selenium deficiency are difficult to produce in rats fed diets adequate in vitamin E. However, one report (McCoy and Weswig, 1969) demonstrated selenium-deficiency signs in rats fed Torula yeast diets with adequate vitamin E. These signs included poor growth, sparse-hair coats, cataracts, and reproductive failure when the diets were fed for two generations. Caution should be exercised when interpreting these results, however. Male rats fed the Torula yeast diet with added selenium only gained about two-thirds as much as rats fed a commercial natural-ingredient diet during the first generation and less than one-half as much during the second. This indicates that the diet may have been lacking in other essential nutrients. Biochemical signs of selenium deficiency include the reduction of selenium-dependent enzyme activities in various tissues and the reduction of T_4 deiodination in liver.

Signs of Selenium Toxicity   Studies by Harr et al. (1967) and Tinsley et al. (1967) showed that 4,000 to 16,000 μg Se/kg diet caused ascites, edema, and poor hair quality in rats fed purified diets for long periods. Many of the animals did not live beyond 100 days. Liver toxicity and hyperplastic hepatocytes were found in rats receiving selenium as selenite or selenate supplemented at 500 to 2,000 μg/kg diet for 30 months.

Zinc

Weanling and adult rats housed individually in wire-bottom stainless steel cages have a dietary zinc requirement on the order of 12 mg/kg when egg white or casein is used as the primary protein source (Williams and Mills, 1970; Pallauf and Kirchgessner, 1971; Wallwork et al., 1981). The requirement is higher (18 mg/kg) when soybean protein is used. The increased requirement for zinc with soybean protein based diets is primarily attributed to the phytic acid content of these diets (Berger and Schneeman, 1988).

For optimal growth and survival of the neonate, the dietary requirement for zinc for the pregnant and lactating rat has been estimated to be on the order of 25 mg/kg, even when a high quality protein such as egg white is used (Rogers et al., 1984). An increased requirement for dietary zinc during pregnancy is also supported by the observation of Fosmire et al. (1977) that, at term, fetal weight and fetal zinc concentrations were higher in litters from dams given water containing 25 mg Zn/L compared to dams given water containing 11 mg Zn/L. Given the above, the recommended concentration of dietary zinc for pregnant and lactating dams is 25 mg Zn/kg diet.

The dietary requirement for zinc can be significantly influenced by an animal's housing conditions. It should be noted that high concentrations of dietary cadmium, iron, phosphorus, and tin have been reported to increase the requirement for dietary zinc (Hambidge et al., 1986; Johnson and Greger, 1984; Sandstrom and Lönnerdal, 1989).

Signs of Zinc Deficiency   The pathologic signs of zinc deficiency depend on the length and severity of the deficiency, the age and sex of the animal, and environmental surroundings. An inadequate intake of zinc can be reflected by marked reductions in plasma zinc concentrations within 24 hours and by mild-to-severe anorexia within 3 days (Hambidge et al., 1986). Prolonged consumption of a zinc-deficient diet can result in continued anorexia, growth retardation/failure, abnormalities in platelet aggregation and hemostasis, alopecia, thickening of the epidermis, increased rates of cell membrane lipid peroxidation, hyperir-

ritability, significant impairment of multiple components of the immune system, and alterations in lipid, carbohydrate, and protein metabolism (Hambidge et al., 1986; Hammermueller et al., 1987; Emery et al., 1990; Keen and Gershwin, 1990; O'Dell and Emery, 1991; Avery and Bettger, 1992). Esophageal lesions can occur in weanling rats within 7 days of the introduction of a zinc-deficient diet (Diamond et al., 1971). For weanling males, the prolonged consumption of a diet containing less than 0.5 mg Zn/kg can result in arrested spermatogenesis, atrophy of the germinal epithelium, and impaired growth of accessory sex organs (Diamond et al., 1971; Hambidge et al., 1986). Zinc deficiency in females results in a disruption of the estrous cycle, a reduced frequency of mating, and a low implantation rate if mating occurs.

The consumption of a zinc-deficient diet after mating can result in severe embryonic and fetal pathologies including prenatal death; a high incidence of central nervous system, soft tissue, and skeletal system defects; and abnormal biochemical development of the lung and pancreas. Zinc-deficient dams are characterized by severe parturition difficulties and the offspring are characterized by lower-than-normal growth rates, a high incidence of early postnatal death, and behavioral abnormalities (Apgar, 1985; Bunce, 1989; Keen and Hurley, 1989).

Signs of Zinc Toxicity   Zinc is often considered to be relatively nontoxic, however, dietary zinc concentrations in excess of 250 mg/kg can induce a copper deficiency if dietary copper is low or marginal (L'Abbé and Fischer, 1984; Keen et al., 1985); and zinc in excess of 5,000 mg/kg diet can result in reduced growth rates, anorexia, anemia, and death even when dietary copper is considered adequate (Hambidge et al., 1986). It should be noted that investigators often use control diets that contain 100 mg Zn/kg. Although this amount of zinc is not thought to represent a "toxic" risk to rats, it is high enough that it may prevent the detection of important nutrient and/or drug-zinc interactions. It may be more accurate to classify 100 mg Zn/kg diet as a zinc-supplemented diet.

NEPHROCALCINOSIS IN RATS FED PURIFIED DIETS

Nephrocalcinosis is histologically demonstrable in the rat as deposits of stainable calcium salts in the kidney, usually in the corticomedullary region. Urolith formation causes increased kidney calcium and phosphorus concentrations and eventually results in renal hypertrophy and heavier kidneys (Woodard and Jee, 1984). Sometimes kidney function is reduced (Ritskes-Hoitinga, 1992).

The etiology is complex. Females are more susceptible than males, and Sprague-Dawley and Wistar strains may be more susceptible than other strains (Ritskes-Hoitinga, 1992). Dietary factors are also important. Generally, rats fed purified diets are more apt to develop nephrocalcinosis than rats fed commercially available stock diets (Ritskes-Hoitinga et al., 1991). As yet no common mechanism has been identified that explains all the dietary factors that have been related to the incidence of nephrocalcinosis. Thus a brief review of potential factors is warranted.

Phosphorus and Calcium

Rats (Sprague-Dawley, Wistar, RIVm:TOX, and Zucker; both males and females) fed purified diets with more than 5.0 g P/kg diet have, in a number of studies, been found to have elevated concentrations of calcium in their kidneys (Hitchman et al., 1979; Schaafsma and Visser, 1980; Woodard and Jee, 1984; Greger et al., 1987a; Ritskes-Hoitinga et al., 1989; Van Camp et al., 1990). Kidney calcification has also been noted in female rats (RIVm:TOX, Sprague-Dawley, and Wistar) fed 4.0 g P/kg diet (Shah et al., 1980; Mars et al., 1988; Schoenmakers et al., 1989; Henskens et al., 1991). For example, Schoenmakers et al. (1989) observed that 5-week-old female RIVm:TOX rats fed diets containing 4.3 g P/kg with 4.8 g Ca/kg and 0.4 g Mg/kg accumulated 25-fold more calcium in their kidneys than rats fed 1.1, 1.9, or 2.8 g P/kg diet. Ritskes-Hoitinga et al. (1993) did long-term studies with three successive generations of rats fed casein-based purified diets containing 5.2 g Ca/kg, 0.6 g Mg/kg, and either 2.0 or 4.0 g P/kg diet. After 4 weeks of age, from 50 to 100 percent of the females fed the diet containing 4 g P/kg developed nephrocalcinosis, while only 2 of 54 rats fed 2.0 g P/kg had measurable nephrocalcinosis. Male rats were not affected.

The amount of calcium consumed and the ratio of dietary calcium to phosphorus are also important. Woodard and Jee (1984) found that ingestion of additional calcium (5.5 versus 3.5 g Ca/kg diet) by Sprague-Dawley rats fed moderately high concentrations of phosphorus (>5.5 g P/kg diet) increased the deposition of calcium in the kidneys. In contrast, Schaafsma and Visser (1980) observed that Zucker rats fed diets with low concentrations of calcium (2.0 g Ca/kg diet) were more sensitive to phosphorus-induced nephrocalcinosis than rats fed 6.0 g Ca/kg diet. Moreover, Hoek et al. (1988) observed that kidney calcium accumulation in RIVm:TOX rats dropped dramatically when dietary calcium was increased to 7.5 g Ca/kg diet. This may reflect the ratio of dietary calcium to phosphorus.

Investigators have noted that a dietary calcium:phosphorus molar ratio below 1.3 is associated with nephrocalcinosis in RIVm:TOX and Wistar strains of rats (Hitchman et al., 1979; Hoek et al., 1988). Ritskes-Hoitinga et al. (1991) compared the responses of Wistar rats to 10 commercial diets and found that the dietary ratios of calcium to phosphorus were inversely correlated to the degree of nephrocalcinosis, as determined by histological score. Reeves et al. (1993a) fed rats purified diets similar to the

AIN-93G diet (Reeves et al., 1993b) but with varying amounts of calcium and phosphorus: 3.3, 5.9, and 6.7 g Ca/kg diet and 2.0, 3.0, and 4.0 g P/kg diet. The Ca:P molar ratio in all diets was 1.3. They found that the concentrations of calcium and phosphorus in the tibia of both male and female rats were similar to those in rats fed a commercial natural-ingredient diet that contained higher concentrations of calcium and phosphorus. In addition, there were no indications of nephrocalcinosis in female rats after they consumed these diets for 16 weeks.

Magnesium

Nephrocalcinosis is a sign of magnesium deficiency in laboratory rats. Several investigators have found that the addition of supplemental magnesium (above required or recommended amounts) reduced the accumulation of calcium in kidneys of rats (Sprague-Dawley and Wistar strains) fed higher concentrations of calcium and/or phosphorus (Goulding and Malthus, 1969; Shah et al., 1980; Ericsson et al., 1986; Shah et al., 1986). Although ingestion of generous amounts of phosphorus and/or calcium have been found to depress magnesium absorption (Greger et al., 1987b; Hoek et al., 1988) and sometimes serum magnesium concentrations (Ericsson et al., 1986) of rats (RIVm:TOX and Sprague-Dawley strains), kidney magnesium concentrations were not reduced. This suggests that the rats were not magnesium deficient per se.

Protein

The ingestion of additional (25 or 30 percent versus 15 percent) protein was found to prevent phosphorus-induced nephrocalcinosis in rats in several studies (Hitchman et al., 1979; Van Camp et al., 1990). Similarly, Shah et al. (1986) observed that Sprague-Dawley rats fed purified diets containing recommended concentrations of phosphorus had less calcium deposited in their kidneys when the protein content of the diets was increased from 10 to 15 percent casein.

The substitution of lactalbumin for casein in semipurified diets, even if dietary phosphorus amounts are similar, also is associated with less accumulation of calcium in kidneys of Sprague-Dawley rats (Greger et al., 1987a,b). Zhang and Beynen (1992) found that an increased intake of protein, provided in the diet by soybean isolate or casein, reduced the incidence of calcinosis in female rats; however, protein from fish meal did not. They concluded that the antinephrocalcinogenic effect of the soybean protein was related to lower urinary phosphorus, and the effect of casein was the result of lower urine pH and elevated urinary magnesium.

Other Dietary Factors

Bergstra et al. (1993) showed that dietary fructose, as opposed to glucose, stimulated nephrocalcinosis in female rats. This was related to fructose stimulating greater concentrations of urinary phosphorus and magnesium and lowering the pH.

Levine et al. (1974) also found that chloride depletion stimulated nephrocalcinosis in Charles River rats but only in the presence of increased dietary phosphate or sulfate. Kootstra et al. (1991) observed that supplementing purified diets that contained 6 mg P/kg with ammonium chloride, but not ammonium sulfate, reduced the accumulation of calcium in the kidneys of female Wistar rats. Supplementation of diets with fluoride has also been observed to decrease the accumulation of calcium in the kidneys of Sprague-Dawley rats in several studies (Shah et al., 1980; Ericsson et al., 1986; Shah et al., 1986; Cerklewski, 1987).

VITAMINS

In the conversion of many of the values to moles from international units or mass that appeared in the original literature, the values reported may not be an exact conversion. The molar values have been rounded to reflect the degree of precision present in the original published estimates. Conversion factors for molar, mass, and IU units of the vitamins are presented in Appendix Tables 3 and 4.

FAT-SOLUBLE VITAMINS

Vitamin A

Vitamin A is essential for many critical functions of the body such as vision, which requires 11-*cis*-retinaldehyde bound to the photoreceptor pigments. Many cellular differentiation processes are mediated by all-*trans*-retinoic acid and 9-*cis*-retinoic acid bound to their respective nuclear

TABLE 2-10   Equivalence of β-Carotene and Retinol at Different Concentrations

Dose of β-Carotene, μmol/kg BW	Relative Molar Biopotency, %	Moles β-Carotene Equivalent to 1 mole Retinol	μg β-Carotene Equivalent to 1 mole Retinol
<0.3	100.0	1.0	1.84
1	50.0	2.0	3.75
2	30.0	3.3	6.25
6	15.4	6.5	12.20
12	9.5	10.5	19.70
48	4.3	23.3	43.60

SOURCE: Adapted from Brubacher and Weiser (1985).

TABLE 2-11   Vitamin A Repletion of Vitamin A-Deficient Rats

Criteria for Repletion	Retinol Required, nmol/kg BW/day	Reference
Normal growth rate	14	K. C. Lewis et al., 1990; Moore, 1957
Maintenance of differentiation of vaginal epithelial cells	19–23	Guilbert et al., 1940
Maintenance of normal testes	20–40	Bieri et al., 1968
Maintenance of a positive vitamin A balance	100	Green, personal communication, 1992
"Natural" hepatic storage[a]	200	Moore, 1957

[a]Defined as liver reserves equal to those found in wild animals.

receptors, RAR and RXR (Zelent et al., 1989; Mangelsdorf et al., 1992). 14-Hydroxy-4,14-*retro*-retinol also has been shown to be involved in signal transduction in B lymphocytes (Buck et al., 1991).

Retinol, the retinyl esters, and β-carotene are the main dietary compounds present in diets with vitamin A activity. Several plant carotenoids are precursor forms of vitamin A, and β-carotene is the most active carotenoid. The concentration of dietary intake influences the biopotency of β-carotene (Brubacher and Weiser, 1985) as shown in Table 2-10. β-Carotene is transformed to retinol in the intestinal mucosa. The retinol, irrespective of its source, is esterified primarily with palmitate or stearate. The esters are transported to the parenchymal cells of the liver as components of the chylomicrons. The esters are either hydrolyzed and transported out of the liver to the target tissues in combination with a specific transport protein, retinol-binding protein, or they may be transferred to the stellate cells of the liver for storage. Vitamin A can be stored in the liver in large amounts.

Rats are born with very low liver stores of vitamin A. As a consequence the vitamin A requirement in weanling rats varies according to the criteria used, overt signs of deficiency (e.g., epithelial keratinization), hepatic storage, or retinol kinetics. Maximum blood concentrations (about 2 μmol/L) were reached when liver deposition was moderate (140 μmol/kg liver) in Holtzman rats (Muto et al., 1972). An elevation in the cerebrospinal fluid pressure occurred when the serum retinol concentration dropped below 0.35 μmol/L in Sprague-Dawley rats (Corey and Hayes, 1972). Some of the different criteria for vitamin A requirements for repletion of deficient animals are presented in Table 2-11.

Guilbert et al. (1940) demonstrated that the need for vitamin A was related to body weight rather than energy intake. This concept is consistent with the vitamin's activity in maintaining integrity of the epithelia, which quantitatively directly correlate with body mass (Mitchell, 1950).

Takahashi et al. (1975) found that 56 nmol retinyl acetate/kg BW/day was sufficient to support gestation in Holzman rats with the delivery of pups of normal weight and with normal brain, liver, and kidney size. However, the mean number of live neonates was reduced to 4.8 compared to 9.5 for the controls, which received 1,400 nmol/kg BW/day. Sixty-three percent of the dams fed the marginal concentration of vitamin A delivered as compared to 100 percent of the dams fed the very high concentration. These low concentrations that supported gestation would not support optimal lactation, however. Davila et al. (1985) found that nursing Sprague-Dawley dams fed 2.1 μmol retinyl acetate/kg diet were healthy and their pups grew as well as the pups of dams fed 52 μmol/kg diet but had lower liver retinyl ester stores.

Vitamin A requirements are sensitive to other nutritional influences. Consumption of protein-deficient diets decreases the serum concentrations of vitamin A and its transport protein, retinol-binding protein (Peterson et al., 1974). Rates of depletion of liver and kidney reserves of vitamin A were linearly related to growth rate, which changed as dietary casein concentration varied from 0 to 18 percent (Rechcigl et al., 1962). In zinc-deficient rats, mobilization of vitamin A from the liver declined (Smith et al., 1973), but this effect was not confirmed by Apgar (1977). Vitamin E deprivation resulted in depletion of liver stores of vitamin A (Moore, 1957).

Age does not appear to have any significant effect on vitamin A requirement other than that related to differences in body weight. Suckling, young adult, and aged rats absorb retinol with about the same efficiency (Hollander and Morgan, 1979; Said et al., 1988).

Green and co-workers (1987) developed a computer model to describe the kinetics of vitamin A metabolism in male Sprague-Dawley rats and have studied the effects of vitamin A intake on vitamin A excretion. With diets containing 2.1 μmol/kg diet or less, the disposal was equal to the vitamin A intake. However, when the diets contained 2.4 μmol/kg diet the disposal rate was essentially the same as observed in the rats receiving 2.1 μmol/kg diet (Green and Green, 1991). The rats had a modest (≈3.5 nmol/day) accumulation of vitamin A in the liver when they were fed 2.4 μmol/kg diet.

The estimated requirement for vitamin A, based on the kinetic studies of Green et al. (1987) and Green and Green (1991) and on the lactation study of Davila et al. (1985) is 2.4 μmol/kg diet (equivalent to 2,300 IU/kg) if retinol or retinyl esters are used. This requirement may be met by retinol at 0.7 mg/kg diet, retinyl acetate at 0.8 mg/kg diet, or retinyl palmitate at 1.3 mg/kg diet. At this low concentration β-carotene is used relatively efficiently, so the requirement would also be met by β-carotene at 12.4 μmol/kg

diet (1.3 mg/kg diet). Because compounds with vitamin A activity are relatively unstable, the use of retinyl acetate or retinyl palmitate in gelatin coated beadlets is strongly recommended.

Several studies have shown that animals exposed to stress respond better at higher dietary concentrations of vitamin A than is recommended above. Gerber and Erdman (1982) reported that the strength of the scar tissue after a surgical incision was about twice as great in animals receiving 21 μmol retinyl acetate/kg diet or 13.4 μmol β-carotene/kg diet as it was in animals receiving 4.2 μmol retinyl acetate/kg diet. Demetriou et al. (1984) reported that rats fed a commercial diet containing 15 μmol retinol or retinyl esters per kg of diet and 12 μmol β-carotene/kg diet had only a 20 percent survival rate 72 hours after intra-abdominal sepsis. In contrast, when the diet was supplemented with an additional 525 μmol retinyl palmitate/kg, survival rose to 70 percent.

Signs of Vitamin A Deficiency   The various procedures used to produce vitamin A-deficient rats have been reviewed in detail (Smith, 1990). The signs of vitamin A deficiency can be divided into six categories.

1. Defect in vision. Because 11-*cis*-retinaldehyde is a necessary part of the visual pigments, a deficiency of vitamin A leads to a loss of vision through the lack of functional visual pigments (Wald, 1968).

2. Bone defects. Vitamin A deficiency leads to improper bone cell differentiation, which causes retardation and disorganization of bone growth and failure of bone resorption during remodeling. The reduced size of the openings in the bones can cause a secondary compression of nerves (Underwood, 1984).

3. Increase in cerebral spinal fluid pressure. The arachnoid villi, which release the fluid, become clogged with fibroblasts (Corey and Hayes, 1972).

4. Reproductive failure. Cessation of spermatogenesis occurs in the male. In the female, severe and lethal deficiency causes cornification of the reproductive tract, which results in loss of reproductive function. With a more moderate deficiency, females will become pregnant but severe lethal fetal malformations and resorption are the most common result of the pregnancy (Wilson et al., 1953).

5. Epithelial metaplasia and keratinization. All epithelia are sensitive to vitamin A deficiency to varying degrees. In early vitamin A deficiency, goblet cells and mucus formation decline in the intestine; squamous metaplasia followed by keratinization takes place in the trachea. Keratinization of the urogenital tract and the corneal epithelium combined with xerophthalmia and porphyrin deposits around the eyelids, and ultimate dissolution of the corneal stroma, takes place in severe vitamin A deficiency (Underwood, 1984).

6. Growth failure. After 5 to 6 weeks of vitamin A defi-

ciency, the weight of a weanling rat plateaus for about a week and then drops rapidly until the animal dies. Under germ-free conditions, the rat can survive at the weight plateau stage for several months (Rogers et al., 1971).

Signs of Vitamin A Toxicity   Acute retinol toxicity occurs with an intake of 180 μmol/kg BW/day, although the long-chain retinyl esters were not toxic at this concentration of intake (Leelaprute et al., 1973). All-*trans*-retinoic acid is much more toxic; signs were found at 47 μmol/kg BW/day (Kurtz et al., 1984). The signs typically associated with vitamin A toxicity are weight loss, fatty liver, hyperlipidemia, calcification of soft tissues, mobilization of bone calcium, bone fractures, increased urinary excretion of 3-methylhistidine, and hemorrhage. Vitamin A is also teratogenic, causing cleft palate in fetuses at a retinyl palmitate intake of 40 μmol/kg BW/day given on days 9 to 12 of gestation (Nanda et al., 1970). Retinoic acid was much more teratogenic when low-protein diets (2.5 to 10 percent) were fed compared to a 20 percent protein diet (Nolen, 1972). Hemorrhage was caused by an interference in vitamin K metabolism and could be corrected by increasing vitamin K intake (McCarthy et al., 1989). α-Tocopherol has been shown to greatly reduce the toxicity of vitamin A (Jenkins and Mitchell, 1975). In contrast to the retinoids, β-carotene was not toxic at doses up to 1,800 μmol/kg BW/day (Heywood et al., 1985).

Vitamin D

Vitamin D is an important precursor of the hormone 1,25-dihydroxycholecalciferol. The hormonal action is mediated by the interaction with a specific nuclear receptor and the corresponding responsive elements in the promotor regions of the genes that code for several proteins, many of which are involved in calcium metabolism. Although 1,25-dihydroxycholecalciferol is best known for its role in the regulation of calcium and phosphorus homeostasis, it may have significant roles in many other processes, including the synthesis of red blood cells, the proliferation of B and T lymphocytes, and the secretion of insulin and prolactin (Reichel et al., 1989). Vitamin D is hydroxylated in two positions before it becomes active. First, cholecalciferol is converted to 25-hydroxycholecalciferol in the liver. In the kidney 25-hydroxycholecalciferol is hydroxylated to form the active 1,25-dihydroxycholecalciferol. The formation of 1,25-dihydroxycholecalciferol is a tightly regulated process. Serum calcium concentration regulates the activity of the kidney 1-α-hydroxylase enzyme via the parathyroid gland. Low concentrations of serum phosphate directly increase the synthesis of this kidney enzyme. High dietary concentrations of both cause a decrease in the formation of 1,25-dihydroxycholecalciferol.

Information about the vitamin D requirements of the rat is surprisingly limited. The actual requirements are difficult to determine because the rat can absorb sufficient calcium and phosphorus to prevent the overt signs of rickets if the dietary ratio of calcium and phosphorus is about equal and they are present in adequate amounts. Also, cholecalciferol can be synthesized in the skin of the rat if it is exposed to ultraviolet light of appropriate wavelength (280 to 320 nm).

Relatively small amounts of vitamin D are needed to produce most of the biological responses, and the amount needed is a function of calcium and phosphorus intakes. A single injection of 0.13 μmol of either ergocalciferol (vitamin D_2) or cholecalciferol (vitamin D_3) promoted maximum active calcium absorption within 48 hours in vitamin D-deficient adult male rats of the Sherman strain fed a diet containing 5 g Ca/kg and 5 g P/kg (Schachter et al., 1961). Vitamin D is clearly required for maximum growth even when calcium and phosphorus are at concentrations typically considered optimal (Coward et al., 1932). However, vitamin D-deficient rats had normal bone mineralization and grew at the same rate as rats supplemented with vitamin D when they were continuously infused with calcium and phosphorus (Underwood and DeLuca, 1984). Male Sprague-Dawley rats fed a vitamin D-deficient diet containing 4.7 g Ca/kg and 3 g P/kg had low reproductive rates, but weekly injections of 5.2 μmol cholecalciferol/rat was sufficient to return reproductive function of deficient rats to normal (Kwiecinski et al., 1989). However, when male rats were fed a vitamin D-deficient diet containing 12 g Ca/kg and 3 g P/kg the fertility rate greatly improved (Uhland et al., 1992). Vitamin D-deficient female rats became pregnant and produced pups; nonetheless, the number of dams giving live births was only one-half that of dams given an oral dose of 1.6 μmol vitamin D/day (Halloran and DeLuca, 1980). Vitamin D-deficient rats were less likely to become pregnant, had more spontaneous abortions, and had a greater risk of death during parturition. In general, suckling pups grow poorly when nursed by vitamin D-deficient dams; however, this was the result of reduced milk production. When the litter size was reduced to two pups, the growth rate was normal (Mathews et al., 1986). Increasing the dietary calcium (16 g/kg) and phosphorus (14 g/kg) concentrations increased milk production so that suckling and weaned pups of vitamin D-deficient rats had normal growth and bone mineral content (Clark et al., 1987).

The rate of active absorption of calcium in the duodenum in middle-aged (12 to 14 months) and elderly rats (18 to 24 months) fed optimal amounts of calcium and phosphorus for a long period was reduced to about 70 percent of the rate observed in young rats (2 to 3 months) (Armbrecht, 1990). In contrast, the older rats actually absorbed cholecalciferol more efficiently than young rats (Hollander and Tarnawski, 1984). The activity of the kidney 1-α-hydroxylase enzyme, however, was reduced in the older rats (Ishida et al., 1987). The concentration of the enzyme can be increased by a 4-month exposure to a low-calcium and vitamin D-deficient diet (Armbrecht and Forte, 1985). In contrast, young rats respond rapidly to this stimulus. There is no evidence that increasing the intake of vitamin D will have a beneficial effect on the absorption of calcium or phosphorus in adult rats.

In the absence of additional data, the estimated requirement of 1,000 IU vitamin D/kg diet recommended in the 1978 edition of this volume (National Research Council, 1978) is retained. This is equivalent to 65 nmol cholecalciferol/kg diet (25 μg/kg). This concentration, although adequate, may not represent the biological minimum.

Signs of Vitamin D Deficiency   Vitamin D deficiency induces rickets. This disease is classically brought about in rats by a diet lacking vitamin D, adequate in calcium, and low in phosphorus. However, a low-calcium diet deficient in vitamin D has a more severe effect on growth rate and results in irritability, tetany, and decreased bone calcification (Steenbock and Herting, 1955). Bones of rachitic rats show decreased or absent calcification with wide areas of uncalcified cartilage at the junction of diaphysis and epiphysis. Bone ash may be less than half normal. A full description of histological changes in bones of vitamin D-deficient rats is given by Jones (1971).

Signs of Vitamin D Toxicity   The first sign of vitamin D toxicity is usually elevated serum calcium followed by calcification of the kidneys (Potvliege, 1962). Soon the arteries become calcified, and then the liver and heart become calcified. The animals usually die from heart failure secondary to the uremia that comes from kidney failure. The animals also have a decreased growth rate and show considerable resorption of bone. Massive arteriosclerotic lesions developed in 100 percent of male rats fed a diet containing 78,800 μmol ergocalciferol/kg, 39 mmol cholesterol/kg, and 12 mmol cholic acid/kg for 6 weeks. Rats receiving this diet but containing only 63,000 μmol ergocalciferol/kg did not develop lesions, and rats fed ergocalciferol without cholesterol did not develop lesions (Bajwa et al., 1971). Vitamin D is also teratogenic. Treatment of pregnant dams with 2,500 μmol ergocalciferol/day reduced the growth rate of the fetuses and retarded the ossification of the long bones (Ornoy et al., 1968). All the pups died shortly after birth. When the dose was lowered to 1,250 μmol ergocalciferol/day the placenta was much smaller but the pups were normal. Nonpregnant rats receiving this lower dose had very high serum calcium concentrations. At 250 μmol/day the nonpregnant animals appeared normal. Shelling and Asher (1932) demonstrated that the toxicity of vitamin D was related to the calcium and phosphorus content of the diet. Rats fed diets containing 4.4 g Ca/kg and 17.8 g P/kg had severe toxicity signs at relatively low vitamin D intakes. However, the animals were able to tolerate high concentrations of vitamin D when they were fed diets with an optimal ratio of calcium to phosphorus.

Vitamin E

"Vitamin E is nature's best fat-soluble antioxidant" (Scott, 1978). In fact the only clearly defined function of vitamin E is as an antioxidant. Most of the signs of vitamin E deficiency can be prevented by feeding the antioxidant *N,N'*-diphenyl-*p*-phenylene diamine (DPPD).

Several compounds have vitamin E activity. The most active naturally occurring compound is *RRR*-α-tocopherol (formerly called D-α-tocopherol). The synthetic all-*rac*-α-tocopherol is a mixture of eight stereoisomers, and the other seven are less active than *RRR*-α-tocopherol. One mole of *RRR*-α-tocopherol has a biopotency equivalent to 1.36 moles of all-*rac*-α-tocopherol (U.S. Pharmacopeia, 1985). Although it does not occur naturally, tocopheryl acetate is frequently used in animal diets. The ester is hydrolyzed in the intestine, and tocopherol is released for absorption. However, having the alcohol group linked to the acetate prevents the tocopherol from being destroyed in the diet before it is consumed by the animal. In this section the concentrations of all compounds having vitamin E activity are expressed as the equivalent concentration of *RRR*-α-tocopherol.

Polyunsaturated fatty acids are labile to autoxidation. Each fatty acid free-radical that is oxidized damages about three other polyunsaturated fatty acid molecules, thus producing a geometrically expanding chain reaction (Chow, 1979). Vitamin E can readily donate hydrogen atoms to the free-radicals to terminate the chain reaction. As a result the requirement for vitamin E is related to the dietary and tissue concentrations of the polyunsaturated fatty acids (Witting and Horwitt, 1964). Adequate dietary selenium will greatly reduce the requirement for vitamin E. The selenium-containing enzyme, glutathione peroxidase, will convert the peroxides that are intermediates in this breakdown process to stable alcohols, thus reducing the requirement for vitamin E (Hoekstra, 1975).

Several criteria have been used to evaluate vitamin E status of the rat including: survival, growth, prevention of nutritional muscular dystrophy, prevention of creatinuria, prevention of fetal resorption, prevention of testicular degeneration, reduction of pentane expiration, reduction of malondialdehyde production, and prevention of the spontaneous hemolysis of red blood cells after they are diluted with saline.

In selenium-deficient rats, Hakkarainen et al. (1986) found that about 5.2 mg α-tocopheryl acetate/kg diet (11 μmol/kg) was required for survival. Gabriel et al. (1980)

used plasma pyruvate kinase and glutamic oxaloacetic transaminase as indicators of myopathy of skeletal muscle. By feeding rats the antioxidant ethoxyquin (which is not stored) instead of vitamin E, they were able to rapidly produce vitamin E deficiency in rats from 12 to 68 weeks old by removing ethoxyquin from the diet. In all age groups the minimum vitamin E requirement to prevent myopathy of muscle was approximately 0.75 mg α-tocopheryl acetate/ kg BW/day (1.6 μmol/kg BW/day). Jager and Houtsmuller (1970) found that preventing the spontaneous hemolysis of red blood cells required 13.2 mg *RRR*-α-tocopherol/kg diet (28 μmol/kg diet) when the diet contained 3.6 percent linoleic acid, but increasing the linoleic acid content to 13 percent increased the requirement to 18 mg/kg diet (38 μmol/kg). In a test of repletion of vitamin E-deficient rats, Bieri (1972) showed that hemolysis was prevented after feeding 20 mg *RRR*-α-tocopheryl acetate/kg diet (42 μmol/ kg). The dietary lipid was a mixture of stripped corn oil and lard that provided 5.2 percent linoleic acid. At this percentage, α-tocopherol in the tissues reached a stable concentration within 8 weeks following the beginning of repletion. Buckingham (1985) studied the requirements of rats fed purified diets containing 20 percent lipid with polyunsaturated to saturated (P:S) ratios of 0.38, 0.82, and 2.30. Pentane expiration in the breath and production of malondialdehyde were reduced to normal levels when rats were fed a diet with 27 mg α-tocopherol/kg diet (62 μmol/ kg). However, 44 percent spontaneous hemolysis was observed when the diet contained a 2.30 P:S ratio even with an α-tocopherol concentration of 67 mg/kg diet (156 μmol/kg).

Evans and Emerson (1943) investigated the rats' requirement for vitamin E during reproduction. As female rats aged, the requirement to maintain pregnancy became higher. To maintain pregnancy and optimal health in the suckling young, 0.57 mg α-tocopheryl acetate/rat/day (1.2 μmol/rat/day) was required. However, after the third pregnancy the suckling pups suffered slight muscular impairment at this concentration of intake. In young male rats 0.18 mg/rat/day (0.39 μmol/rat/day) was adequate to maintain reproduction; but at 9 months and older, the rats required 0.57 mg/rat/day (1.2 μmol/rat/day) to maintain fertility. Ames (1974) determined the amount of vitamin E necessary to maintain pregnancy in female rats. The dose to maintain 50 percent fetal viability increased about sevenfold from the first pregnancy (12 weeks) to the fourth pregnancy (60 weeks). This increase is more than can be explained by an increase in body weight. Gabriel et al. (1980) suggested that the increased requirement may be caused by accumulated toxic products from long-term exposure to very low intakes of antioxidants.

Autofluorescent pigment (lipofuscin) accumulates in the tissues of vitamin E-deficient animals and in old animals. Nonetheless the tissue distribution of the lipofuscin is dif-

ferent in vitamin E-deficient rats than in aging rats (Katz et al., 1984), thus the accumulation in aging rats does not appear to be the result of inadequate vitamin E intake. Although older rats may require more dietary vitamin E to maintain α-tocopherol concentrations in the cerebellum and brain stem (Meydani et al., 1986), the concentration needed has not been established. Hollander and Dadufalza (1989) have demonstrated that older rats absorb α-tocopherol more efficiently than younger rats.

Bendich et al. (1986) examined the vitamin E requirement of spontaneously hypertensive rats, which are more sensitive to vitamin E-deficient diets than the parent Wistar strain. Maintenance of normal growth required 7.5 mg all-*rac*-α-tocopheryl acetate/kg diet (16 μmol/kg diet); 15 mg/kg diet was required to prevent myopathy of muscle; and 50 mg/kg diet was required to prevent the spontaneous hemolysis of red blood cells. Optimal immune responses appeared to require slightly more than 50 mg/kg diet. The immune system response was the most sensitive indicator of vitamin E status.

The vitamin E requirement for most of the frequently used strains of rats is 18 mg *RRR*-α-tocopherol/kg diet (42 μmol/kg) when lipids comprise less than 10 percent of the diet. This corresponds to 27 IU/kg diet. When all-*rac*-α-tocopheryl acetate is used as the dietary source, this would be equivalent to 27 mg/kg diet (57 μmol/kg).

High intakes of either retinyl palmitate (42 μmol/kg diet) or β-carotene (89 μmol/kg diet) depressed plasma and liver concentrations of α-tocopherol to about one-half the normal concentrations (Blakely et al., 1990). It seems probable that high concentrations of vitamin A in the diet interfere with the absorption of vitamin E.

Signs of Vitamin E Deficiency   Red blood cells from vitamin E-deficient rats show an increased hemolysis when diluted with saline or when treated with oxidizing agents (dialuric acid). Other signs are hyaline degeneration of skeletal muscle fibers with infiltration by histocytes, interfibrillar fat cells, and an increase in interstitial cells (Jager, 1972); accumulation of yellow pigment in smooth muscles; in the male, irreversible degeneration of the seminiferous epithelium of the testis, which occurs by age 40 to 50 days; in the female, induction of fetal abnormalities or intrauterine death and resorption; kyphoscoliosis (humpedback) (Machlin et al., 1977); rough coat; skin ulcers; neural lesions; and impaired learning ability (Sarter and Van Der Linde, 1987).

Signs of Vitamin E Toxicity   In general vitamin E is relatively nontoxic. However, 2,000 mg *RRR*-α-tocopheryl acetate/kg BW (4,230 μmol/kg BW) prolonged the process of prothrombin production, and hemorrhagic diathesis developed in rats fed the AIN-76 diet (low vitamin K) (Abdo et al., 1986). The main metabolic product of vitamin E, tocopheryl quinone, inhibits the normal metabolism of vitamin K. In addition, Martin and Hurley (1977) found eye abnormalities in pups from dams receiving 1,600 mg α-tocopheryl acetate/kg BW/day (3,500 μmol/kg BW/day). Yang and Desai (1977) found a decreased implantation index in inseminated females fed a diet containing 7,300 mg α-tocopheryl acetate/kg diet (15,500 μmol/kg).

Vitamin K

The only known function of vitamin K in mammalian systems is the posttranslational conversion of glutamic acid to γ-carboxyglutamic acid (Gla) (Stenflo, 1976). Gla is found only in a limited number of proteins including the blood clotting proteins, prothrombin and factors VII, IX, and X; the anticlotting proteins, protein C and protein S; blood protein Z; the bone proteins, osteocalcin and matrix Gla-containing protein (MGP); and a few other Gla-containing proteins in the kidney and intestine (Suttie, 1991).

Phylloquinone (vitamin K_1) is the metabolically preferred source of vitamin K (Will and Suttie, 1992). Phylloquinone is also actively absorbed in the proximal portion of the small intestine (Hollander, 1973). The menaquinones (vitamin K_2) are a family of compounds synthesized by bacteria. The bacteria in the large intestine produce substantial amounts of menaquinones. Because the rat is coprophagous, excrement provides a substantial source of vitamin K activity. Essentially no absorption of the larger menaquinones, such as menaquinone-9, occurs in the large intestine; but they are effectively absorbed if recycled to the small intestine by coprophagy (Ichihashi et al., 1992). The smaller menaquinone-4 is absorbed to a limited extent in the colon, but menaquinone-4 is a minor product of the intestinal bacteria. The synthetic derivative of vitamin K, menadione, lacks the isoprenoid side chain of the natural compounds with vitamin K activity, and it must be converted to menaquinone-4 in the liver to be functional (Dialameh et al., 1971). Menadione is about one-tenth as active as phylloquinone. Some of the water-soluble derivatives of menadione, such as the menadione sodium bisulfite complex, are much more readily absorbed and are about equally as active as phylloquinone (Griminger, 1966).

The most prominent function of vitamin K is its role in blood clotting. Failure of blood to coagulate occurs in some rats that have been fed a vitamin K-deficient diet for 2 to 3 weeks. However, in most rats overt clotting problems only develop after they have been fed antibiotics or when coprophagy is prevented.

Estimation of the vitamin K requirement of rats is complicated by several factors: (1) the contribution from coprophagy, (2) the difficulty in producing vitamin K-free diets, (3) the large differences in requirements between strains, and (4) the problem of selecting which criteria to use to determine normal status.

The typical contribution from coprophagy is probably on the order of 4 to 9 μmol/rat/day (Mameesh and Johnson, 1960; Wostmann et al., 1963). Rats fed diets with a high nutrient density and high digestibility are less coprophagous than rats fed low-digestibility diets (Giovannetti, 1982; Mathers et al., 1990). Housing rats in a very cold environment (Smith and Borchers, 1972) will almost eliminate coprophagy. The contribution from coprophagy is difficult to estimate and may necessitate the use of germ-free animals for experimental work.

The purified milk proteins, casein and lactalbumin, usually contain substantial vitamin K that is difficult to extract (Matschiner and Doisy, 1965). Therefore, soybean proteins are frequently used as the source of protein when vitamin K metabolism is studied. If the soybean proteins are extracted with ethanol, diets can be formulated containing as little as 9 μg phylloquinone/kg (0.02 μmol/kg) (Kindberg and Suttie, 1989).

Many different criteria have been used to evaluate vitamin K status. Wostmann et al. (1963) used survival of germ-free male Lobund rats as a criterion and estimated that the diet must contain 0.44 μmol phylloquinone/kg to maintain survival. At 0.33 μmol/kg diet most of the rats died from hemorrhages. However, the minimum amount necessary to keep rats from bleeding to death may not be the optimal vitamin K intake. Most frequently, some estimate of clotting time is used to predict the vitamin K requirement. Mameesh and Johnson (1960) reported that male Sprague-Dawley rats required phylloquinone at 0.25 μmol/kg diet to maintain normal prothrombin activity when coprophagy was prevented by tail cups. Matschiner and Doisy (1965) found that male rats of the St. Louis strain required 0.55 μmol/kg diet to maintain normal prothrombin activity when fed a 21 percent soybean protein diet that was not extracted and coprophagy was not prevented.

More sensitive criteria of vitamin K status were used by Kindberg and Suttie (1989) to determine the requirement of male Holtzman and Sprague-Dawley rats. The rats were fed an extracted soybean protein diet and coprophagy was not prevented. A diet containing 1.1 μmol phylloquinone/kg was not adequate to bring liver carboxylase enzyme activity or plasma prothrombin to the concentrations observed in rats fed a 3.3 μmol/kg diet. Based on extrapolations of this data the optimal concentration appears to be about 2.2 μmol/kg diet. Even at higher doses the rats maintained small liver stores of vitamin K that were essentially depleted within 5 days. Consequently, rats depend on a continuous dietary supply of vitamin K to maintain optimal status.

When fed diets with suboptimal vitamin K activity, female rats develop signs of vitamin K deficiency much more slowly than do males (Matschiner and Bell, 1973). The lower requirement seems to reflect differences in the effects of estrogen and testosterone on the production of prothrombin (Matschiner and Willingham, 1974). In pregnancy the concentrations of prothrombin rise as a result of the increased estrogen concentrations. The requirement of female rats appears to be about three-fourths the requirement of males.

Mellette and Leone (1960) reported that rats 15 to 27 weeks old were about twice as likely to have a hemorrhagic death as rats 3 to 5 weeks old when fed irradiated beef diets low in vitamin K.

Warfarin-resistant rats have much higher vitamin K requirements than normal rats because they do not recycle vitamin K epoxide as efficiently as normal rats. Greaves and Ayres (1973) reported that the Wistar strain required 0.13 μmol phylloquinone/kg BW to maintain normal prothrombin activity, the Tolworth HS (Scottish) warfarin-resistant strain required 0.44 μmol/kg BW, and the Tolworth HW (Welsh) warfarin-resistant strain required 1.77 μmol/kg BW. The vitamin K was given by daily subcutaneous injections.

Roebuck et al. (1979) reported that male Wistar/Lewis rats developed hemorrhaging but had normal prothrombin activity when fed a modified AIN-76 diet for 8 to 20 weeks. The bleeding could be corrected by adding phylloquinone, 1.1 μmol/kg, to the diet. The AIN-76 diet (American Institute of Nutrition, 1977) was formulated to contain 0.18 μmol menadione sodium bisulfite/kg plus an undefined amount of vitamin K activity in the dietary casein and corn oil. Bieri (1979) confirmed that the amount of vitamin K activity in the AIN-76 diet was less than optimal for Fischer rats when the casein was replaced with other proteins; the American Institute of Nutrition recommended that the added menadione sodium bisulfite be increased to 1.8 μmol/kg diet (American Institute of Nutrition, 1980) and that the diet be designated AIN-76A.

For the most commonly used strains of rats (Fischer, Sprague-Dawley, and Wistar) a diet containing 1.00 mg phylloquinone/kg diet (2.22 μmol/kg) should satisfy the most sensitive criterion for vitamin K adequacy. Because the rat actively absorbs phylloquinone and gives preference to phylloquinone metabolically, phylloquinone is the recommended form to be used in diets. Menadione and its derivatives are not recommended.

High concentrations of dietary vitamin A and vitamin E have been shown to accelerate the onset of deficiency signs in rats fed vitamin K-deficient diets. As little as 5.2 μmol retinyl acetate/kg diet decreased the prothrombin activity in rats fed a diet low in vitamin K (Doisy, 1961). Supplementation of a low vitamin K diet with 12 μmol α-tocopherol orally twice a week increased hemorrhagic deaths (Mellette and Leone, 1960); hemorrhaging, however, could be corrected by phylloquinone supplementation. Vitamin E metabolites appear to interfere with the normal metabolism of vitamin K. The vitamin E metabolites tocopheryl hydroquinone and tocopheryl quinone are the most proba-

ble causes (Rao and Mason, 1975). Butylated hydroxytoluene (BHT, 1.2 percent of diet) has also been shown to induce hemorrhagic death when included in casein-based diets that were not supplemented with vitamin K. The simultaneous administration of 0.68 μmol phylloquinone/kg BW/day prevented the hemorrhages and maintained normal prothrombin concentrations (Takahashi and Hiraga, 1979).

Signs of Vitamin K Deficiency In vitamin K deficiency thrombin activity is depressed and liver carboxylase activity is greatly increased (Kindberg and Suttie, 1989). The first overt sign of vitamin K deficiency is usually a continuous oozing of blood from a minor injury such as a separated toenail or a small pin-prick on the tail. These injuries would not normally bleed longer than 30 seconds, but the vitamin K-deficient animal may continue to lose blood until death occurs. In an examination of specific-pathogen-free rats not supplemented with vitamin K, hemorrhages were found in the urogenital tract, the central nervous system, the chest cavity, the abdominal cavity, and under the skin (Fritz et al., 1968).

Signs of Vitamin K Toxicity Phylloquinone is essentially nontoxic when given orally. Rats given daily doses of 4,400 μmol phylloquinone/kg BW for 30 days did not show signs of toxicity, whereas 2,000 μmol menadione/kg BW was lethal (Molitor and Robinson, 1940). At 260 μmol/kg diet, menadione depressed the activity of several heme-containing enzymes in vitamin E-deficient rats (Hauswirth and Nair, 1975). This concentration is present in a popular commercial vitamin mix. At higher concentrations both menadione and menadione sodium bisulfite produce liver toxicity.

WATER-SOLUBLE VITAMINS

Vitamin B_6

The vitamin B_6 compounds (pyridoxine, pyridoxal, and pyridoxamine) function as coenzymes for amino acid decarboxylases, racemases, transaminases, and other enzymes in amino acid, glycogen, and fatty acid metabolism (Baker and Frank, 1968). The coenzymes are formed by phosphorylation of the aldehyde and amine; nearly 50 percent of pyridoxal phosphate in the body is stored as coenzyme for muscle glycogen phosphorylase (Anonymous, 1975; Chen and Marlatt, 1975). Pyridoxal phosphate is involved in releasing steroid-hormone-complexes tightly bound to receptors (Compton and Cidlowski, 1986; Bender et al., 1989).

Phosphatase-mediated hydrolysis is the first step in the intestinal absorption of pyridoxal-5'-phosphate. Gastric acid secretions are important for the hydrolysis as intestinal alkaline phosphatase is activated at low pH (3.4) (Middleton, 1986). Products of digestion—amino acids and oligo-peptides—inhibit hydrolysis (Middleton, 1990). Cellulose, pectin, or lignin did not alter the in vitro jejunal absorption rates of pyridoxine, pyridoxal, or pyridoxamine (Nguyen et al., 1983). Weanling Sprague-Dawley rats fed diets containing 1.5 mg vitamin B_6/kg diet or less had lower vitamin B_6 concentrations in the intestinal mucosa than rats fed 3 to 100 mg/kg diet (Roth-Maier et al., 1982). Vitamin B_6 aldehydes can reductively bind to food proteins as epsilon-pyridoxyllysine complexes during processing and storage and, in this form, possess only 50 percent of the vitamin B_6 activity (Gregory, 1980a,b). No differences were observed in urinary pyridoxic acid between germ-free and conventional rats, indicating that coprophagy does not alter requirements (Coburn et al., 1989).

Studies of the dietary requirement have been based on vitamin B_6-dependent enzyme activities, body weight gain, tissue stores of pyridoxal phosphate, or reproductive performance. When male weanling rats were fed 1, 2, 4, or 8 mg vitamin B_6/kg diet, growth was the same in all groups; but liver, serum, and red blood cell alanine-aminotransferase was maintained only at dietary concentrations of 4 mg/kg and above (Chen and Marlatt, 1975). Concentrations of 1.0 to 1.6 mg/kg diet were required to achieve maximum growth (Roth-Maier and Kirchgessner, 1981; Van den Berg et al., 1982; Mercer et al., 1984). Alterations in red blood cell transaminase activity indicated a vitamin B_6 requirement of 6 to 7 mg/kg diet (Beaton and Cheney, 1965). Diets containing 7.0 mg pyridoxine-HCl/kg restored full activity of erythrocyte alanine aminotransferase and aspartate aminotransferase when fed to female Long-Evans rats that had been previously depleted of vitamin B_6 (Skala et al., 1989).

Protein quality had no significant effect on urinary 4-pyridoxic acid, plasma pyridoxal phosphate or total hepatic vitamin B_6 concentrations in rats fed 0.2 and 7.0 mg vitamin B_6/kg diet (Fisher et al., 1984). Erythrocyte alanine aminotransaminase activity decreased with increasing dietary protein concentrations but was not altered when the quality of protein was changed (Dirige and Beaton, 1969).

Vitamin B_6 is required by pregnant rats for normal development of their offspring. Vitamin B_6-deficient offspring had retarded renal differentiation, abnormalities of cerebral lipids, and increased tissue and urinary concentrations of cystathionine (Kurtz et al., 1972; DiPaolo et al., 1974; Pang and Kirksey, 1974). Maternal weight gain and body and brain weight of offspring were normal when the diet contained 3 mg pyridoxine/kg and slightly, but not significantly, lower at 2 mg/kg; 1 mg/kg was clearly inadequate. There was no significant difference between offspring of dams fed 3 mg/kg diet and those fed 6 mg/kg (Driskell et al., 1973). When female rats were reared on diets containing 1.2, 2.4, 4.8, 9.6, or 19.2 mg vitamin B_6/kg diet, maternal and fetal weights were reduced in the group fed

1.2 mg/kg; erythrocyte alanine aminotransferase activity coefficients were maintained at 2.4 mg/kg and above; and tissue vitamin B_6 saturation concentrations indicated that 9.6 mg/kg was required for growth and 4.8 mg/kg for maintenance. However, 19.2 mg/kg diet did not achieve saturation concentrations of maternal liver, fetal brain, or carcass during gestation (Kirksey et al., 1975). Rats fed 8 or 40 mg/kg during pregnancy had similar activities of hepatic aspartate aminotransferase, erythrocyte alanine aminotransferase, and gastrocnemius muscle glycogen phosphorylase (Shibuya et al., 1990).

Female rats fed diets that contained 1.2 to 19.6 mg vitamin B_6/kg diet from weaning through breeding, gestation, and lactation bore offspring of normal weight at 2.4 mg/kg and above. Concentrations of vitamin B_6, protein, and cerebrosides in brains were significantly decreased at 1.2 and 2.4 mg/kg; however, brain protein concentration continued to increase up to the highest dietary vitamin B_6 concentration. Milk pyridoxine and erythrocyte transaminase were decreased when maternal diet was less than 4.8 mg/kg (Moon and Kirksey, 1973; Pang and Kirksey, 1974).

These studies indicate that vitamin B_6 requirements are met by 4 to 7 mg/kg diet, although higher tissue concentrations of pyridoxal phosphate may be achieved by higher dietary concentrations. The estimated vitamin B_6 requirement for maintenance, growth, and reproduction is set at 6 mg pyridoxine-HCl/kg diet.

Signs of Vitamin B_6 Deficiency   Rats fed diets deficient in vitamin B_6 develop symmetrical scaling dermatitis on the tail, paws, face, and ears; microcytic anemia; hyperexcitability; and convulsions (Sherman, 1954). The amount of 3-hydroxykynurenine was increased in neonatal brain at 14 and 18 days of age but not in adult brain (Guilarte and Wagner, 1987). Amplitude of response to both acoustic and tactile stimuli was depressed by vitamin B_6 depletion (Schaeffer, 1987) as well as differences in angle and width of the hind-leg gait (Schaeffer and Kretsch, 1987; Schaeffer et al., 1990). Depleted rats demonstrated a taste preference for NaCl and excreted reduced concentrations of sodium (Mei-Ying and Kare, 1979). They had deficits in active and passive avoidance learning (Stewart et al., 1975). The pyramidal cells of the cerebral cortex of rats fed deficient diets for 2 or 3 months showed partial to nearly complete dendritic loss and axonal swelling in the hippocampus (Root and Longenecker, 1983). Reduced brain concentrations of vitamin B_6, dopamine, homovanillic acid, D-2 dopamine receptors, brain glutamic acid decarboxylase, gamma aminobutyric acid (GABA), and γ-aminobutyric acid transaminase have been measured in vitamin B_6-deficient rats (Driksell and Chuang, 1974; Aycock and Kirksey, 1976; Rajeswari and Radha, 1984; Guilarte, 1989). Cerebroside and ganglioside content and fatty acids (18:2, 20:1, 20:4,

22:6, and 24:0) and n-6 fatty acids (18:2, 20:4, and 22:4) are decreased in brain as a result of vitamin B_6 deficiency (Thomas and Kirksey, 1976a,b).

In deficiency, reproductive performance of both females and males was decreased; deficient production of insulin may occur (Huber et al., 1964). An 8-week deficiency resulted in decreases in hepatic alanine and aspartate aminotransferases, and the ability to use alanine for gluconeogenesis was impaired (Angel, 1980). The turnover of cytosolic, but not mitochondrial aspartate aminotransferase, was altered in vitamin B_6 deficiency (Shibuya and Okada, 1986); and the total activities of both aspartate aminotransferase and alanine transferase were reduced (Ludwig and Kaplowitz, 1980). Pyridoxine deficiency reduced the intestinal uptake of glucose, glycine, alanine, and leucine as well as the activities of mucosal sucrase, lactase, alkaline phosphatase, leucine aminopeptidase, and membrane synthesis (Mahmood et al., 1985). In rats fed a high-protein diet deficient in vitamin B_6, urinary excretion of urea, free ammonia, and free amino acids were altered (Okada and Suzuki, 1974). Female rats fed vitamin B_6-deficient diets excreted less taurine than pair-fed controls (Lewis et al., 1982; Lombardini, 1986). Vitamin B_6-deficient rats had lower plasma homocystine in association with decreased and sporadic food intake (Smolin and Benevenga, 1984). Creatine in skeletal muscle and liver was higher and creatinine excretion lower in deficient rats (Loo et al., 1986). Vitamin B_6 deficiency produced a decrease in glucose-6-phosphate dehydrogenase activity in the periosteum and in the developing callus (Dodds et al., 1986) and a decrease in collagen cross-link formation (Fujii et al., 1979). Liver parenchymal ultrastructural changes after 7 weeks on a vitamin B_6-deficient diet included hyperplasia in the nucleoli and smooth endoplasmic reticulum and hypoplasia in the Golgi, rough endoplasmic reticulum, and mitochondria with reduced numbers of orthoperoxisomes (Riede et al., 1980). Vitamin B_6-deficient male rats had larger and a greater number of urinary calcium oxalate stones than females or castrated males (Gershoff, 1970). Changes in fatty acid oxidation and incorporation of the fatty acids into triglycerides, phospholipids, and cholesterol fractions have been observed (Dussault and Lepage, 1979).

Signs of Vitamin B_6 Toxicity   Excess vitamin B_6 has been shown to have detrimental effects. Twelve-week-old rats fed 6 weeks on diets containing 1,400 mg vitamin B_6/kg had elevated red blood cell alanine aminotransferase activity and pyridoxal-5'-phosphate concentrations, but muscle pyridoxal-5'-phosphate concentrations were lower. Tissue pyridoxal concentrations were higher and muscle glycogen phosphorylase A activity was increased by the consumption of excessive dietary vitamin B_6 (Schaeffer et al., 1989). Rats injected intraperitoneally with 200 mg/kg BW developed gait ataxia with impaired oxidative me-

tabolism in peripheral nerve tissue and an axonal degeneration of the sensory system fibers (Windebank et al., 1985).

Vitamin B_{12}

In mammals, vitamin B_{12} is required as a coenzyme for the transmethylation of homocystine to methionine, in utilizing 5-methyl-tetrahydrofolic acid, and in the conversion of methylmalonyl-CoA to succinyl-CoA (Weissbach and Taylor, 1970). The concentration of vitamin B_{12} needed in the diet of the rat may vary with dietary content of choline, methionine, and folic acid. A number of conditions have been reported that impair vitamin B_{12} absorption. Rats fed a diet with raw kidney bean (*Phaseolus vulgaris*) as 4 percent of dietary protein or 0.5 percent phytohemagglutinin developed vitamin B_{12} malabsorption after only 3 days, and the condition was not correctable by giving intrinsic factor (Banwell et al., 1980). Highly fermentable fibers such as pectin, guar gum, and xylan fed as 5 percent of the diet increased urinary methylmalonic acid and depressed the oxidation of propionate to CO_2 (Cullen and Oace, 1989a). The half-life of vitamin B_{12} was 58 days for rats fed fiber-free diets and 38 days for rats fed a 5 percent pectin diet (Cullen and Oace, 1989b). Vitamin B_{12} deficiency has been induced by diets containing unheated soybean flour, but amino acid deficiencies occur also and unheated soybean flour contains other toxins; therefore, it has been suggested that its use for the study of vitamin B_{12} deficiency is not justified (Edelstein and Guggenheim, 1971; Williams and Spray, 1973). Giardiasis also has been reported to impair vitamin B_{12} absorption (Deka et al., 1981). The oxidation of acetaldehyde, generated from the metabolism of ethanol, by xanthine oxidase inhibited the ability of vitamin B_{12} to bind to intrinsic factor (Shaw et al., 1990). Decreased vitamin B_{12} concentrations have been reported in rats fed liquid diets with ethanol but not natural-ingredient diets with ethanol (Frank and Baker, 1980). Hypothyroidism slowed the rate of depletion of hepatic vitamin B_{12} (Stokstad and Nair, 1988). Vitamin B_{12} deficiency developed rapidly in rats exposed to nitrous oxide (Horne and Briggs, 1980; Muir and Chanarin, 1984).

The minimum requirement for vitamin B_{12} has not been established; however, a dietary concentration of 10 μg vitamin B_{12}/kg seems to be inadequate based on urinary excretion of methylmalonic acid (Thenen, 1989). A concentration of 50 μg vitamin B_{12}/kg diet supported normal growth and reproduction (Woodard and Newberne, 1966). In the absence of more complete information, the requirement is currently set at 50 μg vitamin B_{12}/kg diet.

Signs of Vitamin B_{12} Deficiency   Induction of isolated vitamin B_{12} deficiency in the rat, as well as in other experimental animals, is achieved with difficulty and generally does not reproduce the signs of human vitamin B_{12} deficiency—

megaloblastic blood cells and neurological lesions. Deficiency can be induced in rats fed vegetable rather than animal protein (which contains vitamin B_{12}). Female rats fed a diet that contained soybean protein supplemented with methionine and choline, but not vitamin B_{12}, grew normally or at a slightly decreased rate and bred and littered normally. The average weights of their offspring were decreased, and 10 percent of the litters were hydrocephalic. Hepatic content of vitamin B_{12} was markedly decreased in both mothers and offspring. Deletion of choline from the diet increased the incidence of congenital abnormalities in the neonates. Supplementation of the diet with 50 μg vitamin B_{12}/kg supported normal growth in the mothers and prevented development of hydrocephalic offspring (Woodard and Newberne, 1966). When litters born to vitamin B_{12}-deficient females were fed diets deficient in vitamin B_{12}, growth was retarded by day 30 and the animals remained small throughout the 150 day experiment. Methionine (0.5 percent DL) reduced the growth retardation (Doi et al., 1989). Germ-free females fed soybean protein aborted, bore short-lived pups, or cannibalized their pups. The germ-free condition apparently enhanced vitamin B_{12} deficiency (Valencia and Sacquet, 1968). Vitamin B_{12}-deficient rats accumulated more odd-chained fatty acids in phosphatidylcholine of cerebrum and liver, more 18:2 acids in liver phosphatidylethanolamine, and arachidonate [20:4(n-6)] and 22:5 in liver phosphatidylcholine, but smaller amounts of 20:4(n-6) and 22:6(n-6) in cerebral phosphatidylcholine (Peifer and Lewis, 1979).

Biotin

Biotin is required by four carboxylase enzymes in mammalian systems: acetyl CoA carboxylase (fatty acid synthesis), pyruvate carboxylase (gluconeogenesis), 3-methylcrotonyl CoA carboxylase (leucine catabolism), and propionyl CoA carboxylase (methionine, threonine, and valine catabolism).

Under normal conditions rats do not require biotin in the diet. Adequate biotin is provided by the intestinal microorganisms through coprophagy. Biotin deficiency can be produced in rats fed a biotin-free diet by (1) preventing coprophagy (Barnes et al., 1959b); (2) using germ-free animals (Luckey et al., 1955); (3) feeding sulfa drugs (Daft et al., 1942); or feeding raw egg white, which contains the biotin-binding protein avidin (Nielsen and Elvehjem, 1941).

Klevay (1976) fed 20 percent raw egg white diets to rats and determined that 2 mg d-biotin/kg diet (8.2 μmol/kg) was required to obtain optimal growth for 60 days. Several other investigators have given various amounts of biotin to animals to cure or prevent signs of deficiency, but these experiments do not permit conclusions to be made about requirements because the amounts given were insufficient

to correct fully the deficiency, the animals did not grow at a rate equivalent to controls, controls were not used, or the biotin administration was erratic.

The AIN-76 diet (American Institute of Nutrition, 1977) and AIN-93G and AIN-93M diets (Reeves et al., 1993b) were formulated to contain 0.2 mg d-biotin/kg (0.82 μmol/kg). This concentration appears to be adequate when casein is the dietary protein; however, if spray-dried raw egg white is used as the dietary protein, the concentration of biotin should be increased to 2.0 mg/kg diet.

Signs of Biotin Deficiency Biotin deficiency causes rats to develop a progressive exfoliative dermatitis, "spectacle eye," achromotrichia (in black rats), and general alopecia. With severe deficiency many animals develop a spastic gait or assume a "kangaroo-like" posture. Immune responses were also depressed in biotin-deficient rats (Rabin, 1983).

Signs of Biotin Toxicity Biotin is relatively nontoxic. Paul et al. (1973) reported that 50 mg biotin/kg BW (205 μmol/kg BW) divided between morning and evening subcutaneous injections produced irregularities in the estrous cycle and a massive infiltration of leukocytes in the vagina. Mittelholzer (1976) did not observe problems in the reproduction of females given this dose.

Choline

Choline is a component of lecithin, of sphingomyelin, and of the neurotransmitter acetylcholine. Choline also has an important role in one-carbon metabolism.

Diets are usually formulated with either choline chloride or choline bitartrate. Although all choline compounds tend to be hygroscopic, choline bitartrate is substantially less hygroscopic than choline chloride, and does not add more chloride, so use of choline bitartrate is preferable for most practical diets.

Choline is required by the rat, but the requirement is a function of the methionine and lipid content of the diet. Diets containing about 14 percent casein require more choline than diets containing either more or less casein (Aoyama et al., 1971). Diets that contain 0.8 percent methionine or more do not require choline (Griffith, 1941; Newberne et al., 1969; Aoyama et al., 1971). Rats fed diets containing suboptimal amounts of methionine have been reported to require 0.6 to 2.0 g choline chloride/kg diet (4 to 14 mmol/kg) (Griffith and Wade, 1939; Griffith, 1941; Engel, 1942; Mulford and Griffith, 1942; Hale and Schaefer, 1951).

Using a diet that contained 38 percent sucrose, 20 percent lipid, and 36 percent protein (0.34 percent methionine), Chahl and Kratzing (1966) found that male Wistar rats housed at 21° C required 500 mg choline (free base) per kg diet (4.8 mmol/kg) to prevent lipid accumulation in the liver. However, when the rats were housed at 33° C the requirement was increased to 1,000 mg choline (free base) per kg diet (9.6 mmol/kg). In contrast, housing the rats at 2° C reduced the requirement to 250 mg choline (free base) per kg diet (2.4 mmol/kg).

Based on the studies of Chahl and Kratzing (1966), Mulford and Griffith (1942), and Griffith and Wade (1939), the requirement for choline is set at 750 mg/kg diet (7.2 mmol/kg diet). This is equivalent to 1.8 g choline bitartrate/kg diet. However, if diets with 5 to 15 percent and 20 to 40 percent fat are to be used, if the diets are deficient in folate or vitamin B_{12}, or if the rats are to be housed in a high environmental temperature, then the amount of choline may need to be increased.

Signs of Choline Deficiency   Choline deficiency appears much more rapidly in male than in female rats (Patek et al., 1969). Droplets of triglycerides and abnormalities of the intracellular membranes appear in the livers of weanling male rats within 24 hours after they have ingested a choline-deficient diet. Long-term deficiency leads first to fatty liver, in which triglycerides may compose as much as 50 percent of the total wet weight of the liver and the liver cells are markedly distended with fat vacuoles, and then to cirrhosis, in which there is proliferation of fibrovascular tissue, vascular shunting, and hepatic failure (Zaki et al., 1963; Rogers and MacDonald, 1965). Sucrose was found to cause twice as great an increase in liver triglycerides as dextrin in choline-deficient diets (Chalvardjian and Stephens, 1970).

Choline deficiency has marked effects on both the kidney and the cardiovascular system in young rats. If male rats are fed a choline-deficient diet at weaning, 50 to 90 percent die within 10 days to 2 weeks of hemorrhagic renal necrosis. They may develop myocardial necrosis and atheromatous changes in arteries (Salmon and Newberne, 1962; Monserrat et al., 1974). Supplementation with about one-half the requirement will protect against renal damage and prevent cirrhosis of the liver, but it will not prevent the infiltration of the liver with triglyceride.

Signs of Choline Toxicity   Sahu et al. (1986) reported that 100 mg choline chloride/kg BW/day (700 μmol/kg BW/day) depressed the rate of growth and caused pathological changes in the lungs and lymph nodes. The safety margin between the daily requirement (16 to 42 mg/kg BW or 115 to 300 μmol/kg BW) and the toxic concentration is relatively narrow with choline, so caution should be used in increasing the concentrations. The LD_{50} of choline chloride has been reported to range from 0.28 to 0.75 g/kg BW/day (2,000 to 5,400 μmol/kg BW/day) depending on the concentration of the injected solution (Hodge, 1944; Ho et al., 1979; Sahu et al., 1986). Bell and Slotkin (1985)

reported that a diet containing 5.0 percent soybean lecithin caused alterations in sensorimotor development and brain cell maturation.

Folates

Folates are composed of pteridines linked to *p*-aminobenzoic acid conjugated to one or more glutamic acid residues (Blakley and Benkovic, 1984; Blakley and Whitehead, 1986). Folates in most tissues exist primarily in the pentaglutamyl form. The principal function of folates is to transfer one-carbon units such as in thymidylate synthesis, purine synthesis, serine synthesis from glycine, and methionine synthesis from homocystine (Shane and Stokstad, 1985; Appling, 1991). Measurement of red blood cell, serum, and tissue folate content of the various folate forms as well as measurement of the urinary histidine metabolite, formiminoglutamic acid, is used to assess folate nutritional status (Shane and Stokstad, 1985; Clifford et al., 1989; Ward and Nixon, 1990; Varela-Moreiras and Selhub, 1992). Furthermore, folate biosynthesis by intestinal bacteria may meet much of the folate requirement (Rong et al., 1991). Diets inadequate in choline, methionine, and cobalamin may induce deficiency through the various interactions of these compounds (Thenen and Stokstad, 1973). Dietary concentrations between 0.5 and 10 mg/kg diet (1.1 and 23 μmol/kg) have been used. Studies by Clifford et al. (1989) have demonstrated that 2 mg folic acid/kg diet (4.5 μmol/kg) was no more effective than 1 mg/kg (2.3 μmol/kg) in young Sprague-Dawley rats that had been depleted of folic acid for 52 days. Thus, 1 mg/kg diet (about 2.3 μmol/kg) is recommended.

Signs of Folate Deficiency   Induction of deficiency requires prolonged periods of feeding, usually with an amino acid-based diet and/or with an antibiotic to inhibit intestinal bacteria (Clifford et al., 1989; Ward and Nixon, 1990). Decreased growth rate, leukopenia, anemia, and formiminoglutamate excretion are reported.

Niacin

The only known roles of niacin are as components of the coenzymes nicotinamide adenine dinucleotide (NAD) and nicotinamide adenine dinucleotide phosphate (NADP), which function as electron carriers dehydrogenases in intermediary metabolism.

Both NAD and NADP can be synthesized from tryptophan in the liver of the rat, so pure niacin deficiency does not occur. Harris and Kodicek (1950) found that 24 moles of tryptophan were equivalent to 1 mole of niacin. Hundley (1947) found that rats fed a diet containing 15 percent casein required niacin for optimal growth, but rats fed a 20 percent casein diet did not require additional niacin.

Several investigators have reported that niacin-free diets containing fructose and low concentrations of tryptophan resulted in niacin deficiency. The amount of niacin required to restore normal growth was 15 mg/kg diet (120 μmol/kg) (Krehl et al., 1946; Henderson et al., 1947; Hundley, 1949). The requirement for niacin is much lower with diets containing glucose or starch than with high-sucrose or -fructose diets (Hundley, 1949). A high fat (40 percent peanut oil) diet was found to decrease the urinary excretion of niacin metabolites, apparently because of an inhibition in the conversion of tryptophan to niacin (Shastri et al., 1968). Fleming and Barrows (1982) found that aging did not influence the absorption of nicotinic acid.

The AIN-93G diet (Reeves et al., 1993b) was formulated to contain 30 mg nicotinic acid/kg (244 μmol/kg). This provides an adequate amount for any experimental condition.

Signs of Niacin Deficiency   Niacin deficiency causes reduced growth rate, rough coat, alopecia, and tissue concentrations of NAD and NADP are lowered. The concentrations of myelin in brain tissue are reduced as a result of a reduced synthesis of cerebrosides (Nakashima and Suzue, 1982, 1984).

Signs of Niacin Toxicity   As the main excretory metabolites of nicotinic acid and nicotinamide are methylated, administration of high concentrations of niacin depletes the methyl pools of the body. Handler and Dann (1942) found that 10 g nicotinic acid/kg (80,000 μmol/kg) in a 10 percent casein diet caused an increase in fatty acids in the liver similar to a choline deficiency but without weight loss. A concentration of 2.5 g nicotinamide/kg diet (20,000 μmol/kg) caused the fatty liver and 10 g/kg diet (80,000 μmol/kg) decreased the rate of weight gain. The fatty livers could be prevented by including 1.5 g choline chloride/kg diet (11 mmol/kg), but choline alone would not support growth. Adding 0.6 percent methionine to the diet would correct the fatty liver and support growth. Growth could also be supported by 1.5 g choline chloride/kg and 0.6 percent homocystine in the diet. Jaus et al. (1977) noted that the injection of 0.50 g nicotinamide/kg BW (4,000 μmol/kg BW) twice daily caused enlarged liver cells and increased glycogen deposits. The effect was more pronounced in female rats. The injection of 0.6 g nicotinamide/kg BW (4,900 μmol/kg BW) per day resulted in reduced weight gain, and as little as 0.05 g/kg BW/day (400 μmol/kg BW/day) induced fatty livers (Kang-Lee et al., 1983; Sun et al., 1986). The LD_{50} for nicotinic acid is 4.3 to 4.9 g/kg BW (35,000 to 40,000 μmol/kg BW), but 0.85 g/kg BW/day (6,900 μmol/kg BW/day) was tolerated for 40 days with no abnormal histology. Nicotinamide is more toxic, and the LD_{50} is 1.7 to 3.0 g/kg BW (14,000 to 25,000 μmol/kg BW) (Unna, 1939; Brazda and Coulson, 1946).

Pantothenic Acid

Pantothenic acid functions as a constituent of coenzyme A and as a component of the acyl carrier protein in fatty acid synthesis. Pure pantothenic acid is an unstable viscous oil. Because pantothenic acid is difficult to handle, calcium pantothenate is normally used in diets. Only the *d*-pantothenic acid isomer has biological activity. The *l*-isomer does not have pantothenic acid activity and is actually a pantothenic acid inhibitor when its concentration is 100-fold greater than the *d*-isomer (Kimura et al., 1980). Some commercial preparations are equal molar mixtures of the *d*- and *l*-isomers and are only one-half as active as *d*-pantothenate. In this document the concentrations are expressed as the moles of *d*-pantothenic acid. One mole of calcium pantothenate is equivalent to 2 moles of *d*-pantothenic acid, and sodium pantothenate or hemicalcium pantothenate is only equivalent to 1 mole.

Unna (1940) found that 80 μg *d*-pantothenate/rat/day (0.36 μmol/rat/day) was necessary to maintain optimal growth. Henderson et al. (1942) found that only 40 μg/day (0.17 μmol/day) was needed to prevent the graying of black or hooded rats, that 80 μg/rat/day (0.36 μmol/rat/day) was required for optimal growth, and that about 100 μg/rat/day (0.42 μmol/rat/day) was necessary before pantothenate was excreted in the urine. Barboriak et al. (1957a) found 4 mg calcium pantothenate/kg diet (17 μmol pantothenic acid/kg diet) was needed to obtain optimal growth confirming the earlier values. This concentration would maintain pregnancy, but would not support lactation (Barboriak et al., 1957b). Nelson and Evans (1961) obtained normal growth of the suckling pups from dams receiving 10 mg calcium *d*-pantothenate/kg diet (42 μmol pantothenic acid/kg).

The AIN-76 (American Institute of Nutrition, 1977) and AIN-93G (Reeves et al., 1993b) diets were formulated to contain 15 mg calcium *d*-pantothenate/kg diet (63 μmol pantothenic acid/kg). The diets appear to contain an adequate safety margin because no problems related to pantothenic acid have been reported.

The antibiotics aureomycin, streptomycin, penicillin (Lih and Baumann, 1951; Sauberlich, 1952), and hygromycin (Barboriak and Krehl, 1957) delayed the appearance of the signs of pantothenic acid deficiency. In the later stages of the deficiency the signs develop in spite of the antibiotics. The effect seems to be related to changes in the intestinal microflora as the antibiotics did not delay the onset of deficiency when they were injected.

Signs of Pantothenic Acid Deficiency   Pantothenic acid deficiency induces achromotrichia, exfoliative dermatitis, oral hyperkeratosis, necrosis, and ulceration of the gastrointestinal tract. Focal or generalized hemorrhagic necrosis of the adrenals may occur, and death results after 4 to 6 weeks of deficiency (Ralli and Dumm, 1953). Deficient rats had impaired antibody synthesis, decreased serum globulins, and decreased antibody forming cells in response to antigen. Restoration of antibody synthesis was achieved by parenteral administration of calcium pantothenate beginning 9 days before antigen injection (Lederer et al., 1975).

Signs of Pantothenic Acid Toxicity   Pantothenic acid is relatively nontoxic. When excess amounts are consumed most of the excess is excreted in the urine. Daily administration of 185 mg (840 μmol) of pantothenate for 6 months did not produce any pathological changes in the rats (Unna and Greslin, 1941). The LD_{50} for rats was 3 g/kg BW/day (13,700 μmol/kg BW/day).

Riboflavin

Riboflavin is the precursor of the flavin coenzymes and is stored in the liver primarily as flavin adenine dinucleotide (FAD) (Rivlin, 1970). The coenzymes function with many oxidation-reduction enzymes, e.g., cytochrome *c* reductase, xanthine oxidase, and diaphorase. They are required, as is vitamin B_6, for the conversion of tryptophan to nicotinic acid. Riboflavin is required for normal metabolism of vitamin B_6 and folate coenzymes (Baker and Frank, 1968; Rivlin, 1970; Tamburro et al., 1971). The requirement is influenced by the dietary content of carbohydrate. Starch increases intestinal synthesis of the vitamin and decreases the required amount in the diet; sucrose does the reverse. Rats fed high-fiber diets had lower red blood cell riboflavin concentrations (Brady et al., 1983).

Adult rats fed diets providing 31 kcal ME/day (130 kJ ME/day) and 0 or 12 mg riboflavin/kg diet were compared to groups provided energy supplements containing sucrose:starch:corn oil (10:3:1 by weight). Glutathione reductase activity coefficients in liver, gastrocnemius, and soleus muscles, and erythrocytes were increased by riboflavin deficiency but were unaffected by the supplemental energy intake. The concentration of riboflavin in muscle decreased in the energy-restricted rats (Turkki et al., 1989). In another study, rats were fed 0.6 or 1.8 g casein/day and 30 mg riboflavin and then progressively restricted to 30 percent of energy intake 6.2 kcal ME/day (26 kJ ME/day); muscle riboflavin concentrations decreased during energy deprivation, did not return to normal with riboflavin supplementation at 100 mg/day, and were unaffected by the amount of protein intake (Turkki and Degruccio, 1983). The effect of protein on the riboflavin requirement was shown to be related to the rate of growth and not to the protein intake per se (Turkki and Holtzapple, 1982; Turkki et al., 1986).

Maximum hepatic storage of flavins was found in rats fed 40 mg/day (equivalent to 2.7 mg/kg diet) and the maximum weight gain at 30 mg/day (equivalent to 2 mg/kg diet)

(Bessey et al., 1958). A dietary content of 0.9 or 1.2 mg/kg produced hepatic storage in rats equivalent to that produced by 15.6 or 23.0 mg/kg (Gaudin-Harding et al., 1971). A dietary concentration of 3 mg/kg saturated the pools of riboflavin, flavin mononucleotide (FMN), and FAD in the retina (Batey and Eckhert, 1992). The effect of exercise has been studied in relation to riboflavin status in the rat. It did not increase the requirement in growing rats but did increase the concentration of flavins in the gastrocnemius and soleus muscles (Hunter and Turkki, 1987). Studies to determine the minimal requirement for riboflavin suggested that it may be as low as 3.2 mg/rat/day, the equivalent of 0.2 mg/kg diet; but the dietary concentration recommended was 17 mg/rat/day or about 1.2 mg/kg diet (Anonymous, 1972). The dietary requirement based on growth response and hepatic stores is 2 to 3 mg riboflavin/kg diet. The requirement can be expressed on a caloric basis as 0.6 to 0.8 mg riboflavin/1,000 kcal ME.

Offspring of dams fed a diet containing 1 mg riboflavin/kg after weaning had decreased body weight, brain weight, and brain DNA content compared to offspring of rats fed a diet containing 8 mg/kg. Correction of these defects in the riboflavin-deficient young was achieved by feeding their dams 2.7 mg/kg diet during lactation but not by supplementation of the diet of the offspring after weaning (Fordyce and Driskell, 1975). In another study rats were fed diets containing 0.40, 0.52, 0.65, or 15.4 mg riboflavin/kg diet 4 weeks before mating, throughout pregnancy, and up to day 15 of lactation (Duerden and Bates, 1985). Fetal resorption occurred in rats receiving 0.4 mg/kg, but those receiving 0.52 mg/kg had successful litters. Dams and their offspring receiving 0.52 or 0.65 mg/kg had higher erythrocyte glutathione reductase activity coefficients and their liver riboflavin concentrations were lower than those fed 15 mg/kg. The concentration of milk riboflavin in the rats fed 0.52 or 0.65 mg/kg was about one-eighth that of rats fed 15 mg/kg diet. The offspring of lactating dams fed diets containing 2, 4, 6, 8, 12, or 16 mg riboflavin/kg had blood glutathione reductase activity coefficients of 1.45, 1.12, 1.12, and 1.12, respectively, at 17 days postpartum. At day 34 the activity coefficients for those fed 4 and 8 mg/kg diet were 1.09 and 1.10 (Leclerc and Miller, 1987). In pregnant rats 4 mg/kg diet produced growth and hepatic riboflavin stores equivalent to 100 mg/kg diet (Schumacher et al., 1965). The intestinal transport of riboflavin is decreased during the period between 14 days to 3 months old and remained constant up to 26 months (Said et al., 1985; Said and Hollander, 1985). The requirement for normal reproduction is 3 to 4 mg/kg diet.

Signs of Riboflavin Deficiency   The classical signs of riboflavin deficiency are dermatitis, alopecia, weakness, and decreased growth. Corneal vascularization and ulceration, cataract formation, anemia, and myelin degeneration may occur (Horwitt, 1954). The activity coefficient of lenticular glutathione reductase was increased and gamma crystallin lowered by deficiency (Bhat, 1982, 1987). Riboflavin-deficient rats may have fatty liver, abnormal hepatocyte mitochondria, and metabolic abnormalities of hepatocytes. The complex metabolic effects of riboflavin deficiency have been reviewed and summarized as (1) a decrease in flavoproteins involved in cellular oxidations; (2) increased protein turnover and an increased pool of free amino acids, which result in increased amounts of enzymes associated with amino acid metabolism, particularly enzymes of gluconeogenesis; and (3) a large decrease in mitochondrial respiration and adenosine triphosphate (ATP) synthesis (Garthoff et al., 1973). Reproductive performance is decreased in both males and females; offspring of deficient females may have congenital anomalies. On day 18 of gestation the iron-mobilizing activity in placental mitochondria was reduced. Maternal iron stores were higher and fetal tissue iron was unaffected presumably because of reduced fetal mass that limited maternal iron depletion and maternofetal iron transfer (Powers, 1987).

Riboflavin deficiency leads to a reduction in the storage of liver and spleen iron, transferrin saturation, hemoglobin concentrations, plasma iron, iron absorption, and liver ferritin-Fe (Adelekan and Thurnham, 1986a; Yu and Cho, 1989; Shanghai, 1991). When deficiencies of riboflavin and iron were induced, riboflavin had a sparing effect on iron status because of the reduction in growth rate induced by the vitamin deficiency (Adelekan and Thurnham, 1986b). In rats fed purified diets deficient in riboflavin, both red blood cell and hepatic glutathione reductase was significantly decreased (Bamji and Sharada, 1972). Red blood cells of deficient rats have decreased fluidity and increased membrane bound acetylcholinesterase, increased glutathione peroxidase activity, and higher concentrations of peroxidation products (Levin et al., 1990). Increases in the amount of lipid peroxides were observed in serum and liver after a 5-week deficiency (Taniguchi, 1980). Rats fed riboflavin-deficient diets had decreased activities of hepatic flavokinase and FAD synthetase but not FMN phosphatase and FAD pyrophosphatase (Lee and McCormick, 1983). Rats deficient in riboflavin had depressed hepatic folate stores despite adequate or even increased intake of folate (Tamburro et al., 1971). Riboflavin-deficient rats prevented from coprophagy had lower hepatic methylenetetrahydrofolate reductase activity but not lower dihydrofolate reductase activity (Bates and Fuller, 1986). Deficient rats had lower activities of the mitochondrial FAD-dependent straight-chain acyl-CoA dehydrogenases and the branched-chain acyl-CoA dehydrogenases (Veitch et al., 1988), NADPH-cytochrome c reductase (Taniguchi, 1980; Wang et al., 1985), ethoxycoumarin-0-deethylase and aryl hydrocarbon hydroxylase (Hietanen et al., 1980), and aflatoxin B_1 activation and DNA adduct formation (Prabhu

et al., 1989). Plasma and tissue carnitine concentrations were reduced by deficiency (Khan-Siddiqui and Bamji, 1987). Riboflavin-deficient rats failed to increase their food intake when fed energy-diluted diets even when given insulin. However, cold exposure stimulated their intake (Matsuo and Suzuoki, 1982).

The activity coefficient of glutathione reductase in red blood cells, liver, and skin correlate with chronic marginal riboflavin deficiency (0, 0.5, 1.0, and 1.5 mg/kg diet). Hepatic and renal FAD is conserved at the expense of riboflavin and FMN. ATP:riboflavin 5-phosphotransferase was decreased in proportion to the concentration of dietary riboflavin, but ATP:FMN adenylyltransferase (FAD pyrophosphorylase) was increased in severely deficient rats. A reduction in succinate:(acceptor) oxidoreductase (succinate dehydrogenase) and NADH:(acceptor) oxidoreductase (NADH dehydrogenase) was tissue dependent (Prentice and Bates, 1981).

Signs of Riboflavin Toxicity   Excessive concentrations of dietary riboflavin have been shown to decrease survivability of newborn rat pups. Mortality during the first week of life was increased 19 percent in a strain of Long-Evans rats by increasing the concentration from 6 to 12 mg/kg (Eckhert, 1987), and the percent of newborn Sprague-Dawley rats surviving to weaning was reduced 7 percent by increasing riboflavin from 8 to 80 mg/kg (Shirley, 1982). Chronic intakes of 12 mg/kg have been shown to cause photoreceptor damage (Eckhert et al., 1989, 1991).

Thiamin

Thiamin is the precursor of thiamin pyrophosphate, which is the storage form and the coenzyme for oxidative decarboxylation and other oxidative reactions. The requirement for thiamin in the diet of the rat depends in part on the quantity and source of dietary energy and is increased by increasing carbohydrate. It has been reported to decrease when dietary fat was increased (Scott and Griffith, 1957), but not in all cases (Murdock et al., 1974). Xylitol had a sparing effect on thiamin resulting from an increase in intestinal facultative bacteria that have the ability to synthesize the vitamin (Rofe et al., 1982). Diabetic rats may have an increased requirement for the vitamin as a result of the alterations in glucose turnover (Berant et al., 1988).

Male weanling rats fed a diet containing either 1.25 or 12.5 mg thiamin/kg diet did not differ in their feed efficiency ratio, but rats fed the higher concentration of thiamin grew faster (Mackerer et al., 1973). There was no significant difference in growth of rats fed either 5 or 50 mg thiamin/kg diet (Itokawa and Fujiwara, 1973). Growth curves of young male rats fed diets containing 0, 0.3, 0.6, 1.5, 6.0, 30.0, and 100.0 mg/kg were used to calculate the amount of thiamin required for maximum growth (Mercer et al., 1986). The theoretical maximum growth rate was 7.0 g/day, and a concentration of 3.68 mg/kg was sufficient to achieve 99 percent (6.93 g/day) of this rate. A concentration of 3.3 mg/kg supported retention of carcass thiamin in nonpregnant and pregnant female rats fed an 18.2 percent protein diet and 3.75 kcal ME/kg diet (15.7 kJ ME/kg) (Roth-Maier et al., 1990). The requirement for growth is 4 mg thiamin-HCl/kg diet.

In pregnant rats the concentration of plasma thiamin remained constant throughout the first 18 days of gestation, while erythrocyte thiamin reached a maximum at day 11 and then declined (Chen et al., 1984). Urinary excretion of thiamin in pregnant rats fed diets containing 2, 4, 6, or 8 mg/kg were stable up to day 16 of gestation and then decreased until parturition (Leclerc, 1991). Pregnant rats fed diets containing 0, 0.8, 1.7, 3.3, 6.7, 13.3, 20, 26.7, 100, 1,000, and 10,000 mg/kg were evaluated for organ thiamin retention (Roth-Maier et al., 1990). The amount of thiamin retained by the liver, muscle, brain, and whole carcass increased with each increase in dietary thiamin. Pregnant rats were fed diets that contained either 4 or 100 mg thiamin/kg. The hepatic stores of thiamin were higher at weaning in the offspring of dams fed the higher concentration, but growth was not affected (Schumacher et al., 1965). The requirement for pregnancy and lactation is 4 mg thiamin-HCl/kg diet.

Signs of Thiamin Deficiency Thiamin deficiency can be induced readily and produces anorexia and weight loss with an increase in food spillage (Tagliaferro and Levitsky, 1982) and in coprophagy (Fajardo and Hornicke, 1989). Thiamin-deficient rats avoid eating sucrose (Yudkin, 1979) and are slower to respond to tasks where food pellets are used for reinforcement (Hashimoto, 1981). Evaluations of behavior demonstrated that rats maintained on thiamin-deficient diets showed muricide aggression (Onodera et al., 1981; Onodera, 1987). After 7 days of deficiency there is a decrease in white and red blood cells and a drop in hemoglobin. After 30 days this is reversed and reticulocytes and plasma erythropoietin are increased, but red blood cell 2,3-diphosphoglycerate, membrane cholesterol, and phospholipids decrease (Hobara and Yasuhara, 1981). After 4 weeks there is a decrease in liver thiamin and an elevation in plasma branched-chain amino acids, α-ketoacids, and α-hydroxyacids (Shigematsu et al., 1989). Deficiency during pregnancy resulted in intrauterine growth retardation (Roecklein et al., 1985) and decreased activity of the thiamin-dependent enzymes pyruvate dehydrogenase, α-ketoglutarate dehydrogenase, and transketolase (Fournier and Butterworth, 1990) and gangliosides (Vaswani, 1985) in newborn rat brains. Acetylcholine was reduced in weanling rats of thiamin-deficient mothers (Kulkarni and Gaitonde, 1983).

In the adult, deficiency results in abnormalities of the central and peripheral nervous systems and the heart. Chronic thiamin deprivation leads to selective neuropathological damage in the brain. Rats become ataxic and a decrease in pyruvate dehydrogenase occurs in the midbrain and lateral vestibular nucleus (Butterworth et al., 1985). Conduction velocity in peripheral nerves was reduced, but axonal transport was increased (McLane et al., 1987). Treatment with pyrithiamine, a thiamine phosphokinase inhibitor, in combination with deficiency, resulted in a decrease in brain amino acids and monoamines (Langlais et al., 1988) and in the activities of glutamic acid decarboxylase and GABA-transaminase (Thompson and McGeer, 1985). Thiamin deficiency produced an increase in cardiac weight, a decrease in cardiac and renal ATP and pyruvate carboxylase, but no cardiac ultrastructural abnormalities (Schenker et al., 1969; McCandless et al., 1970). Cytochrome P-450 concentration and drug metabolizing ability are increased (Wade et al., 1983; Yoo et al., 1990). Thiamin deficiency resulted in an increase in liver nicotinamide methyltransferase and in the excretion of N-1-methylnicotinamide (Shibata, 1986).

Thiamin status has been reported to be influenced by folate deficiency. Folate-deficient rats had decreased absorption of low doses of thiamin, but large doses were absorbed normally (Howard et al., 1974). In an earlier study (Thomson et al., 1972), folate-deficient rats fed 22 mg/kg diet had a significant depletion of thiamin in blood and liver but not in brain. In a more recent study (Walzem and Clifford, 1988b), however, no differences were observed in either thiamin absorption or excretion as a result of folate deficiency.

The enzymatic activity of transketolase in blood and tissues of thiamin-deficient animals correlates with thiamin status, and it may or may not be restored in vitro by the addition of thiamin pyrophosphate (TPP). This may depend on the duration of the deficiency and the resultant instability of the apoenzyme (Brin, 1966; Pearson, 1967; Warnock, 1970; Bamji and Sharada, 1972; Walzem and Clifford, 1988a).

POTENTIALLY BENEFICIAL DIETARY CONSTITUENTS

Requirements have not been determined for the nutrients discussed in this section; however, they are ubiquitous and in abundant supply in natural-ingredient diets and are often missing, or their quantities greatly reduced, in purified diets. Although purified diets support growth and reproduction, numerous investigators have noted that animals exposed to stress, whether carcinogens, age, or diet imbalances, survive longer when fed diets composed of natural ingredients (Longnecker et al., 1981). This suggests that dietary substances not recognized as essential may have beneficial effects.

FIBER

Although dietary fiber has not been shown to be required by the rat, its inclusion in diets may be potentially beneficial. The effects elicited by fiber depend on the properties of the fiber source (i.e., viscosity, solubility, fermentability). Feeding rats fiber increases their fecal bulk and decreases gastrointestinal transit time; decreases in transit time are more pronounced with insoluble fibers (Fleming and Lee, 1983). Increases in the weight of the cecum and colon are observed when fiber is included in rat diets. Inclusion of cellulose (insoluble fiber) in the diet led to greater enlargement of the colon; glucomannan (soluble fiber) led to greater enlargement of the cecum (Konishi et al., 1984).

Increases in cecal wall weight occur in rats fed lactulose, a disaccharide fermented in the cecum, suggesting that microbial fermentation plays an important role in stimulating this hypertrophy (Remesy and Demigne, 1989). The viscosity of fiber sources also may be an important factor influencing cecal hypertrophy (Ikegami et al., 1990). The energy value of fiber for rats depends on fermentation in the hindgut. Microbial fermentation of fiber results in volatile fatty acid production, predominantly of acetate, propionate, and butyrate, which are absorbed and can be used as energy sources by the rat. The digestible energy values of cellulose, a relatively unfermentable fiber, and guar, a highly fermentable fiber, were 0 and 2.4 kcal/g (10 kJ/g), respectively, for the rat. Consumption of guar-containing diets, however, increased heat production by rats such that, despite additional energy supply from guar, there was no additional gain of body energy (i.e., NE = 0; Davies et al., 1991). It is unknown if this thermogenic effect of guar applies to other fermentable fibers.

Additions of insoluble, undegradable sources of fiber such as cellulose, oat hulls, wheat bran, and corn bran to rat diets at concentrations up to 20 percent do not affect growth. Because these nonfermentable fiber sources dilute the nutrient density of the diet, feed intake increases and gain:feed decreases as these fiber sources are added to the diet (Schneeman and Gallaher, 1980; Fleming and Lee, 1983; Lopez-Guisa et al., 1988; Nishina et al., 1991). At high concentrations, viscous polysaccharides such as pectin, guar, and carboxymethylcellulose may decrease weight gain. When added at high concentrations, feed intake may decrease, especially during initial adaptation (Davies et al., 1991). The effects of pectin in particular are difficult to assess because its properties can vary greatly among sources depending on molecular weight and degree of esterification. The more viscous pectins (high molecular weight and degree of esterification) tend to cause greater decreases in feed intake than less viscous pectins (Atallah

and Melnik, 1982). Delorme and Gordon (1983) observed a 30 percent decrease in growth of rats when 4.8 percent pectin was added to diets and a 50 percent mortality when 28.6 percent pectin was added. Fleming and Lee (1983) observed a 35 percent decrease in weight gain when 10 percent pectin was added to the diet, but Nishina et al. (1991), Thomsen et al. (1983), and Track et al. (1982) found no differences in growth when 5 to 8 percent pectin was added to purified fiber-free diets. Guar added to diets at 5 percent of dry matter had no effect on body weight (Ikegami et al., 1990), but 8 percent guar depressed gain (Cannon et al., 1980; Track et al., 1982).

Nitrogen metabolism can be altered by dietary additions of fermentable fiber sources. Fecal nitrogen excretion increases and urinary nitrogen excretion decreases as a result of microbial fermentation and growth in the hindgut. Remesy and Demigne (1989) demonstrated that absorption of ammonia from the hindgut increased when fermentable fiber sources (pectin and guar) were added to the diet, but transfer of urea to the gut was stimulated to a greater extent such that net fecal excretion of nitrogen was increased. The addition of fermentable fiber sources to diets deficient in arginine may improve growth by decreasing the need for arginine for hepatic urea synthesis (Ulman and Fisher, 1983).

Many fiber sources have been used in rat diets including soybean fiber (Levrat et al., 1991), carrageenan, xanthan, alginates (Ikegami et al., 1990), and gum arabic (Tulung et al., 1987). The effects of these fibers can generally be predicted based on their physical properties and fermentabilities. Some carbohydrates that cannot be properly called fiber also elicit some responses similar to those observed for true fibers. Lactulose (disaccharide), raffinose (trisaccharide), and fructooligosaccharides are not absorbed in the small intestine but are rapidly fermented in the hindgut (Fleming and Lee, 1983; Remesy and Demigne, 1989; Tokunaga et al., 1989). Some starches, particularly raw potato, escape small intestinal digestion, are fermented in the cecum, and exert effects similar to true fibers (Calvert et al., 1989).

MINERALS

Many of the mineral elements—including chromium, arsenic, boron, nickel, vanadium, silicon, tin, fluorine, lead, and cadmium—are present in very low concentrations in tissues and body fluids. Some of these elements may be essential for metabolic functions. As with other nutrients, the mineral elements are considered nutritionally essential if a dietary deficiency consistently results in a suboptimal response of an *essential* physiological function, and if the suboptimal response is preventable or reversible by providing physiological amounts of the mineral by dietary or parenteral means (Nielsen, 1984; Underwood and Mertz, 1987).

Perhaps this is a simplistic definition of essentiality when nutrients that have similar physiological functions are considered together. For example, in the antioxidant class of nutrients, selenium and vitamin E may interact so that gross signs of selenium deficiency, such as body weight reduction and reproductive failure, may not be evident in the presence of adequate vitamin E (Combs and Combs, 1984). Similar interactions may occur among other minor nutrients when the deficiency of one may not be expressed in the presence of an abundance of another (Nielsen, 1985).

Many of the essential mineral elements known to be required in the diet in very low concentrations are components of enzymes or metabolic cofactors. Although chromium, arsenic, boron, nickel, vanadium, silicon, tin, lithium, fluorine, lead, and cadmium produce some physiological responses when included in the diets of mammals, they have not been found to associate with enzymes or cofactors. Chromium and vanadium, for example, seem to enhance glucose metabolism, but the mechanism is unknown and the physiological significance of their effects has not been demonstrated; thus, their essentiality remains a question.

Requirements cannot be assessed at this time for any of the mineral elements listed above, but they are widespread in natural-ingredient diets. On the other hand, in purified and chemically defined diets they are often at very low concentrations or cannot be detected. Consequently, a selection of the minor elements are sometimes included in purified diets (Reeves et al., 1993b; Table 2-5).

If these minor elements have any positive effect on the metabolic responses of animals, it could be the result of an indirect effect caused by microbial populations in the gut. For example, the amounts of organic nutrients and/ or unknown growth factors produced by microbes may be changed, or changes in microbial populations may affect the utilization of nutrients (Shurson et al., 1990; Rong et al., 1991; Andrieux et al., 1992; Yoshida et al., 1993).

Chromium

Because supplemental trivalent chromium has been reported to have an enhancing effect on insulin and glucose metabolism, it has been suggested that chromium is essential for the rat and that chromium's function is to aid in the utilization of glucose. The work of Schwarz and Mertz (1959), Schroeder et al. (1963), Schroeder (1966), Roginski and Mertz (1967), Mertz et al. (1965), Mertz and Roginski (1969), and Roginski and Mertz (1969), using highly restrictive environmental conditions, often are cited as evidence for the essentiality of chromium for the rat. Whether this constitutes a beneficial physiological function is uncertain. Specificity is questioned because other heavy metals may

initiate similar effects (Fagin et al., 1987; Pederson et al., 1989).

Other studies have failed to show positive effects of chromium on glucose tolerance or glucose utilization by tissues of rats (Woolliscroft and Barbosa, 1977; Flatt et al., 1989; Holdsworth and Neville, 1990). Woolliscroft and Barbosa (1977) fed 6-week-old Sprague-Dawley rats 30 percent torula yeast diets containing low chromium concentrations (30 to 100 μg/kg, estimated) or diets that contained 5,000 μg Cr/kg. After 6 weeks, there was no significant difference in intravenous glucose tolerance between the two groups. Flatt et al. (1989) found no significant difference in food intake, body weight gain, glycosylated hemoglobin, plasma glucose, plasma insulin, glucose tolerance, or insulin sensitivity between two groups of weanling Wistar rats fed either 30 or 1,000 μg Cr/kg diet for 32 days. Differences in chromium concentrations in tissues between the two groups was variable—from no change in skeletal muscle to a 44 percent reduction in the pancreas.

Holdsworth and Neville (1990) found no effect of dietary chromium supplementation on glucose metabolism in rats. They fed weanling Wistar rats Torula yeast diets similar to those designed by Schwarz (1951) but supplemented with L-cystine, L-methionine, and L-histidine. These supplemented diets supported more rapid growth than the original diet and contained 100 (low-chromium diet) or 1,000 μg Cr/kg (high-chromium diet). A control group was fed a commercial natural-ingredient diet. After 5 weeks, the rats fed the Torula yeast diets gained 30 percent less weight than did the control rats, regardless of whether chromium was present. Those fed chromium-supplemented yeast diets did not grow at a significantly higher rate than those without supplemental chromium. The incorporation of glucose carbon into liver glycogen in the rats fed the low-chromium diet was only one-fifth that of the control rats, but was not different from that of rats given the chromium-supplemented yeast diet. Yeast was grown in media with or without chromium. Extracts from this yeast enhanced glucose incorporation into glycogen of hepatocytes isolated from rats fed low- or high-chromium diets regardless of whether chromium was present in the extract.

Others have reported lower sperm counts in rats fed low-chromium diets (<100 μg/kg) for 8 months than in rats fed high-chromium diets (2,000 μg/kg) (Anderson and Polansky, 1981). Effects of chromium supplementation on weight gain in rats are achieved only with restrictive environmental conditions (Schroeder et al., 1963). Under similar conditions, supplementation of the diet with other heavy metals such as cadmium and lead also enhance initial weight gain, suggesting a nonspecific pharmacological response rather than a nutritional response (Schroeder et al., 1963).

Although the earlier studies seemed to indicate that

dietary chromium supplements enhanced glucose metabolism, the more recent studies did not. It could be argued that the duration of the experiments and environmental conditions in the later studies were not sufficient to allow chromium stores to be depleted and deficiency signs to be expressed. Signs of chromium deprivation might have been more evident if longer feeding periods or multiple-generation studies had been used.

Trivalent chromium salts, chromic oxide, and metallic chromium have low orders of toxicity; however, because of their oxidizing properties, chromium trioxide, chromates, and bichromates are potent poisons. A detailed discussion of tolerance concentrations for chromium in animals can be found in *Mineral Tolerance of Domestic Animals* (National Research Council, 1980).

Lithium

Pickett and O'Dell (1992) fed rats a low-lithium diet (5 µg/kg) through five successive generations and found that the weaning weights of the offspring were significantly lower than weaning weights from dams fed diets with 500 µg Li/kg. Other experiments showed that litter size and birth weights were decreased by feeding rats low-lithium diets through three generations. They also showed an interaction between lithium and sodium in that low-lithium effects were exaggerated in rats fed high-sodium diets. Earlier studies by this laboratory (Patt et al., 1978; Pickett, 1983) showed that second- and third-generation females fed low-lithium diets were less fertile than controls. These studies suggest that diets containing less than 10 µg Li/kg fed to rats through multiple generations could impair reproductive performance.

Signs of Lithium Toxicity   Lithium in high doses can be toxic to the kidney. The minimal toxic concentration of dietary lithium is unknown, but rats fed 280 mg Li/kg diet from 0 to 65 weeks of age developed renal failure (Nyengaard et al., 1994). Nephrotoxicity occurred in rats given 14 mg Li/kg BW/day subcutaneously for 8 days (Qureshi et al., 1992). Rat embryos grown in rat serum with a lithium concentration of 0.6 mmol/L showed signs of toxicity. Until more is learned about the minimal toxic concentration of dietary lithium, it is recommended that the dietary concentration not exceed 1 mg/kg diet (Hansen et al., 1990).

Nickel

Nickel biochemistry plays a prominent role in the metabolism of anaerobic bacteria, plants, and tunicates. Many plant species contain the enzyme urease, which is nickel dependent (Cammack, 1988). Although investigators have shown some cause to believe that nickel is essential for animals, the results among experiments are not consistent. Lederer and Lourau (1948) first suggested that nickel was involved in hematopoiesis because it activated an enzyme required for this process. This began a series of nutritional experiments designed to show a relationship between dietary nickel and iron metabolism. Initial studies showed that nickel-deficient rats showed deficiency signs that could be alleviated by ingesting an adequate dietary concentration of iron (Schneggg and Kirchgessner, 1975a,b, 1976a,b, 1978). Subsequent studies by other investigators (Nielsen et al., 1979; Nielsen, 1980a,b, 1984) demonstrated that the apparent nickel-iron interaction in rats depended on the form of iron fed; however, these investigators could not show a consistent effect of nickel deprivation. They suggested that previously observed effects of nickel on iron metabolism were pharmacological rather than physiological because the amount of nickel supplementation was so high.

Studies with nickel have been carried through multiple generations. Nielsen et al. (1975) fed rats low-nickel diets (2 to 15 µg Ni/kg diet) for three generations and reported that this had no effect on growth of the offspring, but thriftiness and coat condition were worse in the nickel-deprived rats than in similar rats fed diets containing added nickel (3,000 µg/kg). They also found lower hematocrits in the deprived rats than in controls. These results strongly suggest that nickel is essential for the well-being of the rat, but the experiments have not been repeated or confirmed in other laboratories. Other studies have shown that higher concentrations of dietary nickel (20 µg Ni/kg diet) increased weight gain in F_1 and F_2 generation neonatal rats (Nielsen et al., 1979). However, growth rate in rats from weanling to 10 weeks old was not affected by this amount of nickel in the diet (Nielsen et al., 1984).

Studies to determine the possible interaction between nickel and other nutrients have not established with certainty that nickel is essential for growth or any known biochemical process in animal tissues (Nielsen et al., 1989). Given the question about pharmacological versus physiological actions (Nielsen et al., 1984), the lack of marked pathological effects with low dietary intakes of nickel, and the lack of defined biochemical functions in animals, it is not certain that nickel is essential.

Signs of Nickel Toxicity   The amount of dietary nickel required to cause a toxic response in rats is relatively high. Numerous studies have demonstrated that rats have no adverse effects when fed 100 to 1,000 mg Ni/kg diet (Phatak and Patwardhan, 1950; Ambrose et al., 1976) or only lose weight (at 1,000 mg/kg diet; Whanger, 1973). Schnegg and Kirchgessner (1976b) found that rats fed 1,000 mg Ni/kg diet developed many abnormal physiological responses such as increased hematocrit, hemoglobin, and serum protein.

Silicon

Earlier work in two separate laboratories suggested that silicon was an essential nutrient for animals. Carlisle (1972) and Schwarz and Milne (1972) described the effects of silicon supplementation on growth of rats and chicks. Schwarz and Milne observed that the addition of silicon to diets at 500 mg/kg led to increased growth rates in rats fed added silicon as opposed those fed diets not supplemented with silicon. These data may be somewhat misleading because the maximal growth rate of the controls was only 25 to 50 percent of the rate expected for the strain of rat used. This suggests that the diets were generally deficient in some other nutrient(s). Carlisle (1972) showed similar results in chicks fed 100 mg Si/kg diet, but the maximal weight gain of the control chicks was relatively small, only 25 percent of the normal rate for chicks of this age. As in the experiments with rats, the diets seemed to be generally inadequate to support rapid growth.

Although these studies have been cited as establishing silicon essentiality, they have not been verified by subsequent studies. Elliot and Edwards (1991) found that weight gains were depressed in chicks fed 250 mg Si/kg diet compared to those fed diets with no added silicon. No significant effects on growth or any other measures were found in the animals fed the higher silicon concentration compared to those not receiving silicon in their diet. In this experiment the chicks grew at a rate expected for the age and strain.

A number of experiments have shown effects of dietary silicon supplementation on various physiological measures in rats, but none has been able to confirm that the effects are nutritional (Carlisle, 1970; Emerick and Kayongo-Male, 1990a,b; Carlisle et al., 1991).

Signs of Silicon Toxicity Large doses of silicon in the form of tetraethylorthosilicate (2 percent: ≈2,600 mg Si/kg diet) cause urolithiasis in the rat. Death occurred in some rats as a result of urethral obstruction. The lesion was enhanced as the dietary calcium concentration was increased (Schreier and Emerick, 1986).

Sulfur

Sulfur is required as an integral part of sulfur-containing amino acids and vitamins. Michells and Smith (1965) showed that dietary sulfate was readily incorporated into cartilage of Wistar rats and spared methionine for other purposes. Bernhart and Tomarelli (1966) reported a positive effect on growth of Sprague-Dawley rats with added sulfate in the diet. When fed an 8.8 percent lactalbumin diet with a mineral mix that met the requirements of the rat (National Research Council, 1962), growth was improved by inclusion of 1,000 mg sulfate/kg diet. No in-creased growth response was observed when sulfate was included in diets with adequate protein.

Jacob and Forbes (1969) found that weanling Sprague-Dawley rats grew slightly better when fed 15 percent casein diets supplemented with methionine and 350 mg S/kg than with a similar diet supplemented with methionine and 40 mg S/kg as sulfate. Smith (1973) reported that 200 mg inorganic sulfate/kg diet was optimal for adult Long-Evans rats because this amount of inorganic sulfate reduced expiration of $^{14}CO_2$ from a test dose of $1-^{14}C$-methionine. On the basis of these limited data, 300 mg S/kg diet as inorganic sulfate may be beneficial.

Vanadium

In the early 1970s there were several reports from different laboratories that led to the conclusion that vanadium was an essential trace mineral for the rat (Schwarz and Milne, 1971; Strasia, 1971; Hopkins and Mohr, 1974); however, this conclusion has not been supported by all studies (Williams, 1973). Later studies suggest that the results of the previous work demonstrated pharmacological rather than nutritional actions of vanadium (Nielsen, 1984; Nechay et al., 1986; Nielsen and Uthus, 1990).

Reports of the pharmacologic effects of vanadium are numerous. One of the most studied effects is on insulin action. Vanadium is an insulinomimetic agent in vitro and may be in vivo as well. Given in the drinking water, vanadium was shown to lower blood glucose and reduce the activity of phosphotyrosyl-protein phosphatase in the liver of mice (Meyerovitch et al., 1991). The concentration of vanadium given was many times higher than that found in a normal rodent diet, however. Others have shown that pervanadate mimics insulin action by activating the insulin receptor kinase (Fantus et al., 1989). Seaborn et al. (1992) found serum glucose significantly lower in male guinea pigs fed 500 μg V/kg diet compared to those fed less than 10 μg/kg. However, plasma cortisol in the vanadium-fed guinea pigs was elevated more than 100 percent over that of guinea pigs not fed vanadium, suggesting that vanadium in the diet might have stressed the animals. Other measurements were enhanced by vanadium in the diet, but because there were no criteria for normalcy in these studies, the results suggest that vanadium may have caused a toxic reaction rather than demonstrating nutritive value.

Because vanadium is known to be required by some haloperoxidases in lower life forms (Yu and Whittaker, 1989), it has been suggested that peroxidases involved in iodine metabolism in animals may be vanadium-dependent. Some studies have attempted to show an interaction between iodine and vanadium, but the findings were inconclusive and more definitive experiments have not been forthcoming (Uthus and Nielsen, 1990). Because of the lack of definitive and repeatable experiments on the nutritive response to vanadium, its essentiality for the rat or other mammals is uncertain.

Signs of Vanadium Toxicity   The toxicity of vanadium is probably manifested through its effect on tissue enzyme activity. Vanadium has been shown to inhibit numerous enzymes that hydrolyze phosphate esters, including ribonuclease (Lindquist et al., 1973), acid and alkaline phosphatases (Lopez et al., 1976), and phosphotyrosyl-protein phosphatase (Swarup et al., 1982). Sodium, K-ATPase also is inactivated by vanadium (Nieder et al., 1979). Vanadium activates other enzymes such as adenylate cyclase (Grupp et al., 1979) and enhances the phosphorylation of the tyrosyl moiety on proteins. The latter is apparently involved in the overstimulation of membrane receptors, as seen when vanadium stimulates insulin action (Fantus et al., 1989). Rau et al. (1987) found that vanadate stimulated NADH oxidation in microsomes with an increased production of hydrogen peroxide and possible superoxide. Earlier work showed that as little as 25 mg V/kg diet caused visible signs of toxicity such as reduced growth and food utilization in the rat (Franke and Moxon, 1937; Hansard, 1975).

VITAMINS

Ascorbic Acid

Rats do not require a dietary source of ascorbic acid. Enzymatic synthesis of this vitamin can occur via glucuronolactone or gulonolactone in the liver. However, ascorbic acid is a potentially beneficial dietary constituent.

Rats fed ascorbic acid store the vitamin as ascorbic acid and ascorbic acid 2-sulfate (Pillai et al., 1990). Ascorbic acid sulfotransferase was increased, whereas ascorbic acid-2-sulfate sulfohydrolase was reduced in activity in ascorbic acid-supplemented rats. When ascorbic acid was withdrawn from the diet, tissue ascorbic acid, ascorbic acid 2-sulfate, and the activity of ascorbic acid sulfotransferase were reduced and ascorbic acid-2-sulfate sulfohydrolase was increased (Pillai et al., 1990).

Ascorbic acid may be potentially beneficial in thiamin and vitamin B_{12} deficiencies. Five percent ascorbic acid in the diet supported normal weight gain in thiamin-deficient rats and increased the fecal content of thiamin (Scott and Griffith, 1957; Murdock et al., 1974). The inclusion of 100 mg ascorbic acid/kg in a vitamin B_{12}-deficient diet raised liver vitamin B_{12} concentrations in rats (Thenen, 1989).

Several interactions of ascorbic acid with minerals have been identified in rats. Magnesium deficiency reduced ascorbic acid concentration in liver and kidney, as well as the enzymatic synthesis of the vitamin from glucuronolac-

tone or gulonolactone in liver (Hsu et al., 1983). Supplementation of the diet with high iron (5 mg/rat/day) decreased tissue, blood, and urinary concentrations of ascorbic acid (Majumder et al., 1975). However, dietary ascorbic acid increased the absorption of nonheme iron in rats but to a lesser extent than in humans (Reddy and Cook, 1991). Ascorbic acid fed at 1 percent of the diet decreased tissue copper and, in the presence of high iron (191 mg/kg), caused severe anemia and reductions in ceruloplasmin in copper-deficient rats (Johnson and Murphy, 1988). A 1 percent ascorbic acid diet decreased the efficiency of intestinal copper absorption. When copper was given intraperitoneally, however, the rate of copper excretion was decreased (Van den Berg et al., 1989). Lead exposure reduced brain ascorbic acid concentrations (Seshadri et al., 1982). Ascorbic acid given orally was as effective on a molar basis as was parenterally administered EDTA in removing lead from the central nervous system (Goyer and Cherian, 1979). In rats fed lead (500 mg/kg diet) the addition of ascorbic acid as 1 percent of the diet and of 400 mg Fe/kg diet decreased the accumulation of lead in the tissues and prevented growth depression, anemia, and food intake decreases (Suzuki and Yoshida, 1979).

Ascorbic acid may help protect against peroxidation and spare vitamin E. It has been shown to decrease expired pentane, used as a marker for lipid peroxidation (Dillard et al., 1984). Ascorbic acid supplementation reduced the elevation of liver thiobarbituric acid values and reversed decreases in hepatic pyruvate kinase, aspartate aminotransferase, plasma creatine phosphokinase, and vitamin E in vitamin E-deficient rats (Chen and Thacker, 1987). However, high concentrations of ascorbic acid (1.5 g/kg diet) increased in vitro erythrocyte hemolysis and liver lipid peroxidation while lowering reduced glutathione in plasma and erythrocytes (Chen, 1981).

A rat mutant (ODS) unable to synthesize ascorbic acid because of a lack of L-gluconolactone oxidase has been identified (Mizushima et al., 1984). Poor growth, muscle and leg joint hemorrhage, decreased cytochrome P-450, elevated serum and adrenal corticosterone, and lowered urinary excretion of hydroxyproline were prevented in the ODS rat by feeding them 300 mg ascorbic acid/kg diet (Horio et al., 1985). Concentrations of 1,000 to 3,000 mg ascorbic acid/kg diet were required to achieve maximum activities of several microsomal drug-metabolizing enzymes in the liver of these rats when exposed to polychlorinated biphenyls (PCBs) (Horio et al., 1986). A dietary concentration of 250 mg/kg increased survival time to at least 36 weeks following exposure to N-butyl-N-(4-hydroxybutyl)-nitrosamine to induce bladder cancer, while unsupplemented rats died within 4 weeks (Mori et al., 1988). Unsupplemented ODS rats developed an increase in ovarian aromatase activity (Tsuji et al., 1989). LDL cholesterol was higher in unsupplemented ODS rats (Horio et al., 1991).

Myo-*inositol*

Myo-inositol is not required by rats in conventional laboratory conditions, but Burton and Wells (1976) reported a requirement in lactating rats fed antibacterial drugs. Lactating rats fed phthalylsulfathiazole, to decrease *myo*-inositol contribution from the microflora, developed fatty livers with increased concentrations of cholesterol esters and triglycerides, but plasma lipoprotein lipid concentrations were depressed. These alterations in the lactating dam were corrected by supplementing the diet with 0.5 percent *myo*-inositol (Wells and Burton, 1978). The free *myo*-inositol content of milk is 80 mg/100 g in rat's milk and 4 mg/100 g in cow's milk. *Myo*-inositol concentration in milk was influenced by dietary intake (Byun and Jenness, 1982).

Galactose, when fed in high concentrations, results in an accumulation of polyol products, galactitol and sorbitol, which alter osmoregulation and deplete *myo*-inositol. The accumulation of polyols and depletion of *myo*-inositol can be prevented by feeding *myo*-inositol or aldose reductase inhibitors (Bondy et al., 1990). In diabetic rats the accumulation of sorbitol and depletion of *myo*-inositol in peripheral nerves reduces axonal transport of choline acetyltransferase, choline-containing lipids, and motor nerve conduction velocity. Maintaining the concentration of *myo*-inositol in tissue, either through ingesting *myo*-inositol or by the inhibition of aldose reductase, can prevent these changes (Greene et al., 1982; Tomlinson et al., 1986). Na^+-K^+-ATPase in diabetic rats was increased in nerve but not kidney tissue by ingesting dietary *myo*-inositol (Finegold and Strychor, 1988). A phospholipid-derived protein kinase C agonist that is *myo*-inositol dependent may be involved (J. Kim et al., 1991).

A comparison of rats fed diets containing 0 or 5 g *myo*-inositol/kg diet demonstrated that after 3 to 4 days, supplementation decreased the activities of liver fatty acid synthetase and acetyl-CoA followed by a return to unsupplemented concentrations (Beach and Flick, 1982). Liver triglyceride accumulation in rats fed diets devoid of *myo*-inositol only occurred in young rats and decreased with age (Andersen and Holub, 1980). Rats fed a liquid formula diet with *myo*-inositol supplementation (114 or 250 mg/100 g) did not exhibit any differences in weight gain, liver fat, or myelination of the brain; but supplemented rats had higher tissue free and lipid-bound *myo*-inositol concentrations (Burton et al., 1976).

REFERENCES

Aaes-Jorgensen, E., E. E. Leppik, H. W. Hayes, and R. T. Holman. 1958. Essential fatty acid deficiency II. In adult rats. J. Nutr. 66:245–259.

Abdo, K. M., G. Rao, C. A. Montgomery, M. Dinowitz, and K. Kanagalingam. 1986. Thirteen-week toxicity study of *d*-α-tocopheryl acetate (vitamin E) in Fischer 344 rats. Chem. Toxicol. 24:1043–1050.

Abrams, G. M., and P. R. Larsen. 1973. Triiodothyronine and thyroxine in the serum and thyroid glands of iodine-deficient rats. J. Clin. Invest. 52:2522–2531.

Abumrad, N. N., A. J. Schneider, D. Steel, and L. S. Rogers. 1981. Amino acid intolerance during prolonged total parenteral nutrition reversed by molybdate therapy. Am. J. Clin. Nutr. 34:2551–2559.

Adelekan, D. A., and D. I. Thurnham. 1986a. The influence of riboflavin deficiency on absorption and liver storage of iron in the growing rat. Br. J. Nutr. 56:171–179.

Adelekan, D. A., and D. I. Thurnham. 1986b. Effects of combined riboflavin and iron deficiency on the hematological status and tissue iron concentrations of the rat. J. Nutr. 116:1257–1265.

Adkins, J. S., J. M. Wertz, and E. L. Hone. 1966. Influence of nonessential L-amino acids on growth of rats fed high levels of essential L-amino acids. Proc. Soc. Exp. Biol. Med. 122:519–523.

Ahlström, A., and M. Jantti. 1969. Effects of various dietary iron levels on rat reproduction and fetal chemical composition. Ann. Acad. Sci. Fenn. A IV. Biologica 152:1–14.

Ahmad, G., and M. A. Rahman. 1975. Effects of undernutrition and protein malnutrition on brain chemistry of rats. J. Nutr. 105:1090–1103.

Ahrens, R. A. 1967. Influence of dietary nitrogen level on heat production in rats of two ages fed casein and amino-acid diets. J. Dairy Sci. 50:237–239.

Akrabawi, S. S., and J. P. Salji. 1973. Influence of meal-feeding on some of the effects of dietary carbohydrate deficiency in rats. Br. J. Nutr. 30:37–43.

Alexander, H. D., and H. E. Sauberlich. 1957. The influence of lipotropic factors on the prevention of nutritional edema in rats. J. Nutr. 61:329–341.

Allen, K. G. D., K. J. Lampi, P. J. Bostwick, and M. M. Mathias. 1991. Increased thromboxane production in recalcified challenged whole blood from copper-deficient rats. Nutr. Res. 11:61–70.

Almquist, H. J. 1949. Amino acid balance at supernormal dietary levels of protein. Proc. Soc. Exp. Biol. Med. 72:179–184.

Ambrose, P., P. S. Larson, J. F. Borzelleca, and G. R. Hennigar, Jr. 1976. Long-term toxicologic assessment of nickel in rats and dogs. J. Food Sci. Technol. 13:181–186.

American Institute of Nutrition. 1977. Report of the American Institute of Nutrition ad hoc committee on standards for nutritional studies. J. Nutr. 107:1340–1348.

American Institute of Nutrition. 1980. Second report of the ad hoc committee on standards for nutritional studies. J. Nutr. 110:1726.

Ames, S. R. 1974. Age, parity, and vitamin A supplementation and the vitamin E requirement of female rats. Am. J. Clin. Nutr. 27:1017–1025.

Ammerman, C. B., L. R. Arrington, A. C. Warnick, J. L. Edwards, R. L. Shirley, and G. K. Davis. 1964. Reproduction and lactation in rats fed excessive iodine. J. Nutr. 84:108–112.

Andersen, D. B., and B. J. Holub. 1980. *Myo*-inositol-responsive liver lipid accumulation in the rat. J. Nutr. 110:488–495.

Anderson, B. M., and H. E. Parker. 1955. The effects of dietary manganese and thiamine levels on growth rate and manganese concentration in tissues of rats. J. Nutr. 57:55–59.

Anderson, R. A., and M. M. Polansky. 1981. Dietary chromium deficiency effect on sperm count and fertility in rats. Biol. Trace Elem. Res. 3:1–5.

Andrieux, C., E. D. Pacheco, B. Bouchet, D. Gallant, and O. Szylit. 1992. Contribution of the digestive tract microflora to amylomaize starch degradation in the rat. Br. J. Nutr. 67:489–499.

Angel, J. F. 1980. Gluconeogenesis in meal-fed, vitamin B_6 deficient rats. J. Nutr. 110:262–269.

Anonymous. 1972. The fate of riboflavin in the mammal. Nutr. Rev. 30:75.

Anonymous. 1975. Regulation of liver metabolism of pyridoxal phosphate. Nutr. Rev. 33:214.

Aoyama, Y., H. Yasui, and K. Ashida. 1971. Effect of dietary protein and amino acids in a choline-deficient diet on lipid accumulation in rat liver. J. Nutr. 101:730–746.

Apgar, J. 1977. Mobilization of vitamin A by the zinc-deficient female rat. Nutr. Rep. Int. 15:553–559.

Apgar, J. 1985. Zinc and reproduction. Annu. Rev. Nutr. 5:43–68.

Apgar, J. L., C. A. Shively, and S. M. Tarka, Jr. 1987. Digestibility of cocoa butter and corn oil and their influence on fatty acid distribution in rats. J. Nutr. 117:660–665.

Appling, D. R. 1991. Compartmentation of folate-mediated one-carbon metabolism in eukaryotes. FASEB J. 5:2645–2651.

Armbrecht, H. J. 1990. Effect of age on calcium and phosphate absorption. Role of 1,25-dihydroxyvitamin D. Miner. Electrolyte Metab. 16:159–166.

Armbrecht, H. J., and L. R. Forte. 1985. Adaptation of middle aged rats to long-term restriction of dietary vitamin D and calcium. Arch. Biochem. Biophys. 242:464–469.

Armbrecht, H. J., and R. H. Wasserman. 1976. Enhancement of Ca^{++} uptake by lactose in the rat small intestine. J. Nutr. 106:1265–1271.

Arthur, J. R., F. Nicol, A. R. Hutchinson, and G. J. Beckett. 1990. The effects of selenium depletion and repletion on the metabolism of thyroid hormones in the rat. J. Inorg. Biochem. 39:101–108.

Arthur, J. R., F. Nicol, E. Grant, and G. F. Beckett. 1991. The effects of selenium deficiency and hepatic type-I iodothyronine deiodinase and protein disulfide-isomerase assessed by activity measurements and affinity labelling. Biochem. J. 274:297–300.

Atallah, M. T., and T. A. Melnik. 1982. Effect of pectin structure on protein utilization by growing rats. J. Nutr. 112:2027–2032.

Avery, R. A., and W. J. Bettger. 1992. Zinc deficiency alters the protein composition of the membrane skeleton but not the extractability or oligomeric form of spectrin in rat erythrocyte membranes. J. Nutr. 122:428–434.

Aycock, J. E., and A. Kirksey. 1976. Influence of different levels of dietary pyridoxine on certain parameters of developing and mature brains in rats. J. Nutr. 106:680–688.

Babu, U., and M. L. Failla. 1990a. Copper status and function of neutrophils are reversibly depressed in marginally and severely copper-deficient rats. J. Nutr. 120:1700–1709.

Babu, U., and M. L. Failla. 1990b. Respiratory burst and candidacidal activity of peritoneal macrophages are impaired in copper-deficient rats. J. Nutr. 120:1692–1699.

Bajwa, G. S., L. M. Morrison, and B. H. Ershoff. 1971. Induction of aortic and coronary athero-arteriosclerosis in rats fed a hypervitaminosis D, cholesterol-containing diet. Proc. Soc. Exp. Biol. Med. 138:975–982.

Baker, D. A. 1986. Problems and pitfalls in animal experiments designed to establish dietary requirements for essential nutrients. J. Nutr. 116:2339–2349.

Baker, D. H., D. E. Becker, A. H. Jensen, and B. G. Harmon. 1967. Response of the weanling rat to alpha- or beta-lactose with or without an excess of dietary phosphorus. J. Dairy Sci. 50:1314–1318.

Baker, H., and O. Frank. 1968. Clinical vitaminology: Methods and interpretation. New York: Interscience, Wiley.

Baldwin, J. K., and P. Griminger. 1985. Nitrogen balance studies in aging rats. Exp. Gerontol. 20:29–34.

Baldwin, R. L., and A. C. Bywater. 1984. Nutritional energetics of animals. Annu. Rev. Nutr. 4:101–114.

Baly, D. L., D. L. Curry, C. L. Keen, and L. S. Hurley. 1985. Dynamics of insulin and glucagon release in rats: Influence of dietary manganese. Endocrinol. Soc. 116:1734–1740.

Baly, D. L., C. L. Keen, and L. S. Hurley. 1986. Effects of manganese deficiency on pyruvate carboxylase and phosphoenol-pyruvate carboxykinase activity and carbohydrate homeostasis in adult rats. Biol. Trace Elem. Res. 11:201–212.

Baly, D. L., J. S. Schneiderman, and A. L. Garcia-Welsh. 1990. Effect of manganese deficiency on insulin binding glucose transport and metabolism in rat adipocytes. J. Nutr. 120:1075–1079.

Bamji, M. S., and D. Sharada. 1972. Hepatic glutathione reductase and riboflavin concentrations in experimental deficiency of thiamin and riboflavin in rats. J. Nutr. 102:443.

Banwell, J., J. Deese, B. Miller, and D. Bolt. 1980. Inhibition of Vitamin B_{12} absorption by a dietary lectin: Effect of phytohemagglutinins in the rat. Am. J. Clin. Nutr. 33:930 (abstr.).

Barboriak, J. J., and W. A. Krehl. 1957. Effect of ascorbic acid in pantothenic acid deficiency. J. Nutr. 63:601–609.

Barboriak, J. J., W. A. Krehl, and G. R. Cowgill. 1957a. Pantothenic acid requirement of the growing and adult rat. J. Nutr. 61:13–21.

Barboriak, J. J., W. A. Krehl, G. R. Cowgill, and A. D. Whedon. 1957b. Effect of partial pantothenic acid deficiency on reproductive performance of the rat. J. Nutr. 63:591–599.

Barki, V. H., H. Nath, E. B. Hart, and C. A. Elvehjem. 1947. Production of essential fatty acid deficiency symptoms in the mature rat. Proc. Soc. Exp. Biol. Med. 66:474–478.

Barnes, R. H., M. J. Bates, and J. E. Maack. 1946. The growth and maintenance utilization of dietary protein. J. Nutr. 32:535–548.

Barnes, R. H., S. Tuthill, E. Kwong, and G. Fiala. 1959a. Effects of the prevention of coprophagy in the rat. V. Essential fatty acid deficiency. J. Nutr. 68:121–130.

Barnes, R. H., E. Kwong, and G. Fiala. 1959b. Effects of the prevention of coprophagy in the rat. IV. Biotin. J. Nutr. 67:599–610.

Bartholmey, S. J., and A. R. Sherman. 1986. Impaired ketogenesis in iron-deficienct rat pups. J. Nutr. 116:2180–2189.

Bates, C. J., and N. J. Fuller. 1986. The effect of riboflavin deficiency on methylenetetrahydrofolate reductase (NADPH) (EC 1.5.1.20) and folate metabolism in the rat. Br. J. Nutr. 55:455–464; erratum 56:683.

Batey, D. B., and C. D. Eckert. 1992. Flavin levels in the rat retina. Exp. Eye Res. 54:605–609.

Bavetta, L. A., and E. J. McClure. 1957. Protein factors and experimental rat caries. J. Nutr. 63:107–117.

Beach, D. C., and P. K. Flick. 1982. Early effect of myo-inositol deficiency on fatty acid synthetic enzymes of rat liver. Biochim. Biophys. Acta 711:452–459.

Beaton, G. H., and M. C. Cheney. 1965. Vitamin B_6 requirement of the male albino rat. J. Nutr. 87:125–132.

Becker, D. E., A. H. Jensen, S. W. Terrill, I. D. Smith, and H. W. Norton. 1957. The isoleucine requirements of weanling swine fed two protein levels. J. Anim. Sci. 16:26–34.

Beckett, G. J., D. A. MacDougall, F. Nicol, and J. R. Arthur. 1989. Inhibition of type I and type II iodothyronine deiodinase activity in rat liver, kidney and brain produced by selenium deficiency. Biochem. J. 259:887–892.

Bell, J. M., and T. A. Slotkin. 1985. Perinatal dietary supplementation with a commercial soy lecithin preparation: Effects on behavior and brain biochemistry in the developing rat. Dev. Psychobiol. 18:383–394.

Bender, D. A., K. Gartery-Sam, and A. Singh. 1989. Effects of vitamin B_6 deficiency and repletion on the uptake of steroid hormones into uterus slices and isolated liver cells of rats. Br. J. Nutr. 61:619–628.

Bendich, A., E. Gabriel, and L. J. Machlin. 1986. Dietary vitamin E requirement for optimum immune responses in the rat. J. Nutr. 116:675–681.

Benditt, E. P., R. L. Woolridge, C. H. Steffe, and L. E. Frazier. 1950. The minimum requirements of the indispensable amino acids for maintenance of the well-nourished male albino rat. J. Nutr. 40:335–350.

Benevenga, N. J., and R. D. Steele. 1984. Adverse effects of excessive consumption of amino acids. Annu. Rev. Nutr. 4:157–181.

Benevenga, N. J., M. J. Gahl, T. D. Crenshawa and M. D. Finke. 1994. Protein and amino acid requirements for maintenance and amino acid requirements for growth of laboratory rats. J. Nutr. 124:451–453.

Berant, M., D. Berkovitz, H. Mandel, O. Zinder, and D. Mordohovich. 1988. Thiamin status of the offspring of diabetic rats. Pediat. Res. 23:574–575.

Berg, B. N. 1965. Dietary restriction and reproduction in the rat. J. Nutr. 87:344–348.

Berger, J., and B. O. Schneeman. 1988. Intestinal zinc and carboxypeptidase A and B activity in response to consumption of test meals containing various proteins by rats. J. Nutr. 118:723–728.

Bergstra, A. E., A. G. Lemmens, and A. C. Beynen. 1993. Dietary fructose vs. glucose stimulates nephrocalcinogenesis in female rats. J. Nutr. 123:1320–1327.

Bernhart, F. W., and R. M. Tomarelli. 1966. A salt mixture supplying the National Research Council estimates of the mineral requirements of rats. J. Nutr. 89:495–500.

Bernhart, F. W., S. Savini, and R. M. Tomarelli. 1969. Calcium and phosphorus requirements for maximal growth and mineralization of the rat. J. Nutr. 98:443–448.

Bernick, S., and R. B. Alfin-Slater. 1963. Pulmonary infiltration of lipid in essential fatty-acid deficiency. Arch. Pathol. 75:13–20.

Bernsohn, J., and F. J. Spitz. 1974. Linoleic and linolenic acid dependency of some brain membrane-bound enzymes after lipid deprivation in rats. Biochem. Biophys. Res. Commun. 57:293–298.

Berry, M. J., L. Banu, and P. R. Larsen. 1991. Type I iodothyronine deiodinase is a selenocysteine-containing enzyme. Nature 349:438–440.

Bessey, O. A., O. H. Lowry, E. B. Davis, and J. L. Dorn. 1958. The riboflavin economy of the rat. J. Nutr. 64:185–202.

Bhat, K. S. 1982. Alterations in lenticular proteins of rats on riboflavin deficient diet. Curr. Eye Res. 83:829–834.

Bhat, K. S. 1987. Changes in lens and erythrocyte glutathione reductase in response to endogenous flavin adenine dinucleotide and liver riboflavin content of rat on riboflavin deficient diet. Nutr. Res. 7:1203–1208.

Bieri, J. G. 1972. Kinetics of tissue α-tocopherol depletion and repletion. Ann. N.Y. Acad. Sci. 203:181–191.

Bieri, J. G. 1979. Letter to the Editor. J. Nutr. 109:925–926.

Bieri, J. G., K. E. Mason, and E. L. Prival. 1968. Testis requirement for vitamin A in the rat and the effect of essential fatty acid. Int. Z. Vit. Forschung 38:312–319.

Bivin, W. S., M. P. Crawford, and N. R. Brewer. 1979. Morphophysiology. Pp. 73–103 in The Laboratory Rat, Vol. 1, H. J. Baker, J. R. Lindsey, and S. H. Wisbroth, eds. New York: Academic Press.

Black, A., K. H. Maddy, and R. W. Swift. 1950. The influence of low levels of protein on heat production. J. Nutr. 42:415–422.

Blakely, S. R., E. Grundel, M. Y. Jenkins, and G. V. Mitchell. 1990. Alterations in β-carotene and vitamin E status in rats fed β-carotene and excess vitamin A. Nutr. Res. 10:1035–1044.

Blakley, R. L., and Benkovic, S. J., eds. 1984. Folates and Pterins, Vol. 1: Chemistry and Biochemistry of Folates. New York: Wiley.

Blakley, R. L., and V. M. Whitehead, eds. 1986. Folates and Pterins, Vol. 3: Nutritional, Pharmacological and Physiological Aspects. New York: Wiley.

Boelter, M. D. D., and D. M. Greenberg. 1941. Severe calcium deficiency in growing rats. I. Symptoms and pathology. J. Nutr. 21:61–74.

Boelter, M. D. D., and D. M. Greenberg. 1943. Effect of severe

calcium deficiency on pregnancy and lactation in the rat. J. Nutr. 26:105–121.

Bondy, C., B. D. Cowley, Jr., S. L. Lightman, and P. F. Kador. 1990. Feedback inhibition of aldose reductase gene expression in rat renal medulla. Galactitol accumulation enzyme reduces mRNA levels and depletes cellular inositor content. J. Clin. Invest. 86:1103–1108.

Booth, A. N., R. H. Wilson, and F. DeEds. 1953. Effects of prolonged ingestion of xylose on rats. J. Nutr. 49:347–355.

Borel, M. J., S. H. Smith, D. E. Brigham, and J. L. Beard. 1991. The impact of varying degrees of iron nutriture on several functional consequences of iron deficiency in rats. J. Nutr. 121:729–736.

Bottino, N. R., G. A. Vandenburg, and R. Reiser. 1967. Resistance of certain long-chain polyunsaturated fatty acids of marine oils to pancreatic lipase hydrolysis. Lipids 2:489–493.

Bourre, J., M. Francois, A. Youyoo, O. Dumont, M. Piciotti, G. Pascal, and G. Durand. 1989. The effects of dietary α-linolenic acid on the composition of nerve membranes, enzymatic activity, amplitude of electrophysiological parameters, resistance to poisons, and performance of learning tasks in rats. J. Nutr. 119:1880–1892.

Bourre, J. M., M. Piciotti, O. Dumont, G. Pascal, and G. Durand. 1990. Dietary linoleic acid and polyunsaturated fatty acids in rat brain and other organs: Minimal requirements of linoleic acid. Lipids 25:465–472.

Boyden, R., V. R. Potter, and C. A. Elvehjem. 1938. Effect of feeding high levels of copper to albino rats. J. Nutr. 15:397–402.

Brady, P. S., C. M. Knoeber, and L. J. Brady. 1983. High fiber breads and breakfast cereals effect on rat growth and selected vitamin bioavailability. Nutr. Rep. Int. 28:295–304.

Brazda, F. G., and R. A. Coulson. 1946. Toxicity of nicotinic acid and some of its derivatives. Proc. Soc. Exp. Biol. Med. 62:19–20.

Bressani, R., and E. T. Mertz. 1958. Relationship of protein level to the minimum lysine requirement of the rat. J. Nutr. 65:481–491.

Breuer, L. H., W. G. Pond, R. G. Warner, and J. K. Loosli. 1963. A comparison of several amino acid and casein diets for the growing rat. J. Nutr. 80:243–250.

Breuer, L. H., W. G. Pond, R. G. Warner, and J. K. Loosli. 1964. The role of dispensable amino acids in the nutrition of the rat. J. Nutr. 82:499–506.

Breuer, L. H., R. G. Warner, D. A. Benton, and J. K. Loosli. 1966. Dietary requirement for asparagine and its metabolism in rats. J. Nutr. 88:143–150.

Bricker, M. L., and H. H. Mitchell. 1947. The protein requirements of the adult rat in terms of the protein contained in egg, milk and soy flour. J. Nutr. 34:491–505.

Brin, M. 1966. Transketolase: Clinical aspects. Methods Enzymol. 9:508–514.

Brinegar, M. J., H. H. Williams, F. H. Ferris, J. K. Loosli, and L. A. Marnard. 1950. The lysine requirement for the growth of swine. J. Nutr. 42:129–138.

Brink, E. J., P. R. Dekker, E. C. H. van Beresteijn, and A. C. Beynen. 1991. Inhibitory effects of dietary soybean protein vs. casein on magnesium absorption in rats. J. Nutr. 121:1374–1381.

Brink, E. J., A. C. Beynen, P. R. Dekker, E. C. H. Van Beresteijn, and R. van der Meer. 1992. Interaction of calcium and phosphate decreases ileal magnesium solubility and apparent magnesium absorption in rats. J. Nutr. 122:580–586.

Britton, R. S., R. O'Neill, and B. R. Bacon. 1991. Chronic dietary iron overload in rats results in impaired calcium sequestration by hepatic mitochondria and microsomes. Gastroenterology 101:806–811.

Brock, A. A., S. A. Chapman, E. A. Ulman, and G. Wu. 1994. Dietary manganese deficiency decreases rat hepatic arginase activity. J. Nutr. 124:340–344.

Brockerhoff, H., R. J. Hoyle, and P. C. Huang. 1966. Positional distribution of fatty acids in the fats of a polar bear and a seal. Can. J. Biochem. 44:1519–1525.

Brody, S. 1945. Bioenergetics and Growth, with Special Reference to the Efficiency Complex in Domestic Animals. New York: Reinhold.

Brody, S., J. Riggs, K. Kaufman, and V. Herring. 1938. Growth and development with special reference to domestic animals. 45. Energy-metabolism levels during gestation, lactation and post-lactation rest. Mo. Agr. Exp. Stn. Res. Bull. 281.

Brommage, R. 1989. Measurement of calcium and phosphorus fluxes during lactation in the rat. J. Nutr. 119:428–438.

Brommage, R., and H. F. DeLuca. 1984. Self-selection of a high calcium diet by vitamin D-deficient lactating rats increases food consumption and milk production. J. Nutr. 114:1377–1385.

Brubacher, G. B., and H. Weiser. 1985. The vitamin A activity of β-carotene. Int. J. Vitam. Nutr. Res. 55:5–15.

Buchowski, M. S., and D. D. Miller. 1991. Lactose, calcium source and age affect calcium bioavailability in rats. J. Nutr. 121:1746–1754.

Buck, J., F. Derguini, E. Levi, K. Nakanishi, and U. Hammerling. 1991. Intracellular signaling by 14-hydroxy-4,14-retro-retinol. Science 254:1654–1656.

Buckingham, K. W. 1985. Effect of dietary polyunsaturated/saturated fatty acid ratio and dietary vitamin E on lipid peroxidation in the rat. J. Nutr. 115:1425–1435.

Bunce, G. E. 1989. Zinc in endocrine function. Pp. 249–258 in Human Biology, C. F. Mills, ed. New York: Springer-Verlag.

Bunce, G. E., H. E. Sauberlich, P. G. Reeves, and T. S. Oba. 1965. Dietary phosphorus and magnesium deficiency in the rat. J. Nutr. 86:406–414.

Burk, R. F., and P. E. Gregory. 1982. Characteristics of ^{75}Se-P, a selenoprotein found in rat liver and plasma, and comparison of it with selenoglutathione peroxidase. Arch. Biochem. Biophys. 213:73–80.

Burk, R. F. 1983. Biological activity of selenium. Annu. Rev. Nutr. 3:53–70.

Burns, M. J., S. M. Hauge, and F. W. Quackenbush. 1951. Utilization of vitamin A and carotene by the rat. I. Effects of tocopherol, tween, and dietary fat. Arch. Biochem. 30:341–346.

Burton, L. E., and W. W. Wells. 1976. Myo-inositol metabolism during lactation and development in the rat: The prevention of lactation-induced fatty liver by myo-inositol. J. Nutr. 106:1617–1628.

Burton, L. E., R. E. Ray, J. R. Bradford, J. P. Orr, J. A. Nickerson, and W. W. Wells. 1976. Myo-inositol metabolism in the neonatal and developing rat fed a myo-inositol-free diet. J. Nutr. 106:1610–1616.

Butterworth, R. F., J. F. Giguere, and A. M. Besnard. 1985. Activities of thiamin-dependent enzymes in the two experimental models of thiamin-deficiency encephalopathy. 1. The pyruvate dehydrogenase complex. Neurochem. Res. 10:1417–1428.

Byun, S. M., and R. Jenness. 1982. Estimation of free myo-inositol in milks of various species and its source in milk of rats (Rattus norvegicus). J. Dairy Sci. 65:531–536.

Calhoun, J. B. 1963. The ecology and sociology of the Norway rat. Baltimore: U.S. Public Health Service.

Calvert, R. J., M. Otsuka, and S. Satchithanandam. 1989. Consumption of raw potato starch alters intestinal function and colonic cell proliferation in the rat. J. Nutr. 119:1610–1616.

Cammack, R. 1988. Nickel in metalloproteins. Adv. Inorg. Chem. 32:297–333.

Cannon, M., A. Flenniken, and N. S. Track. 1980. Demonstration of acute and chronic effects of dietary fibre upon carbohydrate metabolism. Life Sci. 27:1397–1401.

Cannon, P. R. 1948. Some Pathologic Consequences of Protein and Amino Acid Deficiencies. Springfield, Ill.: Charles C Thomas.

Cardot, P., J. Chambaz, G. Thomas, Y. Rayssiguier, and G. Bereziat. 1987. Essential fatty acid deficiency during pregnancy in the rat: Influence of dietary carbohydrates. J. Nutr. 117:1504–1513.

Carey, M. C., D. M. Small, and C. M. Bliss. 1983. Lipid digestion and absorption. Annu. Rev. Physiol. 45:651–678.

Carlisle, E. M. 1970. Silicon: A possible factor in bone calcification. Science 167:279–280.

Carlisle, E. M. 1972. Silicon: An essential element for the chick. Science 178:619–621.

Carlisle, E. M., M. J. Curran, and T. Duong. 1991. The effect of interrelationships between silicon, aluminum, and the thyroid on zinc content in brain. Pp. 12.16–12.17 in Trace Elements in Man and Animals, B. Momcilovic, ed. Zagreb, Yugoslavia: Institute for Medical Research and Occupational Health, University of Zagreb.

Carmel, N., A. M. Konijn, N. A. Kaufmann, and K. Guggenheim. 1975. Effects of carbohydrate-free diets on the insulin-carbohydrate relationships in rats. J. Nutr. 105:1141–1149.

Caster, W. O., and P. Ahn. 1963. Electrocardiographic notching in rats deficient in essential fatty acids. Science 139:1213.

Cerklewski, F. 1987. Influence of dietary magnesium on fluoride bioavailabilty in the rat. J. Nutr. 117:496–500.

Cerklewski, F. L. 1979. Determination of a copper requirement to support gestation and lactation for the female albino rat. J. Nutr. 109:1529–1533.

Cha, C.-J. M., and H. T. Randall. 1982. Effects of substitution of glucose-oligosaccharides by sucrose in a defined formula diet on intestinal disaccharidases, hepatic lipogenic enzymes and carbohydrate metabolism in young rats. Metabolism 31:57–66.

Chahl, J. S., and C. C. Kratzing. 1966. Environmental temperature and choline requirements in rats. II. Choline and methionine requirements for lipotropic activity. J. Lipid Res. 7:22–26.

Chalvardjian, A., and S. Stephens. 1970. Lipotropic effect of dextrin versus sucrose in choline-deficient rats. J. Nutr. 100:397–403.

Champigny, O. 1963. Respiratory exchange and nitrogen balance in the pregnant rat: Energy cost of anabolic processes of gestation. C. R. Acad. Sci. Paris 256:4755.

Chang, S., P. M. Brannon, and M. Korc. 1990. Effects of dietary manganese deficiency on rat pancreatic amylase mRNA levels. J. Nutr. 120:1228–1234.

Chen, I. S., S. Subramanian, M. M. Cassidy, A. J. Sheppard, and G. V. Vahouny. 1985. Intestinal absorption and lipoprotein transport of (φ-3) eicosapentanoic acid. J. Nutr. 115:219–225.

Chen, I. S., L. Truc, S. Subramanian, M. M. Cassidy, A. J. Sheppard, and G. V. Vahouny. 1987a. Comparison of the clearances of serum chylomicron triglycerides enriched with eicosapentanoic acid or oleic acid. Lipids 22:318–321.

Chen, I. S., S. Hotta, I. Ikeda, M. M. Cassidy, A. J. Sheppard, and G. V. Vahouny. 1987b. Digestion, absorption and effects on cholesterol absorption of Menhaden oil, fish oil concentrate and corn oil by rats. J. Nutr. 117:1676–1680.

Chen, I. S., S. Subramanian, G. V. Vahouny, M. M. Cassidy, I. Ikeda, and D. Kritchevsky. 1989. A comparison of the digestion and absorption of cocoa butter and palm kernel oil and their effects on cholesterol absorption in rats. J. Nutr. 119:1569–1573.

Chen, L. H. 1981. An increase in vitamin E requirement induced by high supplementation of vitamin C in rats. Am. J. Clin. Nutr. 34:1036–1041.

Chen, L. H., and A. L. Marlatt. 1975. Effects of dietary vitamin B_6 levels and exercise on glutamic-pyruvic transaminase activity in rat tissues. J. Nutr. 105:401.

Chen, L. H., and R. R. Thacker. 1987. Effect of ascorbic acid and vitamin E on biochemical changes associated with vitamin E deficiency in rats. Int. J. Vitam. Nutr. Res. 57:385–390.

Chen, L. T., C. H. Bowen, and M. F. Chen. 1984. Vitamin B_1 and B_{12} blood levels in rats during pregnacy. Nutr. Rep. Int. 30:433–437.

Chow, C. K. 1979. Nutritional influence on cellular antioxidant defense systems. Am. J. Clin. Nutr. 32:1066–1081.

Clandinin, M. T., S. Cheema, C. J. Field, M. L. Garg, J. Venkatraman, and T. R. Clandinin. 1991. Dietary fat: Exogenous determination of membrane structure and cell function. FASEB J. 5:2761–2769.

Clark, I., and L. F. Belanger. 1967. Effects on alteration of dietary magnesium on calcium, phosphate and skeletal metabolism. Calcif. Tissue Res. 1:204–218.

Clark, S. A., A. Boass, and S. U. Toverud. 1987. Effects of high dietary contents of calcium and phosphorus on mineral metabolism and growth of vitamin D-deficient suckling and weaned rats. Bone Mineral 2:257–270.

Clifford, A. J., D. S. Wilson, and N. D. Bills. 1989. Repletion of folate-depleted rats with an amino acid-based diet supplemented with folic acid. J. Nutr. 119:1956–1961.

Coburn, S. P., J. D. Mahuren, B. S. Wostmann, D. L. Snyder, and D. W. Townsend. 1989. Role of intestinal microflora in the metabolism of vitamin B_6 and 4'-deoxypyridoxine examined using germ-free guinea pigs and rats. J. Nutr. 119:181–188.

Cohen, N. L., C. L. Keen, and B. Lönnerdal. 1985. Effects of varying iron on the expression of copper deficiency in the growing rat: Anemia and ferroxidase I and II, tissue trace elements, ascorbic acid, and xanthine dehydrogenase. J. Nutr. 115:633–649.

Cole, A. S., and W. Robson. 1951. Tryptophan deficiency and requirements in the adult rat. Br. J. Nutr. 5:306–320.

Combs, G. F., Jr., and S. B. Combs. 1984. The nutritional biochemistry of selenium. Annu. Rev. Nutr. 4:257–280.

Compton, M. M., and J. A. Cidlowski. 1986. Vitamin B_6 and glucocorticoid action. Endocrine Rev. 7:140–148.

Conrad, H. E., W. R. Watts, J. M. Iacino, H. F. Kraybill, and T. E. Friedmann. 1958. Digestibility of uniformly labeled carbon-14 soybean cellulose in the rat. Science 127:1293.

Corey, J. E., and K. C. Hayes. 1972. Cerebrospinal fluid pressure, growth, and hematology in relation to retinol status of the rat in acute vitamin A deficiency. J. Nutr. 102:1585–1593.

Coward, K. H., K. M. Key, and B. G. E. Morgan. 1932. The quantitative determination of vitamin D by means of its growth-promoting property. Biochem. J. 26:1585–1592.

Crawford, M. A., N. M. Casperd, and A. J. Sinclair. 1976. The long chain metabolites of linoleic and linolenic acids in liver and brain in herbivores and carnivores. Comp. Biochem. Physiol. 54B:395–401.

Crockett, M. E., and H. J. Deuel, Jr. 1947. A comparison of the coefficient of digestibility and rate of absorption of several natural and artificial fats as influenced by melting point. J. Nutr. 33:187–194.

Crosby, L. O., and T. R. Cline. 1973. Effect of asparagine on growth and protein synthesis in weanling rats. J. Anim. Sci. 37:713–717.

Cullen, R. W., and S. M. Oace. 1989a. Fermentable dietary fibers elevate urinary methylmalonate and decrease propionate oxidation in rats deprived of vitamin B_{12}. J. Nutr. 119:1115–1120.

Cullen, R. W., and S. M. Oace. 1989b. Neomycin has no persistent sparing effect on vitamin B_{12} status in pectin-fed rats. J. Nutr. 119:1399–1403.

Cuthbertson, E. M., and D. M. Greenberg. 1945. Chemical and pathological changes in dietary chloride deficiency in the rat. J. Biol. Chem. 160:83–94.

Czarnecki, G. L., M. S. Edmonds, O. A. Izquierdo, and D. H. Baker. 1984. Effect of 3-nitro-4'-hydroxyphenylarsenic acid on copper utilization by the pig, rat, and chick. J. Anim. Sci. 59:997–1002.

Daft, F. S., L. L. Ashburn, and W. H. Sebrell. 1942. Biotin deficiency and other changes in rats given sulfanilylguanidine or succinyl sulfathiazole in purified diets. Science 96:321–322.

Davies, I. R., J. C. Brown, and G. Livesey. 1991. Energy values and energy balance in rats fed on supplements of guar gum or cellulose. Br. J. Nutr. 65:415–433.

Davila, M. E., L. Norris, M. P. Cleary, and A. C. Ross. 1985. Vitamin A during lactation: Relationship of maternal diet to milk vitamin A content and to the vitamin A status of lactating rats and their pups. J. Nutr. 115:1033–1041.

Davis, C. D., D. M. Ney, and J. L. Greger. 1990. Manganese, iron and lipid interactions in rats. J. Nutr. 120:507–513.

Davis, G. K., and W. Mertz. 1987. Copper. Pp. 301–364 in Trace Elements in Human and Animal Nutrition, W. Mertz, ed. Orlando, Fla.: Academic Press.

Day, H. G., and E. V. McCollum. 1939. Mineral metabolism, growth and symptomatology of rats and a diet extremely deficient in phosphorus. J. Biol. Chem. 130:269–283.

Day, H. G., and W. Pigman. 1957. Carbohydrates in nutrition. Pp. 779–806 in The Carbohydrates, W. Pigman, ed. New York: Academic Press.

Deb, S., R. J. Martin, and T. V. Hershberger. 1976. Maintenance requirement and energetic efficiency of lean and obese Zucker rats. J. Nutr. 106:191–197.

Deka, N. C., A. K. Sehgal, and P. N. Chhuttani. 1981. Absorption and transport of radioactive cobalt-57 labeled vitamin B_{12} in experimental giardiasis in rats. Indian J. Med. Res. 74:675–679.

Delorme, C. B., and C. I. Gordon. 1983. The effect of pectin on the utilization of marginal levels of dietary protein by weanling rats. J. Nutr. 113:2432–2441.

Demetriou, A. A., I. Franco, S. Bark, G. Rettura, E. Seifter, and S. M. Levenson. 1984. Effects of vitamin A and beta carotene on intra-abdominal sepsis. Arch. Surg. 119:161–165.

Deuel, H. J., Jr. 1950. Non-caloric functions of fat in the diet. J. Am. Diet. Assoc. 26:255–259.

Deuel, H. J., Jr. 1955. Fat as a required nutrient of the diet. Fed. Proc. 14:639–649.

Deuel, H. J., E. R. Meserve, E. Straub, C. Hendrick, and B. T. Scheer. 1947. The effect of fat level of the diet on general nutrition. I. Growth, reproduction, and physical capacity of rats receiving diets containing various levels of cottonseed oil or margarine fat ad libitum. J. Nutr. 33:569–582.

Deuel, H. J., Jr., C. R. Martin, and R. B. Alfin-Slater. 1954. The effect of fat level of the diet on general nutrition. XII. The requirement of essential fatty acids for pregnancy and lactation. J. Nutr. 54:193–199.

Dialameh, G. H., W. V. Taggart, J. T. Matschiner, and R. E. Olson. 1971. Isolation and characterization of menaquinone-4 as a product of menadione metabolism in chicks and rats. Int. J. Vitam. Nutr. Res. 41:391–400.

Diamond, I., H. Swenerton, and L. S. Hurley. 1971. Testicular and esophageal lesions in zinc-deficient rats and their reversibility. J. Nutr. 101:77–84.

Dibak, O., M. Krajcovicova-Kudlackova, E. Grancicova, and M. Jankovicova. 1986. Body composition and physiological casein and wheat gluten protein requirements of 180-day-old rats. Physiol. Bohemoslov. 35:71–80.

Dillard, C. J., J. E. Downey, and A. L. Tappel. 1984. Effect of antioxidants on lipid peroxidation in iron-loaded rats. Lipids 19:127–133.

DiPaolo, R. V., V. S. Caviness, Jr., and J. N. Kanfer. 1974. Delayed maturation of the renal cortex in the vitamin B_6-deficient newborn rat. Pediat. Res. 8:546.

Dirige, O. V., and J. R. Beaton. 1969. Factors affecting vitamin B_6

requirement in the rat as determined by erythrocyte transaminase activity. J. Nutr. 97:109–116.

Dodds, R. A., A. Catterall, L. Bitensky, and J. Chayen. 1986. Abnormalities in fracture healing induced by vitamin B_6 deficiency in rats. Bone 7:489–495.

Doi, T., T. Kawata, N. Tadano, T. Iijima, and A. Maekawa. 1989. Effect of vitamin B_{12} deficiency on the activity of hepatic cystathionine beta-synthase in rats. J. Nutr. Sci. Vitaminol. 35:101–110.

Doisy, E. A., Jr. 1961. Nutritional hypoprothrombinemia and metabolism of vitamin K. Fed. Proc. 20:989–994.

Draper, H. H., T.-L. Sie, and J. G. Bergan. 1972. Osteoporosis in aging rats induced by high phosphorus diets. J. Nutr. 102:1133–1142.

Drescher, A. N., N. B. Talbot, P. A. Meara, M. Terry, and J. D. Crawford. 1958. A study of the effects of excessive potassium intake upon body potassium stores. J. Clin. Invest. 37:1316–1322.

Driskell, J. A., and S.-L. L. Chuang. 1974. Relation between glutamate decarboxylase activities in brains and the vitamin B_6 requirement of male rats. J. Nutr. 104:1657–1661.

Driskell, J. A., L. A. Strickland, C. H. Poon, and D. P. Foshee. 1973. The vitamin B_6 requirement of the male rat as determined by behavioral patterns, brain pyridoxal phosphate and nucleic acid compositin and erythrocyte alanine aminotransferase activity. J. Nutr. 103:670–680.

Dubick, M. A., G. S. M. Yu, and A. P. N. Majumdar. 1989. Morphological and biochemical changes in the pancreas of the copper-deficient female rat. J. Nutr. 119:1165–1172.

Duerden, J. M., and C. J. Bates. 1985. Effect of riboflavin deficiency on reproductive performance and on biochemical indices of riboflavin status in the rat. Br. J. Nutr. 53:97–105.

Dupont, J. 1990. Lipids. Pp. 56–66 in Present Knowledge in Nutrition, M. L. Brown, ed. Washington, D.C.: International Life Sciences Institute.

Dussault, P. E., and M. Lepage. 1979. In vitro studies of fatty acid metabolism in vitamin B_6 deficient rats. J. Nutr. 109:138–141.

Eckhert, C. D. 1987. Differential effects of riboflavin and *RRR*-tocopheryl acetate on the survival of newborn RCS rats with inheritable retinal degeneration. J. Nutr. 117:208–211.

Eckhert, C. D., M. H. Hsu, and D. W. Batey. 1989. Effect of dietary riboflavin on retinal density and flavin concentrations in normal and dystrophic RCS rats: Inherited and environmentally induced retinal degenerations. Prog. Clin. Biol. Res. 314:331–341.

Eckhert, C. D., M. H. Hsu, and N. Pang. 1991. Photoreceptor damage following exposure to excess riboflavin. Experientia 49:1084–1087.

Eckhert, C. D., M. K. Lockwood, and B. Shen. 1993. Influence of selenium on the microvasculature of the retina. Microvas. Res. 45:74–78.

Edelstein, S., and K. Guggenheim. 1971. Effects of sulfur-amino acids and choline on vitamin B_{12}-deficient rats. Nutr. Metab. 13:339–343.

Elliot, M. A., and H. M. Edwards, Jr. 1991. Effect of dietary silicon on growth and skeletal development in chickens. J. Nutr. 121:201–207.

Emerick, R. J., and H. Kayongo-Male. 1990a. Interactive effects of dietary silicon, copper and zinc in the rat. J. Nutr. Biochem. 1:35–40.

Emerick, R. J., and H. Kayongo-Male. 1990b. Silicon facilitation of copper utilization in the rat. J. Nutr. Biochem. 1:487–492.

Emery, M. P., J. D. Browning, and B. L. O'Dell. 1990. Impaired hemostasis and platelet function in rats fed low zinc diets based on egg white protein. J. Nutr. 120:1062–1067.

Engel, R. W. 1942. The relation of B vitamins and dietary fat to the lipotropic action of choline. J. Nutr. 24:175–185.

Enslen, M., H. Milon, and A. Malnoe. 1991. Effect of low intake of n-3 fatty acids during development on brain phospholipid fatty acid composition and exploratory behavior in rats. Lipids 26:203–208.

Ericsson, Y., H. Luoma, and O. Ekberg. 1986. Effects of calcium, fluoride and magnesium supplementations on tissue mineralization in calcium- and magnesium-deficient rats. J. Nutr. 116:1018–1027.

Ertel, N. H., S. Akgun, F. W. Kemp, and J. C. Mittler. 1983. The metabolic fate of exogenous sorbitol in the rat. J. Nutr. 113:566–573.

Evans, H. M., and G. A. Emerson. 1943. The prophylactic requirement of the rat for alpha-tocopherol. J. Nutr. 26:555–568.

Evenson, J. K., K. M. Thompson, S. L. Weiss, and R. A. Sunde. 1992. Dietary selenium regulation of ^{75}Se—Implications for the selenium requirement. FASEB J. 6:(Part 1)A1398.

Fagin, J. A., K. Ikejiri, and S. R. Levin. 1987. Insulinotropic effects of vanadate. Diabetes 36:1248–1252.

Failla, M. L., U. Babu, and K. E. Seidel. 1988. Use of immunoresponsiveness to demonstrate that the dietary requirement for copper in young rats is greater with dietary fructose than dietary starch. J. Nutr. 118:487–496.

Fajardo, G., and H. Hornicke. 1989. Problems in estimating the extent of coprophagy in the rat. Br. J. Nutr. 62:551–561.

Fantus, G. I., S. Kadota, G. Deragon, B. Foster, and B. I. Posner. 1989. Pervanadate [peroxide(s) of vanadate] mimics insulin action in rat adipocytes via activation of the insulin receptor tyrosine kinase. Biochemistry 28:8864–8871.

Feldmann, J. D. 1960. Iodine deficiency in newborn rats. Am. J. Physiol. 199:1081–1083.

Fergusson, M. A., and K. G. Koski. 1990. Comparison of effects of dietary glucose versus fructose during pregnancy on fetal growth and development in rats. J. Nutr. 120:1312–1319.

Fields, M., J. Holbrook, D. Scholfield, A. Rose, J. C. Smith, and S. Reiser. 1986. Development of copper deficiency in rats fed fructose or starch: Weekly measurements of copper indices in blood. Proc. Soc. Exp. Biol. Med. 181:120–124.

Finegold, D. N., and S. Strychor. 1988. Renal ouabain inhibitable Na^+-K^+-ATPase activity and myoinositol supplementation in experimental diabetes mellitus. Metabolism 37:557–561.

Finke, M. D., G. R. DeFoliart, and N. J. Benevenga. 1987a. Use of a four-parameter logistic model to evaluate the protein quality of mixtures of mormon cricket meal and corn gluten meal. J. Nutr. 117:1740–1750.

Finke, M. D., G. R. DeFoliart, and N. J. Benevenga. 1987b. Use of simultaneous curve fitting and a four-parameter logistic model to evaluate the nutritional quality of protein sources at growth rates of rats from maintenance to maximum gain. J. Nutr. 117:1681–1688.

Finke, M. D., G. R. DeFoliart, and N. J. Benevenga. 1989. Use of a four-parameter logistic model to evaluate the quality of the protein from three insect species when fed to rats. J. Nutr. 119:864–871.

Fisher, J. H., R. A. Willis, and B. E. Haskell. 1984. Effect of protein quality on vitamin B_6 status in the rat. J. Nutr. 114:786–791.

Flatt, P. R., L. Juntti-Berggren, P. O. Berggren, B. J. Gould, and S. K. Swanston-Flatt. 1989. Effects of dietary inorganic trivalent chromium (Cr^{3+}) on the development of glucose homeostasis in rats. Diabetes Metab. 15:93–97.

Fleming, B. B., and C. H. Barrows, Jr. 1982. The influence of aging on intestinal absorption of vitamin B_{12} and niacin in rats. Exp. Gerontol. 17:121–126.

Fleming, S. E., and B. Lee. 1983. Growth performance and intestinal transit time of rats fed purified and natural dietary fibers. J. Nutr. 113:592–601.

Follis, R. H. 1958. Deficiency disease. Springfield, Ill.: Charles C Thomas.

Forbes, E. B., R. W. Swift, R. R. Elliott, and W. H. James. 1946a. Relation of fat to economy of food utilization. I. By the growing albino rat. J. Nutr. 31:203–212.

Forbes, E. B., R. W. Swift, R. R. Elliott, and W. H. James. 1946b.

Relation of fat to economy of food utilization. II. By the mature albino rat. J. Nutr. 31:213–227.

Forbes, R. M. 1966. Effects of magnesium, potassium and sodium nutriture on the mineral composition of selected tissues of the albino rat. J. Nutr. 88:403–410.

Forbes, R. M., and L. Vaughan. 1954. Nitrogen balance of young albino rats force-fed methionine or histidine deficient diets. J. Nutr. 52:25–37.

Forbes, R. M., L. Vaughan, and H. W. Norton. 1955. Studies on the utilization of dietary isoleucine by the growing albino rat. I. Isoleucine requirements determined with amino acid mixtures. J. Nutr. 57:593–598.

Forbes, R. M., and T. Rao. 1959. The effect of age on the net requirements for nitrogen, lysine and tryptophan by the well-fed rat. Arch. Biochem. Biophys. 82:348–354.

Fordyce, M. K., and J. A. Driskell. 1975. Effects of riboflavin repletion during different developmental phases on behavioral patterns, brain nucleic acid and protein contents, and erythrocyte glutathione reductase activity of male rats. J. Nutr. 105:1150–1156.

Fosmire, A. J., S. Greeley, and H. H. Sandstead. 1977. Maternal and fetal response to various suboptimal levels of zinc intake during gestation in the rat. J. Nutr. 107:1543–1550.

Fournier, H., and R. F. Butterworth. 1990. Effects of thiamin deficiency on thiamin-dependent enzymes in regions of the brain of pregnant rats and their offspring. Metab. Brain Dis. 5:77–84.

Frank, O., and H. Baker. 1980. Vitamin profile in rats fed stock or liquid ethanolic diets. Am. J. Clin. Nutr. 33:221–226.

Franke, K. W., and A. L. Moxon. 1937. The toxicity of orally ingested arsenic, selenium, tellurium, vanadium and molybdenum. J. Pharmacol. Exp. Ther. 61:89–102.

French, C. E., R. H. Ingram, J. A. Uram, G. P. Barron, and R. W. Swift. 1953. The influence of dietary fat and carbohydrate on growth and longevity in rats. J. Nutr. 51:329–339.

Fritz, T. E., D. V. Tolle, and R. J. Flynn. 1968. Hemorrhagic diathesis in laboratory rodents. Proc. Soc. Exp. Biol. Med. 128:228–234.

Fujii, K., T. Kajiwara, and H. Kurosu. 1979. Effect of vitamin B_6 deficiency on the crosslink formation of collagen. FEBS Lett. 97:193–195.

Furuse, M., Y. Tamura, S. Matsuda, T. Shimizu, and J. Okumura. 1989. Lower fat deposition and energy utilization of growing rats fed diets containing sorbose. Comp. Biochem. Physiol. 94A:813–817.

Gabriel, E., L. J. Machlin, R. Filipski, and J. Nelson. 1980. Influence of age on the vitamin E requirement for resolution of necrotizing myopathy. J. Nutr. 110:1372–1379.

Gahl, M. J., M. D. Finke, T. D. Crenshaw, and N. J. Benevenga. 1991. Use of a four-parameter logistic equation to evaluate the response of growing rats to ten levels of each indispensable amino acid. J. Nutr. 121:1720–1729.

Gander, J. E., and M. O. Schultze. 1955. Concerning the alleged occurrence of an "animal protein factor" required for the survival of young rats. I. Studies with unpurified rations. J. Nutr. 55:543–557.

Ganguli, M. C., J. D. Smith, and L. E. Hanson. 1969a. Sodium metabolism and its requirement during reproduction in female rats. J. Nutr. 99:225–234.

Ganguli, M. C., J. D. Smith, and L. E. Hanson. 1969b. Sodium metabolism and requirements in lactating rats. J. Nutr. 99:395–400.

Garthoff, L. H., S. K. Garthoff, R. B. Tobin, and M. A. Mehlman. 1973. The effect of riboflavin deficiency on key gluconeogenic enzyme activities in rat liver. Proc. Soc. Exp. Biol. Med. 143:693–697.

Gaudin-Harding, F., S. Griglio, B. Bois-Joyeuz, P. de Gasquet, and R. Karlin. 1971. Réserves en vitamines du group B chez des rats Westar soumis a des régimes témoins ou hyperlipidiques à deux niveaux vitaminiques. J. Int. Vitaminol. Nutr. 42:21–32.

Gerber, L. E., and J. W. Erdman. 1982. Effect of dietary retinyl acetate, β-carotene and retinoic acid on wound healing in rats. J. Nutr. 112:1555–1564.

Gershoff, S. N. 1970. Production of urinary calculi in vitamin B_6-deficient male, female, and castrated male rats. J. Nutr. 100:117–122.

Giovanetti, P. M. 1982. Effect of coprophagy on nutrition. Nutr. Res. 2:335–349.

Goettsch, M. 1949. Minimal protein requirement of the rat for reproduction and lactation. Arch. Biochem. 21:289–300.

Goldberg, A. 1971. Carbohydrate metabolism in rats fed carbohydrate-free diets. J. Nutr. 101:693–698.

Goulding, A., and R. S. Malthus. 1969. Effect of dietary magnesium on the development of nephrocalcinosis in rats. J. Nutr. 97:353–358.

Goyer, R. A., and M. G. Cherian. 1979. Ascorbic acid and EDTA treatment of lead toxicity in rats. Life Sci. 24:433–438.

Grau, C. R. 1948. Effect of protein level on the lysine requirement of the chick. J. Nutr. 36:99–108.

Gray, G. M., R. J. White, and J. R. Majer. 1978. 1–(3′-0-acyl)-β-glucosyl-N-dihydroxypentatriacontadienoylsph igosine, a major component of the glucosyl ceramides of pig and human epidermis. Biochim. Biophys. Acta 528:127–137.

Gray, L. F., and L. J. Daniel. 1964. Effect of the copper status of the rat on the copper-molybdenum-sulfate interaction. J. Nutr. 84:3137.

Greaves, J. H., and P. Ayres. 1973. Warfarin resistance and vitamin K requirement in the rat. Lab. Anim. 7:141–148.

Green, M. H., and J. B. Green. 1991. Influence of vitamin A intake on retinol (ROH) balance, utliization and dynamics. FASEB J. 5:A718.

Green, M. H., J. B. Green, and K. C. Lewis. 1987. Variation in retinol utilization rate with vitamin A status in the rat. J. Nutr. 117:694–703.

Greenberg, D. M., and E. M. Cuthbertson. 1942. Dietary chloride deficiency and alkalosis in the rat. J. Biol. Chem. 145:179–187.

Greenberg, S. M., C. E. Calbert, E. E. Savage, and H. J. Deuel, Jr. 1950. The effect of fat level of the diet on general nutrition. J. Nutr. 41:473–486.

Greene, D. A., R. A. Lewis, S. A. Lattimer, and M. J. Brown. 1982. Selective effects of *myo*-inositol administration on sciatic and tibial motor nerve conduction parameters in the streptozocin-diabetic rat. Diabetes 31:573–578.

Greger, J. L. 1982. Effect of phosphorus-containing compounds on iron and zinc utilization. Pp. 107–120 in Nutritional Bioavailability of Iron, ACS Symposia Series 203, C. Kies, ed. Washington, D.C.: American Chemical Society.

Greger, J. L. 1989. Effect of dietary protein and minerals on calcium and zinc utilization. Crit. Rev. Food Sci. Nutr. 28:249–271.

Greger, J. L., C. E. Krzykowski, R. R. Khazen, and C. L. Krashoc. 1987a. Mineral utilization by rats fed various commercially available calcium supplements or milk. J. Nutr. 117:717–724.

Greger, J. L, C. L. Krashoc, and C. E. Krzykowski. 1987b. Calcium, sodium and chloride interactions in rats. Nutr. Res. 7:401–412.

Gregory, J. F., III. 1980a. Effects of epsilon-pyridoxyllysine bound to dietary protein on the vitamin B_6 status of rats. J. Nutr. 110:995–1005.

Gregory, J. F., III. 1980b. Effects of epsilon pyridoxyl lysine and related compounds on liver and brain pyridoxal kinase EC 2.7.1.35 and liver pyridoxamine pyridoxine 5′-phosphate oxidase EC 1.4.3.5. J. Biol. Chem. 255:2355–2359.

Grenby, T. H., and A. Phillips. 1989. Dental and metabolic effects of lactitol in the diet of laboratory rats. Br. J. Nutr. 61:17–24.

Griffith, W. H. 1941. Choline metabolism. V. The effect of supplementary choline, methionine, and cystine and of casein, lactalbumin, fibrin, edestin and gelatin in hemorrhagic degeneration in young rats. J. Nutr. 21:291–306.

Griffith, W. H., and N. J. Wade. 1939. Choline metabolism. I. The occurrence and prevention of hemorrhagic degeneration in young rats on a low choline diet. J. Biol. Chem. 131:567–577.

Grigor, M. R., M. J. Sneyd, A. Geursen, and U. R. Gain. 1984. Effects of litter size at mid-lactation on lactation in rats. J. Endocrinol. 101:69–73.

Griminger, P. 1966. Biological activity of the various vitamin K forms. Vitamin. Horm. 24:605–618.

Gross, J. 1962. Iodine and bromine. In Mineral Metabolism, C. L. Comar and F. Bronner, eds. New York: Academic Press.

Gross, K. L., W. J. Hartman, A. Ronnenberg, and R. L. Prior. 1991. Arginine-deficient diets alter the plasma and tissue amino acids in young and aged rats. J. Nutr. 121:1591–1599.

Grunert, R. R., J. H. Meyer, and P. H. Phillips. 1950. The sodium and potassium requirements of the rat for growth. J. Nutr. 42:609–618.

Grupp, G., I. Grupp, C. L. Johnson, E. T. Wallick, and A. Schwartz. 1979. Effect of vanadate on cardiac contraction and adenylate cyclase. Biochem. Biophys. Res. Commun. 88:440–447.

Guggenheim, M., and R. Jurgens. 1944. Das schachtelhalm schwanzsymptom bei saugenden jungratten die beziehungen der ungesättigten fettsäuren zu den vitaminen des B-komplexes. Helv. Physiol. Pharmacol. Acta 2:417–433.

Guilarte, T. R. 1989. Effect of vitamin B_6 nutrition on the levels of dopamine, dopamine metabolites, dopa decarboxylase activity, tyrosine, and GABA in the developing rat corpus striatum. Neurochem. Res. 14:571–578.

Guilarte, T. R., and H. N. Wagner, Jr. 1987. Increased concentrations of 3–hydroxykynurenine in vitamin B_6 deficient neonatal rat brain. J. Neurochem. 49:1918–1926.

Guilbert, H. R., C. E. Howell, and G. H. Hart. 1940. Minimum vitamin A and carotene requirements of mammalian species. J. Nutr. 19:91–103.

Guthneck, B. T., B. A. Bennett, and B. S. Schweigert. 1953. Utilization of amino acids from foods by the rat. II. Lysine. J. Nutr. 49:289–294.

Hafeman, D. G., R. A. Sunde, and W. G. Hoekstra. 1974. Effect of dietary selenium on erythrocyte and liver glutathione peroxidase in the rat. J. Nutr. 104:580–587.

Hakkarainen, J., J. Tyopponen, and L. Jonsson. 1986. Vitamin E requirement of the growing rat during selenium deficiency with special reference to selenium-dependent and selenium-independent glutathione peroxidase. J. Vet. Med. 33:247–258.

Hale, O. M., and A. E. Schaefer. 1951. Choline requirement of rats as influenced by age, strain, vitamin B_{12} and folacin. Proc. Soc. Exp. Biol. Med. 77:633–636.

Halloran, B. P., and H. F. DeLuca. 1980. Effect of vitamin D deficiency on fertility and reproductive capacity in the female rat. J. Nutr. 110:1573–1580.

Hallquist, N. A., L. K. McNeil, J. F. Lockwood, and A. R. Sherman. 1992. Maternal-iron-deficiency effects on peritoneal macrophage and peritoneal natural-killer-cell cytotoxicity in rat pups. Am. J. Clin. Nutr. 55:741–746.

Halverson, A. W., J. H. Shaw, and E. B. Hart. 1945. Goiter studies in the rat. J. Nutr. 30:59–65.

Hambidge, K. M., C. E. Casey, and N. F. Krebs. 1986. Zinc. Pp. 1–137 in Trace Elements in Human and Animal Nutrition, W. Mertz, ed. Orlando, Fla.: Academic Press.

Hamilton, T. S. 1939. The growth, activity, and composition of rats fed diets balanced and unbalanced with respect to protein. J. Nutr. 17:565–582.

Hammermueller, J. D., T. M. Bray, and W. J. Bettger. 1987. Effect of zinc and copper deficiency on microsomal NADPH-dependent active oxygen generation in rat lung and liver. J. Nutr. 117:894–901.

Handler, P., and W. J. Dann. 1942. The inhibition of rat growth by nicotinamide. J. Biol. Chem. 146:357–368.

Hansard, S. L., II. 1975. Toxicity and physiological movement of vanadium in the sheep and rat. Ph.D. dissertation. University of Florida, Gainesville, Fla.

Hansen, D. K., R. C. Walker, T. F. Grafton. 1990. Effect of lithium carbonate on mouse and rat embryos in vitro. Teratology 41:155–160.

Hansen, H. S., and B. Hensen. 1985. Essential function of linoleic acid esterified in acylglucosylceramide and acylceramide in maintaining the epidermal water permeability barrier. Evidence from feeding studies with oleate, linoleate, arachidonate, columbinate, and α-linolenate. Biochim. Biophys. Acta 834:357–363.

Hardwick, L. L., R. A. Clemens, and M. R. Jones. 1987. Effects of calcium phosphate supplementation on calcium, phosphorus, and magnesium metabolism in the Wistar rat. Nutr. Res. 7:787–796.

Hardwick, L. L., M. R. Jones, N. Bratubar, and K. B. N. Lee. 1991. Magnesium absorption: Mechanisms and the influence of vitamin D, calcium, and phosphate. J. Nutr. 121:13–23.

Harper, A. E., and C. A. Elvehjem. 1957. Dietary carbohydrates. A review of the effects of different carbohydrates on vitamin and amino acid requirements. J. Agr. Food Chem. 5:754–758.

Harper, A. E., and H. E. Spivey. 1958. Relationship between food intake and osmotic effect of dietary carbohydrate. Am. J. Physiol. 193:483–487.

Harper, A. E., N. J. Benevenga, and R. M. Wohlhueter. 1970. Effects of ingestion of disproportionate amounts of amino acids. Physiol. Rev. 50:428–458.

Harr, J. R., J. F. Bone, I. J. Tinsley, P. H. Weswig, and R. S. Yamamoto. 1967. Selenium toxicity in rats. II. Histopathology. Pp. 153–178 in Symposium: Selenium in Biomedicine, O. H. Muth, ed. Westport, Conn.: AVI Publishing.

Harris, H. A., A. Neuberger, and F. Sanger. 1943. Lysine deficiency in young rats. Biochem. J. 37:508–513.

Harris, L. J., and E. Kodicek. 1950. Quantitative studies and dose-response curves in nicotinamide deficiency in rats. Br. J. Nutr. 4:xiii-xiv.

Hartroft, P. M., and E. Sowa. 1964. Effect of potassium on juxtaglomerular cells and adrenal zona glomerulosa of rats. J. Nutr. 82:439–442.

Hartsook, E. W., and H. H. Mitchell. 1956. The effect of age on the protein and methionine requirements of the rat. J. Nutr. 60:173–195.

Hartsook, E. W., T. V. Hershberger, and J. C. M. Nee. 1973. Effects of dietary protein content and ratio of fat to carbohydrate calories on energy metabolism and body composition of growing rats. J. Nutr. 103:167–178.

Hashimoto, K. 1981. Effects of repeated administration of pyrithiamin and oxythiamin on timing behavior in rats. Nippon Yakurigaku Zasshi 78:521–528.

Hauswirth, G. W., and P. P. Nair. 1975. Effects of different vitamin E deficient basal diets on hepatic catalase and microsomal cytochromes P-450 and B_5 in rats. Am. J. Clin. Nutr. 28:1087–1094.

Haywood, S. 1979. The effect of the sex of weaned rats on the accumulation of dietary copper in their livers. J. Comp. Pathol. 89:481–487.

Haywood, S. 1985. Copper toxicosis and tolerance in the rat. I. Changes in copper content of the liver and kidney. J. Pathol. 145:149–158.

Heger, J., and Z. Frydrych. 1985. Efficiency of utilization of essential amino acids in growing rats at different levels of intake. Br. J. Nutr. 54:499–508.

Hegsted, D. M., and Y. Chang. 1965. Protein utilization in growing rats. I. Relative growth index as a bioassay procedure. J. Nutr. 85:159–168.

Henderson, L. M., J. M. McIntire, H. A. Waisman, and C. A. Elvehjem. 1942. Pantothenic acid in the nutrition of the rat. J. Nutr. 23:47–58.

Henderson, L. M., T. Deodhar, W. A. Krehl, and C. A. Elvehjem. 1947. Factors affecting the growth of rats receiving niacin-tryptophan-deficient diets. J. Biol. Chem. 170:261–268.

Henskens, Y. M. C., J. Ritskes-Hoitinga, J. N. J. J. Mathot, I. Van Camp, and A. C. Beynen. 1991. The influence of dietary lactose on phosphorus-induced nephrocalcinosis in female rats. Int. J. Vit. Nutr. Res. 61:77–86.

Hepburn, F. N., and W. B. Bradley. 1964. The glutamic acid and arginine requirement for high growth rate of rats fed amino acid diets. J. Nutr. 84:305–312.

Heppel, L. A., and C. A. A. Schmidt. 1949. Studies on the potassium metabolism of the rat during pregnancy, lactation and growth. Univ. Calif. Publ. Physiol. 8:189–195.

Herzberg, G. R., and M. Rogerson. 1988a. Hepatic fatty acid synthesis and triglyceride secretion in rats fed fructose- or glucose-based diets containing corn oil, tallow or marine oil. J. Nutr. 118:1061–1067.

Herzberg, G. R., and M. Rogerson. 1988b. Interaction of dietary carbohydrate and fat in the regulation of hepatic and extrahepatic lipogenesis in the rat. Br. I. Nutr. 59:233–241.

Heywood, R., A. K. Palmer, R. L. Gregson, and H. Hummler. 1985. The toxicity of beta-carotene. Toxicology 36:91–100.

Hietanen, E., U. Koivusaari, M. Laitinen, and A. Norling. 1980. Hepatic drug metabolism during ethanol ingestion in riboflavin deficient rats. Toxicology 16:103–111.

Higgings, E. S., D. Richert, and W. W. Westerfeld. 1956. Molybdenum deficiency and tungstate inhibition studies. J. Nutr. 59:539–559.

Hill, E. G., S. B. Johnson, and R. T. Holman. 1979. Intensification of essential fatty acid deficiency in the rat by dietary *trans* fatty acids. J. Nutr. 109:1759–1765.

Hirano, T., J. Mamo, M. Poapst, and G. Steiner. 1988. Very-low-density lipoprotein triglyceride kinetics in acute and chronic carbohydrate-fed rats. Am. J. Physiol. 255:E236–E240.

Hitchman, A. J., S. A. Hasany, A. Hitchman, J. E. Harrison, and C. Tam. 1979. Phosphate-induced renal calcification in the rat. Can. J. Physiol. Pharmacol. 57:92–97.

Ho, I. K., H. H. Loh, and E. L. Way. 1979. Toxic interaction between choline and morphine. Toxicol. Appl. Pharmacol. 51:203–208.

Hoagland, R., and G. G. Snider. 1943. Digestibility of some animal and vegetable fats. J. Nutr. 25:295–302.

Hoagland, R., N. R. Ellis, O. G. Hankins, and G. G. Snider. 1948. Supplemental value of certain amino acids for beef protein. J. Nutr. 35:167–176.

Hobara, R., and H. Yasuhara. 1981. Erythrocytosis in thiamine deficient rats. Jpn. J. Pharmacol. 31:985–993.

Hodge, H. C. 1944. Acute toxicity of choline hydrochloride administered intraperitoneally to rats. Proc. Soc. Exp. Biol. Med. 57:26–28.

Hoek, A. C., A. G. Lemmens, J. W. M. A. Mullink, and A. C. Beynen. 1988. Influence of dietary calcium:phosphorus ratio on mineral excretion and nephrocalcinosis in female rats. J. Nutr. 118:1210–1216.

Hoekstra, W. G. 1975. Biochemical function of selenium and its relation to vitamin E. Fed. Proc. 34:2083–2089.

Holdsworth, E. S., and E. Neville. 1990. Effects of extracts of high- and low-chromium brewer's yeast on metabolism of glucose by hepatocytes from rats fed on high- or low-Cr diets. Br. J. Nutr. 63:623–630.

Hollander, D. 1973. Vitamin K_1 absorption by everted intestinal sacs of the rat. Am. J. Physiol. 225:360–364.

Hollander, D., and D. Morgan. 1979. Aging: Its influence on vitamin A intestinal absorption in vivo by the rat. Exp. Geront. 14:301–305.

Hollander, D., and V. Dadufalza. 1989. Lymphatic and portal absorption of vitamin E in aging rats. Digest. Dis. Sci. 34:768–772.

Hollander, D., and H. Tarnawski. 1984. Influence of aging on vitamin D absorption and unstirred water layer dimensions in the rat. J. Lab. Clin. Med. 103:462–468.

Holman, R. T. 1960. The ratio of trienoic:tetraenoic acids in tissue lipids as a measure of essential fatty acid requirement. J. Nutr. 70:405–410.

Holman, R. T. 1968. Essential fatty acid deficiency. Prog. Chem. Fats Other Lipids 9:275–348.

Holman, R. T. 1970. Biological activities of and requirements for polyunsaturated acids. Prog. Chem. Fats Other Lipids 9:607–682.

Holman, R. T., and J. J. Peiffer. 1960. Acceleration of essential fatty acid deficiency by dietary cholesterol. J. Nutr. 70:411–417.

Holtkamp, D. E., and R. M. Hill. 1950. The effect on growth of the level of manganese in the diet of rats, with some observations on the manganese-thiamine relationship. J. Nutr. 41:307–316.

Hopkins, L. L., Jr., and H. E. Mohr. 1974. Vanadium as an essential nutrient. Fed. Proc. 33:1773–1775.

Horio, F., K. Ozaki, A. Yoshida, S. Makino, and Y. Hayashi. 1985. Requirement for ascorbic acid in a rat mutant unable to synthesize ascorbic acid. J. Nutr. 115:1630–1640.

Horio, F., K. Ozaki, M. Kohmura, A. Yoshida, S. Makino, and Y. Hayashi. 1986. Ascorbic acid requirement for the induction of microsomal drug-metabolizing enzymes in a rat mutant unable to synthesize ascorbic acid. J. Nutr. 116:2278–2289.

Horio, F., N. Takahashi, S. Makino, Y. Hayashi, and A. Yoshida. 1991. Ascorbic acid deficiency elevates serum level of LDL-cholesterol in a rat mutant unable to synthesize ascorbic acid. J. Nutr. Sci. Vitaminol. 37:63–71.

Horne, D. W., and W. T. Briggs. 1980. Effect of dietary and nitrous oxide-induced vitamin B_{12} deficiency on uptake of 5-methyltetrahydrofolate by isolated rat hepatocytes. J. Nutr. 110:223–230.

Horwitt, M. K. 1954. Riboflavin. Pp. 380–391 in The Vitamins, Vol. III, W. H. Sebrell, Jr., and R. S. Harris, eds. New York: Academic Press.

Houglum, K., M. Filip, J. L. Witztum, and M. Chojkier. 1990. Malondialdehyde and 4-hydroxynoneal protein adducts in plasma and liver of rats with iron overload. J. Clin. Invest. 86:1191–1198.

Howard, L., C. Wagner, and S. Schenker. 1974. Malabsorption of thiamin in folate-deficient rats. J. Nutr. 104:1024–1032.

Hsu, J. M., J. C. Smith, Jr., A. A. Yunice, and G. Kepford. 1983. Impairment of ascorbic acid synthesis in liver extracts of magnesium-deficient rats. J. Nutr. 113:2041–2047.

Hsueh, A. M., C. E. Agustin, and B. F. Chow. 1967. Growth of young rats after differential manipulation of maternal diets. J. Nutr. 91:195–200.

Huber, A. M., S. N. Gershoff, and D. M. Hegsted. 1964. Carbohydrate and fat metabolism and response to insulin in vitamin B_6-deficient rats. J. Nutr. 82:371.

Hundley, J. M. 1947. Production of niacin deficiency in rats. J. Nutr. 34:253–262.

Hundley, J. M. 1949. Influence of fructose and other carbohydrates on the niacin requirement of the rat. J. Biol. Chem. 181:1–9.

Hunter, K. E., and P. R. Turkki. 1987. Effect of exercise on riboflavin status of rats. J. Nutr. 117:298–304.

Hurley, L. S., and L. T. Bell. 1974. Genetic influence on response to dietary manganese deficiency. J. Nutr. 104:133–137.

Hurley, L. S., and C. L. Keen. 1987. Manganese. Pp. 185–223 in Trace Elements in Human and Animal Nutrition, W. Mertz, ed. Orlando, Fla.: Academic Press.

Hurley, L. S., G. Cosens, and L. L. Therialut. 1976a. Teratogenic effects of magnesium deficiency in rats. J. Nutr. 106:1254–1260.

Hurley, L. S., G. Cosens, and L. L. Therialut. 1976b. Magnesium, calcium, and zinc levels of maternal and fetal tissues in magnesium-deficient rats. J. Nutr. 106:1261–1264.

Ichihashi, T., Y. Takagishi, K. Uchida, and H. Yamada. 1992. Colonic absorption of menaquinone-4 and menaquinone-9 in rats. J. Nutr. 122:506–512.

Ikegami, S., F. Tsuchihashi, H. Harada, N. Tsuchihashi, E. Nishide, and S. Innami. 1990. Effect of viscous indigestible polysaccharides on pancreatic-biliary secretion and digestive organs in rats. J. Nutr. 120:353–360.

Ishida, M., B. Bulos, S. Takamoto, and B. Sacktor. 1987. Hydroxylation of 25-hydroxyvitamin D_3 by renal mitochondria from rats of different ages. Endocrinology 121:443–448.

Itokawa, Y., and M. Fujiwara. 1973. Lead and vitamin effect on heme synthesis. Arch. Environ. Health 27:31–35.

Jacob, M., and R. M. Forbes. 1969. Effects of magnesium deficiency, dietary sulfate and thryoxine treatment on kidney calcification and tissue protein-bound carbohydrate in the rat. J. Nutr. 99:51–57.

Jager, F. C. 1972. Long-term dose-response effects of vitamin E in rats. Significance of the in vitro haemolysis test. Nutr. Metab. 14:1–7.

Jager, F. C., and U. M. T. Houtsmuller. 1970. Effect of dietary linoleic acid on vitamin E requirement and fatty acid composition of erythrocyte lipids in rats. Nutr. Metab. 12:3–12.

Jaus, H., G. Sturm, B. Grassle, W. Romen, and G. Siebert. 1977. Metabolic effects and liver damage after prolonged administration of high doses of nicotinamide to rats. Nutr. Metab. 21(Suppl. 1):38–41.

Jenkins, M. Y., and G. V. Mitchell. 1975. Influence of excess vitamin E on vitamin A toxicity in rats. J. Nutr. 105:1600–1606.

Johnson, J. L., W. R. Waud, K. V. Rajagopalan, M. Duran, F. A. Beemer, and S. K. Wadman. 1980. Inborn errors of molybdenum metabolism: Combined deficiencies of sulfite oxidase and xanthine dehydrogenase in a patient lacking the molybdenum cofactor. Proc. Natl. Acad. Sci. USA 77:3715–3719.

Johnson, M., and J. L. Greger. 1984. Absorption, distribution and endogenous excretion of zinc by rats fed various dietary levels of inorganic tin and zinc. J. Nutr. 114:1843–1852.

Johnson, M. A., and C. L. Murphy. 1988. Adverse effects of high dietary iron and ascorbic acid on copper status in copper-deficient and copper-adequate rats. Am. J. Clin. Nutr. 47:96–101.

Johnson, P. E., and E. D. Korynta. 1992. Effects of copper, iron and ascorbic acid on manganese availability to rats. Proc. Soc. Exp. Biol. Med. 199:470–480.

Johnson, W. T., and S. N. Dufault. 1989. Altered cytoskeletal organization and secretory response of thrombin-activated platelets from copper-deficient rats. J. Nutr. 119:1404–1410.

Johnson, W. T., S. N. Dufault, and A. C. Thomas. 1993. Platelet cytochrome c oxidase activity is an indicator of copper status in rats. Nutr. Res. 13:1153–1162.

Johnston, P. V. 1985. Dietary fat, eicosanoids, and immunity. Adv. Lipid Res. 21:103–141.

Jones, J. H. 1971. Vitamin D: Requirement of animals. Pp. 285–289 in The Vitamins, Vol. III, W. H. Sebrell, Jr., and R. S. Harris, eds. New York: Academic Press.

Kahlenberg, O. J., A. Black, and E. B. Forbes. 1937. The utilization of energy producing nutriment and protein as affected by sodium deficiency. J. Nutr. 13:97–108.

Kametaka, M., J. Inaba, and R. Ishikawa. 1974. Estimation of the daily milk intake of the suckling rat using the turnover rate of potassium. J. Nutr. Sci. Vitamin. 20:421–429.

Kane, G. G., F. E. Lovelace, and C. M. McCay. 1949. Dietary fat and calcium wastage in old age. J. Gerontol. 4:185–192.

Kang, S. S., R. G. Price, J. Yudkin, N. A. Worcester, and K. R. Bruckdorfer. 1979. The influence of dietary carbohydrate and fat on kidney calcification and the urinary excretion of N-acetyl-β-glucosaminidase (EC 3.2.1.30). Br. J. Nutr. 41:65–71.

Kang-Lee, Y. A., R. W. McKee, S. M. Wright, M. E. Swendseid, D. J. Jenden, and R. S. Jope. 1983. Metabolic effects of nicotinamide administration in rats. J. Nutr. 113:215–221.

Katz, M. L., W. G. Robison, Jr., R. K. Herrmann, A. B. Groome, and J. G. Bieri. 1984. Lipofuscin accumulation resulting from senescence and vitamin E deficiency: Spectral properties and tissue distribution. Mech. Ageing Dev. 25:149–159.

Kaup, S. N., A. R. Behling, L. Choquette, and J. L. Greger. 1990. Calcium and magnesium utilization in rats: Effect of dietary butterfat and calcium and of age. J. Nutr. 120:226–273.

Kaup, S. M., A. R. Behling, and J. L. Greger. 1991a. Sodium, potassium and chloride utilization by rats given various inorganic anions. Br. J. Nutr. 66:523–532.

Kaup, S. M., J. L. Greger, and K. Lee. 1991b. Nutritional evaluation with an animal model of cottage cheese fortified with calcium and guar gum. J. Food. Sci. 56:692–695.

Kaup, S. M., J. L. Greger, M. S. K. Marcus, and N. M. Lewis. 1991c. Blood pressure, fluid compartments and utilization of chloride in rats fed various chloride diets. J. Nutr. 121:330–337.

Kawano, J., D. N., Ney, C. L. Keen, and B. O. Schneeman. 1987. Altered high density lipoprotein composition in manganese-deficient Sprague-Dawley and Wistar rats. J. Nutr. 117:902–906.

Keen, C. L., and L. S. Hurley. 1989. Zinc and reproduction: Effects of deficiency on foetal and postnatal development. Pp. 183–220 in Zinc in Human Biology, C. F. Mills, ed. New York: Springer-Verlag.

Keen, C. L., and M. E. Gershwin. 1990. Zinc deficiency and immune function. Annu. Rev. Nutr. 10:415–431.

Keen, C. L., B. Lönnerdal, and L. S. Hurley. 1982. Teratogenic effects of copper deficiency and excess. Pp. 109–122 in Inflammatory Diseases and Copper, J. R. J. Sorenson, ed. New Jersey: Humana Press.

Keen, C. L., N. H. Reinstein, J. Goudey-Lefevre, M. Lefevre, B. Lönnerdal, B. O. Schneeman, and L. S. Hurley. 1985. Effect of dietary copper and zinc levels on tissue copper, zinc, and iron in male rats. Biol. Trace Elem. Res. 8:123–136.

Kellerman, J. H. 1934. A well-balanced ration for stock rats. Onderstepoort J. Vet. Sci. Anim. Ind. 2:649–654.

Keymer, A., D. W. T. Crompton, and A. Singhvi. 1983. Mannose and the "crowding effect" of hymenolepis in rats. Int. J. Parasitol. 13:561–570.

Khan, M. A., and B. Munira. 1978. Biological utilization of protein as influenced by dietary carbohydrates. Acta Agr. Scand. 28:282–284.

Khan-Siddiqui, L., and M. S. Bamji. 1987. Effect of riboflavin or pyridoxine deficiency on tissue carnitine levels in rats. Nutr. Res. 7:445–448.

Kielanowski, J. 1965. Estimates of the Energy Costs of Protein Deposition in Growing Animals: Animal Production (EAPP) Pub. No. 11. London: Butterworths.

Kielanowski, J. 1976. Energy cost of protein deposition. Pp. 207–217 in Protein Metabolism and Nutrition, Animal Production (EAPP) Pub. No. 16, D. J. A. Cole, K. N. Boorman, P. J. Buttery, D. Lewis, R. J. Neale, and H. Swan, eds. London: Butterworths.

Kim, H. Y., M. F. Picciano, M. A. Wallig, and J. A. Milner. 1991. The role of selenium nutrition in the development of neonatal rat lung. Pediat. Res. 29:440–445.

Kim, J., H. Kyriazi, and D. A. Green. 1991. Normalization of Na$^+$-K$^+$-ATPase activity in isolated membrane fraction from sciatic nerves of streptozocin-induced diabetic rats by dietary *myo*-inositol supplementation in vivo or protein kinase C agonist in vitro. Diabetes 40:558–567.

Kimura, S., Y. Furukawa, J. Wakasugi, Y. Ishihara, and A. Nakayama. 1980. Antagonism of L(-)pantothenic acid on lipid metabolism in animals. J. Nutr. Sci. Vitaminol. 26:113–117.

Kindberg, C. G., and J. W. Suttie. 1989. Effect of various intakes of phylloquinone on signs of vitamin K deficiency and serum and liver phylloquinone concentrations in the rat. J. Nutr. 119:175–180.

Kinsella, J. E., D. H. Hwang, P. Yu, J. Mai, and J. Shimp. 1979. Prostaglandins and their precursors in tissues from rats fed on *trans,trans*-linoleate. Biochem. J. 184:701–704.

Kirksey, A., and R. L. Pike. 1962. Some effects of high and low sodium intakes during pregnancy in the rat. I. Food consumption, weight gain, reproductive performance, electrolyte balances, plasma total protein and protein fractions in normal pregnancy. J. Nutr. 77:33–42.

Kirksey A., R. L. Pang, and W.-J. Lin. 1975. Effects of different levels of pyridoxine fed during pregnancy superimposed upon growth in the rat. J. Nutr. 105:607–615.

Kleiber, M. 1975. The Fire of Life: An Introduction to Animal Energetics. New York: Krieger.

Kleiber, M., A. H. Smith, and T. N. Chernikoff. 1956. Metabolic rate of female rats as a function of age and body size. Am. J. Physiol. 186:9–12.

Kleiber, M., and H. H. Cole. 1945. Body size and energy metabolism during pregnancy of normal and precocious rats. Fed. Proc. 4:40 (abstr.).

Klevay, L. M. 1976. The biotin requirement of rats fed 20% egg white. J. Nutr. 106:1643–1646.

Klevay, L. M., and J. T. Saari. 1993. Comparative responses of rats to different copper intakes and modes of supplementation. Proc. Soc. Exp. Biol. Med. 203:214–220.

Kligler, D., and W. A. Krehl. 1952. Lysine deficiency in rats. II. Studies with amino acid diets. J. Nutr. 46:61–74.

Knapka, J. J. 1983. Nutrition. Pp. 51–67 in The Mouse in Biomedical Research, Vol. 3, H. L. Foster, J. D. Small, J. G. Fox, eds. New York: Academic Press.

Knapka, J. J., K. P. Smith, and F. J. Judge. 1974. Effect of open and closed formula rations on the performance of three strains of laboratory mice. Lab. Anim. Sci. 24:480–487.

Ko, K. W., F. X. Fellers, and J. M. Craig. 1962. Observations on magnesium deficiency in the rat. Lab. Invest. 11:294–305.

Kochanowski, B. A., and A. R. Sherman. 1983. Iron status of suckling rats as influenced by maternal diet during gestation and lactation. Br. J. Nutr. 49:51–57.

Kochanowski, B. A., and A. R. Sherman. 1984. Phagocytosis and lysozyme activity in granulocytes from iron-deficient rat dams and pups. Nutr. Res. 4:511–520.

Kochanowski, B. A., and A. R. Sherman. 1985. Decreased antibody formation in iron-deficient rat pups—effect of iron repletion. Am. J. Clin. Nutr. 41:278–284.

Koh, E. T. 1990. Comparison of copper status in rats when dietary fructose is replaced by either cornstarch or glucose. Proc. Soc. Exp. Biol. Med. 194:108–113.

Koh, E. T., S. Reiser, and M. Fields. 1989. Dietary fructose as compared to glucose and starch increases the calcium content of kidney of magnesium-deficient rats. J. Nutr. 119:1173–1178.

Koller, L. D., S. A. Mulhern, N. C. Frankel, M. G. Steven, and J. R. Williams. 1987. Immune dysfunction in rats fed a diet deficient in copper. Am. J. Clin. Nutr. 45:997–1006.

Kollmorgen, G. M., M. M. King, S. D. Kosanke, and C. Do. 1983. Influence of dietary fat and indomethacin on the growth of transplantable mammary tumors in rats. Cancer Res. 43:4714–4719.

Konijn, A. M., D. N. C. Muogbo, and K. Guggenheim. 1970. Metabolic effects of carbohydrate-free diets. Isr. J. Med. Sci. 6:498–505.

Konishi, F., T. Oku, and N. Hosoya. 1984. Hypertrophic effect of unavailable carbohydrate on cecum and colon in rats. J. Nutr. Sci. Vitaminol. 30:373–379.

Koong, L. J., C. L. Ferrell, and J. A. Neinaber. 1985. Assessment of interrelationships among levels of intake and production, organ size and fasting heat production in growing animals. J. Nutr. 115:1383–1390.

Kootstra, Y., J. Ritskes-Hoitinga, A. G. Lemmens, and A. C. Beynen. 1991. Diet-induced calciuria and nephrocalcinosis in female rats. Int. J. Vitam. Nutr. Res. 61:100–101.

Kornberg, A., and K. M. Endicott. 1946. Potassium deficiency in the rat. Am. J. Physiol. 145:291–298.

Koski, K. G., and F. W. Hill. 1986. Effect of low carbohydrate diets during pregnancy on parturition and postnatal survival of the newborn rat pup. J. Nutr. 116:1938–1948.

Koski, K. G., and F. W. Hill. 1990. Evidence for a critical period during late gestation when maternal dietary carbohydrate is essential for survival of newborn rats. J. Nutr. 120:1016–1027.

Koski, K. G., F. W. Hill, and L. S. Hurley. 1986. Effect of low carbohydrate diets during pregnancy on embryogenesis and fetal growth and development in rats. J. Nutr. 116:1922–1937.

Koski, K. G., F. W. Hill, and B. Lönnerdal. 1990. Altered lactational performance in rats fed low carbohydrate diets and its effect on growth of neonatal rat pups. J. Nutr. 120:1028–1036.

Kotchen, T. A., R. G. Luke, C. E. Ott, J. H. Galla, and S. Whitescarver. 1983. Effect of chloride on renin and blood pressure responses to sodium chloride. Ann. Intern. Med. 98:817–822.

Kramar, J., and V. L. Levine. 1953. Influence of fats and fatty acids on the capillaries. J. Nutr. 50:149–160.

Kramer, T. R., W. T. Johnson, and M. Briske-Anderson. 1988. Influence of iron and the sex of rats on hematological, biochemical and immunological changes during copper deficiency. J. Nutr. 118:214–221.

Krehl, W. A., P. S. Sarma, L. J. Teply, and C. A. Elvehjem. 1946. Factors affecting the dietary niacin and tryptophane requirement of the growing rat. J. Nutr. 31:85–106.

Kulkarni, A. B., and B. B. Gaitonde. 1983. Effects of early thiamin deficiency and subsequent rehabilitation on the cholinergic system in developing rat brain. J. Nutr. Sci. Vitaminol. 29:217–225.

Kummerow, F. A., H. P. Pan, and H. Hickman. 1952. The effect of dietary fat on the reproductive performance and the mixed fatty acid composition of fat-deficient rats. J. Nutr. 46:489–498.

Kumta, U. S., and A. E. Harper. 1962. Amino acid balance and imbalance. Effect of amino acid imbalance on blood amino acid pattern. Proc. Soc. Exp. Biol. Med. 110:512–517.

Kunkel, H. O., and P. B. Pearson. 1948. The quantitative requirements of the rat for magnesium. Arch. Biochem. 18:461–465.

Kurtz, D. J., H. Levy, and J. N. Kanfer. 1972. Cerebral lipids and amino acids in the vitamin B$_6$-deficient suckling rat. J. Nutr. 102:291.

Kurtz, P. J., D. C. Enunerling, and D. J. Donofrio. 1984. Subchronic toxicity of all-*trans*-retinoic acid and retinylidene dimedone in Sprague-Dawley rats. Toxicology 30:115–124.

Kwiecinski, G. G., G. I. Petrie, and H. F. DeLuca. 1989. Vitamin D is necessary for reproductive functions of the male rat. J. Nutr. 119:741–744.

L'Abbe, M. R., and P. W. F. Fischer. 1984. The effects of high dietary zinc and copper deficiency on the activity of copper-requiring metalloenzymes in the growing rat. J. Nutr. 114:813–822.

L'Abbe, M. R., P. W. F. Fischer, D. D. Trick, J. S. Campbell, and

E. R. Chavez. 1991. Dietary Se and tumor glutathione peroxidase and superoxide dismutase activity. J. Nutr. Biochem. 2:430–436.

Lamptey, M. S., and B. L. Walker. 1976. A possible essential role for dietary linolenic acid in the development of the young rat. J. Nutr. 106:86–93.

Lane, H. W., R. Strength, J. Johnson, and M. White. 1991. Effect of chemical form of selenium on tissue glutathione peroxidase activity in developing rats. J. Nutr. 121:80–86.

Langlais, P. J., R. G. Mair, C. D. Anderson, and W. J. McEntee. 1988. Long-lasting changes in regional brain amino acids and monoamines in recovered pyrithiamine treated rats. Neurochem. Res. 13:1199–1206.

Laskey, J. W., G. L. Rehnberg, J. F. Hein, and S. D. Carter. 1982. Effects of chronic manganese exposure on selected reproductive parameters in rats. J. Toxicol. Environ. Health 9:677–687.

Leclerc, J. 1991. Study of vitamin B_1 nutrition in the pregnant rat and its litter as a function of dietary thiamin supply. Int. J. Vitam. Nutr. Res. 48:333–340.

Leclerc, J., and M. L. Miller. 1987. Relationship between riboflavin intake and excretion in female rats after weaning of the litter. Int. J. Vitam. Nutr. Res. 57:45–51.

Lederer, E., and M. Lourau. 1948. Action antagoniste de certains cations devalents sur un enzyme secrété par la muquerise gastrique et jouant un rôle dans l'hematopoiese. Biochim. Biophys. Acta 2:278–285.

Lederer, W. H., M. Kumar, and A. E. Axelrod. 1975. Effects of pantothenic acid deficiency on cellular antibody synthesis in rats. J. Nutr. 105:17–25.

Lee, J. H., M. Fukumoto, H. Nishida, I. Ikeda, and M. Sugano. 1989. The interrelated effects of n-6/n-3 and polyunsaturated/saturated ratios of dietary fats on the regulation of lipid metabolism in rats. J. Nutr. 119:1893–1899.

Lee, S. S., and D. B. McCormick. 1983. Effect of riboflavin status on hepatic activities of flavin-metabolizing enzymes in rats. J. Nutr. 113:2274–2279.

Leelaprute, V., V. Boonpucknavig, N. Shamarapravati, and W. Weerapradist. 1973. Hypervitaminosis A in rats. Arch. Pathol. 96:5–9.

Levander, O. A., and R. F. Burk. 1990. Selenium. Pp. 268–273 in Present Knowledge in Nutrition, M. L. Brown, ed. Washington, D.C.: International Life Sciences Institute, Nutrition Foundation.

Levin, G., U. Cogan, Y. Levy, and S. Mokady. 1990. Riboflavin deficiency and the function and fluidity of rat erythrocyte membranes. J. Nutr. 120:857–861.

Levine, D. Z., D. Roy, G. Tolnai, L. Nash, and B. G. Shah. 1974. Chloride depletion and nephrocalcinosis. Am. J. Physiol. 227:878–883.

Levine, H., R. E. Remington, and H. von Kilnitz. 1933. Studies on the relation of diet to goiter. II. The iodine requirement of the rat. J. Nutr. 6:347–354.

Levrat, M.-A., S. R. Behr, C. Remesy, and C. Demigne. 1991. Effects of soybean fiber on cecal digestion in rats previously adapted to a fiber-free diet. J. Nutr. 121:672–678.

Lewis, C. J., B. Shannon, and B. Kleeman. 1982. Taurine and cystathionine excretion in female rats as influenced by dietary vitamin B_6 and estradiol administration. Nutr. Rep. Int. 25:269–276.

Lewis, C. G., M. Fields, and T. Beal. 1990. Effect of changing the type of dietary carbohydrate or copper level of copper-deficient, fructose-fed rats on tissue sorbitol concentrations. J. Nutr. Biochem. 1:160–166.

Lewis, K. C., M. H. Green, J. B. Green, and L. A. Zech. 1990. Retinol metabolism in rats with low vitamin A status: A compartmental model. J. Lipid Res. 31:1535–1548.

Lih, H., and C. A. Baumann. 1951. Effects of certain antibiotics on the growth of rats fed diets limiting in thiamine, riboflavin, or pantothenic acid. J. Nutr. 45:143–152.

Likins, R. C., L. A. Bavetta, and A. S. Posner. 1957. Calcification in lysine deficiency. Arch. Biochem. Biophys. 70:401–412.

Lin, W., and A. Kirksey. 1976. Effects of different levels of dietary iron on pregnancy superimposed upon growth in the rat. J. Nutr. 106:543–554.

Lindquist, R. N., J. L. Lynn, Jr., and G. E. Lienhard. 1973. Possible transition state analogs for ribonuclease. The complexes of uridine with oxovanadium (IV) and vanadium (V) ion. J. Am. Chem. Soc. 95:8762–8768.

Lindsey, J. R. 1979. Historical foundations. In The Laboratory Rat, Vol. 1, H. J. Baker, J. R. Lindsey, and S. H. Wisbroth, eds. New York: Academic Press.

Linscher, W. G., and A. J. Vergroeson. 1988. Lipids. Pp. 72–107 in Modern Nutrition in Health and Disease, M. E. Shils and V. R. Young, eds. Washington, D.C.: Lea & Febiger.

Lockwood, M. K., and C. D. Eckhert. 1992. Sucrose induced lipid, glucose and insulin elevations, microvascular injury and Se. Am. J. Physiol. 262:R144–R149.

Loeb, H. G., and G. O. Burr. 1947. A study of sex differences in the composition of rats, with emphasis on the lipid component. J. Nutr. 33:541–551.

Lojkin, M. L. 1967. Effect of levels of nitrogen intake on tryptophan metabolism and requirement for pregnancy of the rat. J. Nutr. 91:89–98.

Lombardini, J. B. 1986. Taurine levels in blood and urine of vitamin B_6 deficient and estrogen-treated rats. Biochem. Med. Metab. Biol. 35:125–131.

Longnecker, D. S., B. D. Roebuck, J. D. Yager, Jr., H. S. Lilja, and B. Siegmund. 1981. Pancreatic carcinoma in azaserine-treated rats; induction, classification and dietary modulation of incidence. Cancer 47:1561–1572.

Longnecker, J. B., and N. L. Hause. 1959. Relationship between plasma amino acids and composition of the ingested protein. Arch. Biochem. Biophys. 84:46–59.

Loo, G., P. J. Goodman, K. A. Hill, and J. T. Smith. 1986. Creatine metabolism in the pyridoxine-deficient rat. J. Nutr. 116:2403–2408.

Loosli, J. K., J. F. Lingenfelter, J. W. Thomas, and L. A. Maynard. 1944. The role of dietary fat and linoleic acid in the lactation of the rat. J. Nutr. 28:81–88.

Lopez, V., T. Stevens, and R. N. Lindquist. 1976. Vanadium ion inhibition of alkaline phosphatase-catalyzed phosphate ester hydrolysis. Arch. Biochem. Biophys. 175:31–38.

Lopez-Guisa, J. M., M. C. Harned, R. Dubielzig, S. C. Rao, and J. A. Marlett. 1988. Processed oat hulls as potential dietary fiber sources in rats. J. Nutr. 118:953–962.

Lowry, R. R., and I. J. Tinsley. 1966. Oleic and linoleic acid interaction in polyunsaturated fatty acid metabolism in the rat. J. Nutr. 88:26–32.

Luckey, T. D., T. J. Mende, and J. Pleasants. 1954. The physical and chemical characterization of rat's milk. J. Nutr. 54:345–359.

Luckey, T. D., J. R. Pleasants, M. Wagner, H. A. Gordon, and J. A. Reyniers. 1955. Some observations on vitamin metabolism in germ-free rats. J. Nutr. 57:169–182.

Ludwig, S., and N. Kaplowitz. 1980. Effect of pyridoxine deficiency on serum and liver transaminases in experimental liver injury in the rat. Gastroenterology 79:545–549.

Lushbough, C. H., T. Porter, and B. S. Schweigert. 1957. Utilization of amino acids from foods by the rat. J. Nutr. 62:513–526.

Machlin, L. J., R. Filipski, J. Nelson, L. R. Horn, and M. Brin. 1977. Effects of a prolonged vitamin E deficiency in the rat. J. Nutr. 107:1200–1208.

Mackerer, C. R., M. A. Mehlman, and R. B. Tobin. 1973. Effects of chronic acetylsalicylate administration on several nutrition and biochemical parameters in rats fed diets of varied thiamin content. Biochem. Med. 8:51–60.

Magour, S., H. Maser, and I. Steffen. 1983. Effect of daily oral intake of manganese on free polysomal protein synthesis of rat brain. Acta Pharmacol. Toxicol. 53:88–91.

Mahmood, S., H. M. Dani, and A. Mahmood. 1985. Effect of dietary pyridoxine deficiency on intestinal functions in rats. Nutr. Res. 5:299–304.

Majumder, A. K., K. B. Nandi, N. Subramanian, and I. B. Chatterjee. 1975. Nutrient interrelation of ascorbic acid and iron in rats and guinea pigs fed cereal diets. J. Nutr. 105:240–244.

Mameesh, M. S., and B. C. Johnson. 1960. Dietary vitamin K requirement of the rat. Proc. Soc. Exp. Biol. Med. 103:378–380.

Mangelsdorf, D. J., U. Borgmeyer, R. A. Heyman, J. Y. Zhou, E. S. Ong, A. E. Oro, A. Kakizuka, and R. M. Evans. 1992. Characterization of three RXR genes that mediate the action of 9-cis retinoic acid. Genes Dev. 6:329–344.

Mars, Y. W. H. M., A. G. Lemmens, and A. C. Beynen. 1988. Dietary phosphorus and nephrocalcinosis in female rats. Nutr. Rep. Int. 38:249–258.

Martin, M. M., and L. S. Hurley. 1977. Effect of large amounts of vitamin E during pregnancy and lactation. Am. J. Clin. Nutr. 30:1629–1637.

Martindale, L., and F. W. Heaton. 1964. Magnesium deficiency in the adult rat. Biochem. J. 92:119–126.

Mathers, J. C., F. Fernandez, M. J. Hill, P. T. McCarthy, M. J. Shearer, and A. Oxley. 1990. Dietary modification of potential vitamin K supply from enteric bacterial menaquinones in rats. Br. J. Nutr. 63:639–652.

Mathews, C. H. E., R. Brommage, and H. F. DeLuca. 1986. Role of vitamin D in neonatal skeletal developments in rats. Am. J. Physiol. 250:E725–E730.

Matschiner, J. T., and A. K. Willingham. 1974. Influence of sex hormones on vitamin K deficiency and epoxidation of vitamin K in the rat. J. Nutr. 104:660–665.

Matschiner, J. T., and R. G. Bell. 1973. Effect of sex and sex hormones on plasma prothrombin and vitamin K deficiency. Proc. Soc. Exp. Biol. Med. 144:316–320.

Matschiner, J. T., and E. A. Doisy, Jr. 1965. Effect of dietary protein on the development of vitamin K deficiency in the rat. J. Nutr. 86:93–99.

Matsuo, T., and Z. Suzuoki. 1982. Feeding responses of riboflavin-deficient rats to energy dilution, cold exposure and gluoprivation. J. Nutr. 112:1052–1056.

Mattson, F. H. 1959. The absorbability of stearic acid when fed as a simple or mixed triglyceride. J. Nutr. 69:338–342.

Mayer, J., N. B. Marshall, J. J. Vitale, J. H. Christensen, M. B. Mashay, Mashekhi, and F. J. Store. 1954. Exercise, food intake and body weight in normal rats and genetically obese adult mice. Am. J. Physiol. 177:544–548.

McAleese, D. M., and R. M. Forbes. 1961. The requirement and tissue distribution of magnesium in the rat as influenced by environmental temperature and dietary calcium. J. Nutr. 73:94–106.

McCall, M. G., G. E. Newman, J. R. P. Obrien, L. S. Valberg, and L. J. Witts. 1962a. Studies in iron metabolism. 1. The experimental production of iron metabolism. Br. J. Nutr. 16:297–304.

McCall, M. G., G. E. Newman, J. R. P. Obrien, and L. J. Witts. 1962b. Studies in iron metabolism. 2. The effects of experimental iron deficiency in the growing rat. Br. J. Nutr. 16:305–323.

McCandless, D. W., C. Hanson, K. V. Speeg, Jr., and S. Schenker. 1970. Cardiac metabolism in thiamin deficiency in rats. J. Nutr. 100:991–1002.

McCarthy, D. J., C. Lindamood III, C. M. Gundberg, and D. L. Hill. 1989. Retinoid-induced hemorrhaging and bone toxicity in rats fed diets deficient in vitamin K. Toxicol. Appl. Pharmacol. 97:300–310.

McCoy, K. E. M., and P. H. Weswig. 1969. Some selenium responses in the rat not related to vitamin E. J. Nutr. 98:383–389.

McCraken, K. J. 1975. Effect of feeding pattern on the energy metabolism of rats given low-protein diets. Br. J. Nutr. 33:277–289.

McLane, J. A., T. Khan, and I. R. Held. 1987. Increased axonal transport in peripheral nerves of thiamin-deficient rats. Exp. Neurol. 95:482–491.

Mead, J. F. 1984. The noneicosanoid functions of the essential fatty acids. J. Lipid Res. 25:1517–1521.

Medeiros, D. M., Z. Liao, and R. L. Hamlin. 1992. Electrocardiographic activity and cardiac function in copper-restricted rats. Proc. Soc. Exp. Biol. Med. 200:78–84.

Mei-Ying, C. M., and M. R. Kare. 1979. Effect of vitamin B_6 deficiency on preference for several taste solutions in the rat. J. Nutr. 109:339–344.

Meister, A. 1957. Biochemistry of the Amino Acids. New York: Academic Press.

Mellette, S. J., and L. A. Leone. 1960. Influence of age, sex, strain of rat and fat soluble vitamins on hemorrhagic syndromes in rats fed irradiated beef. Fed. Proc. 19:1045–1049.

Menaker, L., and J. M. Navia. 1973. Appetite regulation in the rat under various physiological conditions: The role of dietary protein and calories. J. Nutr. 103:347–352.

Mercer, L. P., J. M. Gustafson, P. T. Higbee, C. E. Geno, M. R. Schweisthal, and T. B. Cole. 1984. Control of physiological response in the rat by nutrient concentration. J. Nutr. 114:144–152.

Mercer, L. P., S. J. Dodds, and J. M. Gustafson. 1986. The determination of nutritional requirements: A modeling approach. Nutr. Rep. Int. 34:337–350.

Mercer, L. P., H. E. May, and S. J. Dodds. 1989. The determination of nutritional requirements in rats: Mathematical modeling of sigmoidal, inhibited nutrient-response curves. J. Nutr. 119:1465–1471.

Mertz, W., and E. E. Roginski. 1969. Effect of chromium (III) supplementation on growth and survival under stress in rats fed low protein diets. J. Nutr. 97:531–536.

Mertz, W., E. E. Roginski, and H. A. Schroeder. 1965. Some aspects of glucose metabolism of chromium deficient rats raised in the strictly controlled environment. J. Nutr. 86:107–112.

Meydani, M., J. B. Macauley, and J. B. Blumberg. 1986. Influence of dietary vitamin E, selenium and age on regional distribution of α-tocopherol in the rat brain. Lipids 21:786–791.

Meyer, J. H. 1956. Influence of dietary fiber on metabolic and endogenous nitrogen excretion. J. Nutr. 58:407–413.

Meyer, J. H. 1958. Interactions of dietary fiber and protein on food intake and body composition of growing rats. Am. J. Physiol. 193:488–494.

Meyerovitch, J., P. Rothenberg, Y. Shechter, S. Bonner-Weir, and C. R. Kahn. 1991. Vanadate normalizes hyperglycemia in two mouse models of non-insulin-dependent diabetes mellitus. J. Clin. Invest. 87:1286–1294.

Michaelis, O. E., IV, R. E. Martin, L. B. Gardner, and K. C. Ellwood. 1981. Effect of simple and complex carbohydrate on lipogenic parameters of spontaneously hypertensive rats. Nutr. Rep. Int. 24:313–321.

Michells, F. G., and J. T. Smith. 1965. A comparison of the utilization of organic and inorganic sulfur by the rat. J. Nutr. 87:217–220.

Middleton, H. M. 1986. Intestinal hydrolysis of pyridoxal 5'-phosphate in vivo in the rat: Effect of ethanol. Am. J. Clin. Nutr. 43:374–381.

Middleton, H. M. 1990. Intestinal hydrolysis of pyridoxal 5'-phosphate in vitro and in vivo in the rat: Effect of amino acids and oligopeptides. Dig. Dis. Sci. 35:113–120.

Miller, H. G. 1926. Sodium deficiency in a corn ration. J. Biol. Chem. 70:759–762.

Miller, R. F., N. O. Price, and R. W. Engel. 1956. Added dietary inorganic sulfate and its effect upon rats fed molybdenum. J. Nutr. 60:539–545.

Miller, S. C., M. A. Miller, and T. H. Omura. 1988. Dietary lactose improves endochondral growth and bone development and mineralization in rats fed a vitamin D-deficient diet. J. Nutr. 118:72–77.

Mills, C. F., and G. K Davis. 1989. Molybdenum. Pp. 429–463 in Trace Elements in Human and Animal Nutrition, Vol. 1, W. Mertz, ed. San Diego: Academic Press.

Milner, J. A., A. E. Wakeling, and W. J. Visek. 1974. Effect of arginine deficiency on growth and intermediary metabolism in rats. J. Nutr. 104:1681–1689.

Mitchell, H. H. 1950. Nutrient requirements as related to body size and body function. Scientia 85:165–175.

Mitchell, H. H., and J. R. Beadles. 1952. The determination of the protein requirement of the rat for maxiumum growth under conditions of restricted consumption of food. J. Nutr. 47:133–145.

Mittelholzer, E. 1976. Absence of influence of high doses of biotin on reproductive performance in female rats. Int. J. Vitam. Nutr. Res. 46:33–39.

Mizushima, Y., T. Harauchi, T. Yoshizaki, and S. Makino. 1984. A rat mutant unable to synthesize vitamin C. Experientia 40:359–361.

Molitor, H., and H. J. Robinson. 1940. Oral and parenteral toxicity of vitamin K_1, phthiocol, and 2-methyl-1,4-naphthoquinone. Proc. Soc. Exp. Biol. Med. 43:125–128.

Monserrat, A. J., E. A. Porta, A. K. Ghoshal, and S. B. Hartman. 1974. Sequential renal lipid changes in weanling rats fed a choline-deficient diet. J. Nutr. 104:1496–1502.

Moon, W.-H., and A. Kirksey. 1973. Cellular growth during prenatal and early postnatal periods in progeny of pyridoxine-deficient rats. J. Nutr. 103:123–133.

Moore, R. J., P. G. Reeves, and T. L. Veum. 1984. Influence of dietary phosphorus and sulphaguanidine levels on P utilization in rats. Br. J. Nutr. 51:453–465.

Moore, T. 1957. Vitamin A. New York: Elsevier.

Morhauer, H., and R. T. Holman. 1963. The effect of dose level of essential fatty acids upon fatty acid composition of the rat liver. J. Lipid Res. 4:151–159.

Mori, S., Y. Takeuchi, M. Toyama, S. Makino, T. Harauchi, Y. Kurata, and S. Fukushima. 1988. Assessment of L-ascorbic acid requirement for prolonged survival in ODS rats and their susceptibility to urinary bladder cancinogenesis by N-butyl-N-(4–hydroxybutyl)-nitrosamine. Cancer Lett. 38:275–282.

Morrison, S. D. 1956. The total energy and water metabolism during pregnancy in the rat. J. Physiol. 134:650–664.

Morrison, S. D. 1968. The constancy of the energy expended by rats on spontaneous activity, and the distribution of activity between feeding and non-feeding. J. Physiol. 197:305–323.

Mudd, S. H., F. Irrverre, and L. Laster. 1967. Sulfite oxidase deficiency in man: Demonstration of the enzymatic defect. Science 156:1599–1602.

Muir, M., and I. Chanarin. 1984. Conversion of endogenous cobalamins into microbiologically-inactive cobalamin analogues in rats exposed to nitrous oxide. Br. J. Haematol. 58:517–523.

Mulford, D. J., and W. H. Griffith. 1942. Choline metabolism. VIII. The relation of cystine and of methionine to the requirement of choline in young rats. J. Nutr. 23:91–100.

Murdock, D. S., M. L. Donaldson, and C. J. Gubler. 1974. Studies on the mechanism of the "thiamin-sparing" effect on ascorbic acid in rats. Am. J. Clin. Nutr. 27:696–699.

Muto, Y., J. E. Smith, P. O. Milch, and D. S. Goodman. 1972. Regulation of retinol-binding protein metabolism by vitamin A status in the rat. J. Biol. Chem. 247:2542–2550.

Nakashima, Y., and R. Suzue. 1982. Effect of nicotinic acid on myelin lipids in the brain of developing rats. J. Nutr. Sci. Vitaminol. 28:491–500.

Nakashima, Y., and R. Suzue. 1984. Influence of nicotinic acid on cerebroside synthesis in the brain of developing rats. J. Nutr. Sci. Vitaminol. 30:525–534.

Naismith, D. J., D. P. Richardson, and A. E. Pritchard. 1982. The utilization of protein and energy during lactation in the rat, with particular regard to the use of fat accumulated during pregnancy. Br. J. Nutr. 48:433–441.

Nanda, R., F. P. G. M. Van Der Linden, and H. W. B. Jansen. 1970. Production of cleft palate with dexamethasone and hypervitaminosis A in rat embryos. Experientia 26:1111–1112.

Narayan, K. A., and J. J. McMullen. 1980. Accelerated induction of fatty livers in rats fed fat-free diets containing sucrose or glycerol. Nutr. Rep. Int. 21:689–697.

National Institutes of Health. 1982. NIH Rodents, 1980 Catalogue: Strains and Stocks of Laboratory Rodents Provided by the NIH Genetic Resource. NIH Publication 83-606. Bethesda, Md.: U.S. Public Health Service.

National Research Council. 1962. Nutrient Requirements of Domestic Animals, Number 10. Pub. 990. Washington, D.C.: National Academy of Sciences.

National Research Council. 1972. Nutrient Requirements of Laboratory Animals, Second Edition. Washington, D.C.: National Academy Press.

National Research Council. 1978. Nutrient requirements of the laboratory rat. Pp. 7–37 in Nutrient Requirements of Laboratory Animals, Third Revised Ed. Washington, D.C.: National Academy Press.

National Research Council. 1980. Chromium. Pp. 142–153 in Mineral Tolerance of Domestic Animals. Washington, D.C.: National Academy Press.

National Research Council. 1983. Selenium in Nutrition. Washington, D.C.: National Academy Press.

National Research Council. 1988. Nutrient Requirements of Swine, Ninth Revised Ed. Washington, D.C.: National Academy Press.

National Research Council. 1994. Nutrient Requirements of Poultry, Ninth Revised Ed. Washington, D.C.: National Academy Press.

Nechay, B. R., L. B. Nanninga, P. S. E. Nechay, R. L. Post, J. J. Grantham, I. G. Macara, L. F. Kubena, T. D. Phillips, and F. H. Nielsen. 1986. Role of vanadium in biology. Fed. Proc. 45:123–132.

Nelson, G. J., and R. G. Ackman. 1988. Absorption and transport of fat in mammals with emphasis on n-3 polyunsaturated fatty acids. Lipids 23:1005–1014.

Nelson, M. M., and H. M. Evans. 1953. Relation of dietary protein levels to reproduction in the rat. J. Nutr. 51:71–84.

Nelson, M. M., and H. M. Evans. 1958. Sulfur amino acid requirement for lactation in the rat. Proc. Soc. Exp. Biol. Med. 99:723–725.

Nelson, M. M., and H. M. Evans. 1961. Dietary requirements for lactation in rats and other laboratory animals. Pp. 137–191 in Milk: The Mammary Gland and Its Secretion, Vol. II, S. K. Kon and A. T. Cowie, eds. New York: Academic Press.

Newberne, P. M. 1964. Cardiorenal lesions of potassium depletion or steroid therapy in the rat. Am. J. Vet. Res. 25:1256–1266.

Newberne, P. M., A. E. Rogers, C. Bailey, and V. R. Young. 1969. The induction of liver cirrhosis in rats by purified amino acid diets. Cancer Res. 29:230–235.

Newburg, D. S., and L. C. Fillios. 1979. A requirement for dietary asparagine in pregnant rats. J. Nutr. 109:2190–2197.

Newburg, D. S., D. L. Frankel, and L. C. Fillios. 1975. An asparagine requirement in young rats fed the dietary combinations of aspartic acid, glutamine, and glutamic acid. J. Nutr. 105:356–363.

Nguyen, L. B., J. F. Gregory III, and J. J. Cerda. 1983. Effect of dietary fiber on absorption of B-6 vitamins in a rat jejunal perfusion study. Proc. Soc. Exp. Biol. Med. 173:568–573.

Nieder, G. L., C. N. Corder, and P. A. Culp. 1979. The effect of vanadate on human kidney potassium dependent phosphatase. Arch. Pharmacol. 307:191–198.

Nielsen, E., and C. A. Elvehjem. 1941. Cure of spectacle eye condition in rats with biotin concentrate. Proc. Soc. Exp. Biol. Med. 48:349–352.

Nielsen, F. H. 1980a. Effect of form of iron on the interaction between nickel and iron in rats: Growth and blood parameters. J. Nutr. 110:965–973.

Nielsen, F. H. 1980b. Evidence of the essentiality of arsenic, nickel, and vanadium and their possible nutritional significance. Pp. 157–172 in Advances in Nutritional Research, Vol. 3, H. H. Draper, ed. New York: Plenum Press.

Nielsen, F. H. 1984. Ultra-trace elements in nutrition. Annu. Rev. Nutr. 4:21–41.

Nielsen, F. H. 1985. The importance of diet composition in ultratrace element research. J. Nutr. 115:1239–1247.

Nielsen, F. H., and E. O. Uthus. 1990. The essentiality and metabolism of vanadium. Pp. 51–62 in Vanadium in Biological Systems, N. D. Chasteen, ed. The Netherlands: Klumer Academic.

Nielsen, F. H., D. R. Myron, S. H. Givand, T. J. Zimmerman, and D. A. Ollerich. 1975. Nickel deficiency in rats. J. Nutr. 105:1620–1630.

Nielsen, F. H., T. J. Zimmerman, M. E. Collings, and D. R. Myron. 1979. Nickel deprivation in rats: Nickel-iron interactions. J. Nutr. 109:1623–1632.

Nielsen, F. H., T. R. Shuler, T. G. McLeod, and T. J. Zimmerman. 1984. Nickel influences iron metabolism through physiologic, pharmacologic and toxicologic mechanisms in the rat. J. Nutr. 114:1280–1288.

Nielsen, F. H., T. J. Zimmerman, T. R. Shuler, B. Brossart, and E. O. Uthus. 1989. Evidence for a cooperative metabolic relationship between nickel and vitamin B_{12} in rats. J. Trace Elem. Exp. Med. 2:21–29.

Nishina, P. M., B. O. Schneeman, and R. A. Freedland. 1991. Effects of dietary fibers on nonfasting plasma lipoprotein levels in rats. J. Nutr. 121:431–442.

Nolen, G. A. 1972. The effects of various levels of dietary protein on retinoic acid-induced teratogenicity in rats. Teratology 5:143–152.

Nowak, R. M. 1991. Walker's Mammals of the World, 5th Ed. Baltimore, Md.: Johns Hopkins University Press.

Nyengaard, J. R., T. F. Bendtsen, S. Christensen, and P. D. Ottosen. 1994. The number and size of glomeruli in long-term lithium-induced nephropathy in rats. APMIS 102:59–66.

O'Connor, D. L., M. F. Picciano, T. Tamura and B. Shane. 1990. Impaired milk folate secretion is not corrected by supplemental folate during iron deficiency in rats. J. Nutr. 120:499–506.

O'Dell, B. L., and M. Emery. 1991. Compromised zinc status in rats adversely affects calcium metabolism in platelets. J. Nutr. 121:1763–1768.

O'Dell, B. L., E. R. Morris, and W. O. Regan. 1960. Magnesium requirement of guinea pigs and rats: Effect of calcium and phosphorus and symptoms of magnesium deficiency. J. Nutr. 70:103–110.

Oftedal, O. T. 1984. Milk composition, milk yield, and energy output at peak lactation: A comparative review. Symp. Zool. Soc. London 51:33–85.

Okada, M., and K. Suzuki. 1974. Amino acid metabolism in rats fed a high protein diet with pyridoxine. J. Nutr. 104:287–293.

Okuyama, H., M. Saitoh, Y. Naito, T. Hori, A. Hashimoto, A. Moriuchi, and N. Yamamoto. 1987. Re-evaluation of the essentiality of alpha-linolenic acid in rats. Pp. 296–300 in Polyunsaturated Fatty Acids and Eicosanoids, W. E. M. Lands, ed. Champaign, Ill.: American Oil Chemists Society.

Onodera, K. 1987. Effects of decarboxylase inhibitors on muricidal suppression by L-dopa in thiamine deficient rats. Arch. Int. Pharmacodyn. Ther. 285:263–276.

Onodera, K., Y. Ogura, and K. Kisara. 1981. Characteristics of muricide induced by thiamine deficiency and its suppression by antidepressants or intraventricular serotonin. Physiol. Behav. 27:847–853.

Orent-Keiles, E., A. Robinson, and E. V. McCollum. 1937. The effects of sodium deprivation on the animal organism. Am. J. Physiol. 119:651–661.

Ornoy, A., J. Menczel, and L. Nebel. 1968. Alterations in the mineral composition and metabolism of rat fetuses and their placentas induced by maternal hypervitaminosis D_2. Isr. J. Med. Sci. 4:827–831.

Pallauf, J., and M. Kirchgessner. 1971. Zum Zinkbedarf wachsender Ratten. Int. J. Vitam. Nutr. Res. 41:543–553.

Pang, R. L., and A. Kirksey. 1974. Early postnatal changes in brain composition in progeny of rats fed different levels of dietary pyridoxine. J. Nutr. 104:111–117.

Parker, H. E., F. N. Andrews, S. M. Hauge, and F. W. Quackenbush. 1951. Studies on the iodine requirement of white rats during growth, pregnancy and lactation. J. Nutr. 44:501–511.

Patek, A. J., E. Erenoglu, N. M. O'Brian, and R. L. Hirsch. 1969. Sex hormones and susceptibility of the rat to dietary cirrhosis. Arch. Pathol. 87:52–56.

Patt, E. L., E. E. Pickett, and B. L. O'Dell. 1978. Effect of dietary lithium levels on tissue lithium concentration, growth rate, and reproduction in the rat. Bioinorg. Chem. 9:299–310.

Paul, P. K., P. N. Duttagupta, and H. C. Agarwal. 1973. Effects of an acute dose of biotin on the reproductive organs of the female rat. Curr. Sci. 42:206–208.

Pazos-Moura, C. C., E. G. Moura, M. L. Dorris, S. Rehnmark, L. Melendez, J. E. Silva, and A. Taurog. 1991. Effect of iodine deficiency on thyroxine 5'-deiodinase activity in various rat tissues. Am. J. Physiol. 260:E175–E182.

Pearson, P. B. 1948. High levels of dietary potassium and magnesium and growth of rats. Am. J. Physiol. 153:432–435.

Pearson, W. N. 1967. Blood and urinary vitamin levels as potential indices of body stores. Am. J. Clin. Nutr. 20:514–525.

Pederson, R. A., S. Ramanadham, A. M. J. Buchan, and J. H. McNeill. 1989. Long-term effects of vanadyl treatment on streptozocin-induced diabetes in rats. Diabetes 38:1390–1395.

Peifer, J. J., and R. D. Lewis. 1979. Effects of vitamin B_{12} deprivation on phospholipid fatty acid patterns in liver and brain of rats fed high and low levels of linoleate in low methionine diets. J. Nutr. 109:2160–2172.

Pence, B. C. 1991. Dietary selenium and antioxidant status: Toxic effects of 1,2-dimethylhydrazine in rats. J. Nutr. 121:138–144.

Perisse, J., and E. Salmon-Legagneur. 1960. Influence of plane of nutrition during pregnancy and lactation on milk production in the rat. Arch. Sci. Physiol. 14:105.

Peters, J. C., and A. E. Harper. 1985. Adaptation of rats to diets containing different levels of protein: Effects on food intake, plasma and brain amino acid concentrations and brain neurotransmitter metabolism. J. Nutr. 115:382–398.

Peterson, A. D., and B. R. Baumgardt. 1971a. Food and energy intake of rats fed diets varying in energy concentration and density. J. Nutr. 101:1057–1067.

Peterson, A. D., and B. R. Baumgardt. 1971b. Influence of level of energy demand on the ability of rats to compensate for diet dilution. J. Nutr. 101:1069–1074.

Peterson, C. A., J. A. C. Eurell, and J. W. Erdman, Jr. 1992. Bone composition and histology of young growing rats fed diets of varied calcium bioavailability: Spinach, nonfat dry milk, or calcium carbonate added to casein. J. Nutr. 122:137–144.

Peterson, P. A., S. F. Nilsson, L. Ostberg, L. Rask, and A. Vahlquist. 1974. Aspects of the metabolism of retinol-binding protein and retinol. Vitam. Horm. 32:181–214.

Pfaller, W., W. M. Fischer, N. Streider, H. Wurnig, and P. Deetjen. 1974. Morphologic changes of cortical nephron cells in potassium-adapted rats. Lab. Invest. 31:678–684.

Phatak, S. S., and V. N. Patwardhan. 1950. Toxicity of nickel. J. Sci. Ind. Res. 9B:70–75.

Phillips, R. D. 1981. Linear and nonlinear models for measuring protein nutritional quality. J. Nutr. 111:1058–1066.

Picciano, M. F. 1970. Chloride nutriture of the growing rat. M.S. thesis. Pennsylvania State University, University Park, Pa.

Pickett, E. E. 1983. Evidence for the essentiality of lithium in the rat. Pp. 66–70 in Fourth Trace Element Symposium: Lithium, M. Anke, W. Baumann, H. Bräunlich, and Chr. Brückner, eds. Jena, Germany: Friedrich Schiller Universität.

Pickett, E. E., and B. L. O'Dell. 1992. Evidence for the dietary essentiality of lithium in the rat. Biol. Trace Elem. Res. 334(3):299–319.

Pillai, G. R., M. Indira, and P. L. Vijayammal. 1990. Ascorbic acid 2-sulfate storage form of ascorbic acid in rats. Curr. Sci. 59:800–804.

Poiley, S. M. 1972. Growth tables for 66 strains and stocks of laboratory animals. Lab. Anim. Sci. 22:759–779.

Potvliege, P. R. 1962. Hypervitaminosis D_2 in gravid rats. Arch. Pathol. 73:371–382.

Powers, H. J. 1987. A study of maternofetal iron transfer in the riboflavin-deficient rat. J. Nutr. 117:852–856.

Prabhu, A. L., V. S. Aboobaker, and R. K. Bhattacharya. 1989. In vivo effect of dietary factors on the molecular action of aflatoxin B_1: Role of riboflavin on the catalytic activity of liver fractions. Cancer Lett. 48:89–94.

Prentice, A. M., and C. J. Bates. 1981. A biochemical evaluation of the erythrocyte glutathione reductase (EC 1.6.4.2) test for riboflavin status. 2. Dose-response relationships in chronic marginal deficiency. Br. J. Nutr. 45:53–65.

Pudelkewicz, C., J. Seuffert, and R. T. Holman. 1968. Requirements of the female rat for linoleic and linolenic acids. J. Nutr. 94:138–146.

Pullar, J. D., and A. J. F. Webster. 1974. Heat loss and energy retention during growth in congenitally obese and lean rats. Br. J. Nutr. 31:377–392.

Pullar, J. D., and A. J. F. Webster. 1977. The energy cost of fat and protein deposition in the rat. Br. J. Nutr. 37:355–363.

Qureshi, S. A., H. S. Buttar, and I. J. McGilveray. 1992. Lithium-induced nephrotoxicity in rats following subcutaneous multiple injections and infusion using miniosmotic pumps. Fund. Appl. Toxicol. 18:616–620.

Rabin, B. S. 1983. Inhibition of experimentally induced autoimmunity in rats by biotin deficiency. J. Nutr. 113:2316–2322.

Rahm, J. J., and R. T. Holman. 1964. The relationship of single dietary polyunsaturated fatty acids to fatty acid composition of lipids from subcellular particles of liver. J. Lipid Res. 5:169–176.

Rajagopalan, K. V. 1988. Molybdenum: An essential trace element in human nutrition. Annu. Rev. Nutr. 8:401–427.

Rajeswari, T. S., and E. Radha. 1984. Age-related effects of nutritional vitamin B_6-dependent enzymes of glutamate, gamma-aminobutyrate and glutamine systems in the rat brain. Exp. Gerontol. 19:87–93.

Ralli, E. P., and M. E. Dumm. 1953. Relation of pantothenic acid to adrenal cortical function. Vitam. Horm. 11:133–154.

Ranhotra, G. S., and B. C. Johnson. 1965. Effects of feeding different amino acids on growth rate and nitrogen retention of weanling rats. Proc. Soc. Exp. Biol. Med. 118:1197–1201.

Rao, G. H., and K. E. Mason. 1975. Antisterility and antivitamin K activity of d-α-tocopheryl hydroquinone in the vitamin E-deficient female rat. J. Nutr. 105:495–498.

Rau, M., M. S. Patole, S. Vijaya, C. K. R. Kurup, and T. Ramasarma. 1987. Vanadate-stimulated NADH oxidation in microsomes. Mol. Cell. Biochem. 75:151–159.

Rechcigl, M., Jr., S. Berger, J. K. Loosli, and H. H. Williams. 1962. Dietary protein and utilization of vitamin A. J. Nutr. 76:435–440.

Reddy, M. B., and J. D. Cook. 1991. Assessment of dietary determinants of nonheme-iron absorption in humans and rats. Am. J. Clin. Nutr. 54:723–728.

Reeves, P. G. 1989. AIN-76 diet: Should we change the formulation? J. Nutr. 119:1081–1082.

Reeves, P. G., and R. M. Forbes. 1972. Prevention by thyroxine of nephrocalcinosis in early magnesium deficiency in the rat. Am. J. Physiol. 222:220–224.

Reeves, P. G., K. L. Rossow, and J. Lindlauf. 1993a. Development and testing of the AIN-93 purified diets for rodents: Results on growth, kidney calcification and bone mineralization in rats and mice. J. Nutr. 123:1923–1931.

Reeves, P. G., F. H. Nielsen, and G. C. Fahey, Jr. 1993b. AIN-93 purified diets for laboratory rodents: Final report of the American Institute of Nutrition Ad Hoc Writing Committee on the reformulation of the AIN-76A rodent diet. J. Nutr. 123:1939–1951.

Reeves, P. G., K. L. Rossow, and L. Johnson. 1994. Maintenance nutrient requirements for copper in adult male mice fed AIN-93 rodent diet. Nutr. Res. 14:1219–1226.

Rehnberg, G. L., J. F. Hein, S. D. Carter, R. S. Linko, and J. W. Laskey. 1982. Chronic ingestion of Mn304 by rats: Tissue accumulation and distribution of manganese in two generations. J. Toxicol. Environ. Health 9:175–188.

Reichel, H., H. P. Koeffler, and A. W. Norman. 1989. The role of the vitamin D endocrine system in health and disease. N. Engl. J. Med. 320:980–991.

Remesy, C., and C. Demigne. 1989. Specific effects of fermentable carbohydrates on blood urea flux and ammonia absorption in rat cecum. J. Nutr. 119:560–565.

Remington, R. E., and J. W. Remington. 1938. The effect of enhanced iodine intake on growth and on the thyroid glands of normal goitrous rats. J. Nutr. 15:539–545.

Richardson, L. R., J. Godwin, S. Wilkes, and M. Cannon. 1964. Reproductive performance of rats receiving various levels of dietary protein and fat. J. Nutr. 82:257–262.

Riede, U. N., W. Sandritter, A. Pietzsch, and R. Rohrbach. 1980. Reaction patterns of cell organelles in vitamin B_6 deficiency: Ultrastructural-morphometric analysis of the liver parenchymal cell. Pathol. Res. Prac. 170:376–387.

Ritskes-Hoitinga, J. 1992. Diet and Nephrocalcinoisis in the Laboratory Rat. Ph.D. dissertation. State University, Wageningen, The Netherlands.

Ritskes-Hoitinga, J., A. G. Lemmens, L. H. J. C. Danse, and A. C. Beynen. 1989. Phosphorus-induced nephrocalcinosis and kidney function in female rats. J. Nutr. 119:1423–1431.

Ritskes-Hoitinga, J., J. N. J. J. Mathot, L. H. J. C. Danse, and C. C. Beynen. 1991. Commercial rodent diets and nephrocalcinosis in weanling female rats. Lab. Anim. 25:126–132.

Ritskes-Hoitinga, J., J. N. J. J. Mathot, A. G. Lemmens, L. H. J. C. Danse, G. W. Meijer, G. Van Tintelen, and A. C. Beynen. 1993. Long-term phosphorus restriction prevents corticomedullary nephrocalcinosis and sustains reproductive performance but delays bone mineralization in rats. J. Nutr. 123:754–763.

Rivlin, R. S. 1970. Riboflavin metabolism. N. Engl. J. Med. 283:463–472.

Robbins, J. D., R. R. Oltjen, C. A. Cabell, and E. H. Dolnick. 1965. Influence of varying levels of dietary minerals on the development of urolithiosis, hair growth and weight gains in rats. J. Nutr. 85:355–361.

Roebuck, B. D., S. A. Wilpone, D. S. Fifield, and J. D. Yager, Jr. 1979. Letter to the Editor. J. Nutr. 109:924–925.

Roecklein, B., S. W. Levin, M. Comly, and A. B. Mukherjee. 1985. Intrauterine growth retardation induced by thiamin deficiency and pyrithiamine during pregnancy in the rat. Am. J. Obstet. Gynecol. 151:455–460.

Rofe, A. M., R. Krishnan, R. Bais, J. B. Edwards, and R. A. Conyers. 1982. A mechanism for the thiamin-sparing action of dietary xylitol in the rat. Aust. J. Exp. Biol. Med. Sci. 60:101–111.

Rogers, A. E., and R. A. MacDonald. 1965. Hepatic vasculature and cell proliferation in experimental cirrhosis. Lab. Invest. 14:1710–1724.

Rogers, J. M., C. L. Keen, N. Reinstein, and L. S. Hurley. 1984. Maternal zinc nutriture during pregnancy and lactation in the rat: Survivability and growth of offspring. Fed. Proc. 43:1052 (abstr.).

Rogers, Q. R., and A. E. Harper. 1965. Amino acid diets and maximum growth in the rat. J. Nutr. 87:267–273.

Rogers, W. E., Jr., J. G. Bieri, and E. G. McDaniel. 1971. Vitamin A deficiency in the germfree state. Fed. Proc. 30:1773–1778.

Roginski, E. E., and W. Mertz. 1967. An eye lesion in rats fed low chromium diets. J. Nutr. 93:249–251.

Roginski, E. E., and W. Mertz. 1969. Effects of chromium (III) supplementation on glucose and amino acid metabolism in rats fed a low protein diet. J. Nutr. 97:525–530.

Rolls, B. J., and E. A. Rowe. 1982. Pregnancy and lactation in the obese rat: Effects on maternal and pup weights. Physiol. Behav. 28:393–400.

Rong, N., J. Selhub, B. R. Goldin, and I. H. Rosenberg. 1991. Bacterially synthesized folate in rat large intestine is incorporated into host tissue folyl polyglutamates. J. Nutr. 121:1955–1959.

Root, E. J., and J. B. Longnecker. 1983. Brain cell alterations suggesting premature aging induced by dietary deficiency of vitamin B_6 and/or copper. Am. J. Clin. Nutr. 37:540–552.

Rose, W. C., M. J. Oesterling, and M. Womack. 1948. Comparative growth on diets containing ten and nineteen amino acids, with further observation upon the role of glutamic and aspartic acid. J. Biol. Chem. 176:753–762.

Rosenberg, H. F., and R. Culik. 1955. Lysine requirement of the growing rat as a function of the productive energy level of the diet. J. Anim. Sci. 14:1221 (abstr.).

Roth-Maier, D. A., and M. Kirchgessner. 1981. Homeostasis and requirement of vitamin B_6 growing rats. Z. Tierphysiol. Tierernaehr. Futtermittelkde 46:247–254.

Roth-Maier, D. A., P. M. Zinner, and M. Kirchgessner. 1982. Effect of varying dietary vitamin B_6 supply on intestinal absorption of vitamin B_6. Int. J. Vitam. Nutr. Res. 52:272–279.

Roth-Maier, D. A., M. Kirchgessner, and S. Rajtek. 1990. Retention and utilization of thiamin by gravid and non gravid rats with varying dietary thiamin supply. Int. J. Vitam. Nutr. Res. 60:343–350.

Rotruck, J. T., A. L. Pope, H. E. Ganther, and W. G. Hoekstra. 1973. Selenium: Biochemical role as a component of glutathione peroxidase. Science 179:588–590.

Saari, J. T. 1992. Dietary copper deficiency and endothelium-dependent relaxation of rat aorta. Proc. Soc. Exp. Biol. Med. 200:19–24.

Sahu, A. P., A. K. Saxena, K. P. Singh, and R. Shanker. 1986. Effect of chronic choline administration in rats. Indian J. Exp. Biol. 24:91–96.

Said, A. K., and D. M. Hegsted. 1970. Response of adult rats to low dietary levels of essential amino acids. J. Nutr. 100:1363–1375.

Said, H. M., and D. Hollander. 1985. Does aging affect the intestinal transport of riboflavin? Life Sci. 36:69–73.

Said, H. M., F. K. Ghishan, H. L. Greene, and D. Hollander. 1985. Development maturation of riboflavin intestinal transport in the rat. Pediat. Res. 19:1175–1178.

Said, H. M., D. Ong, and R. Redha. 1988. Intestinal uptake of retinol in suckling rats: Characteristics and ontogeny. Pediat. Res. 24:481–485.

Sainz, R. D., C. C. Calvert, and R. L. Baldwin. 1986. Relationships among dietary protein feed intake and changes in body composition of lactating rats. J. Nutr. 116:1529–1539.

Salmon, W. D., and P. M. Newberne. 1962. Cardiovascular disease in choline-deficient rats. Arch. Pathol. 73:190–208.

Sampson, D. A., and G. R. Janson. 1984. Protein and energy nutrition during lactation. Annu. Rev. Nutr. 4:43–67.

Sandstrom, B., and B. Lönnerdal. 1989. Promoters and antagonists of zinc absorption. Pp. 57–78 in Zinc in Human Biology, C. F. Mills, ed. New York: Springer-Verlag.

Sarter, M., and A. Van Der Linde. 1987. Vitamin E deprivation in rats: Some behavioral and histochemical observations. Neurobiol. Aging 8:297–307.

Sato, Y., K. Ando, E. Ogata, and T. Fujita. 1991. High-potassium diet attenuates salt-induced acceleration of hypertension in SHR. Am. J. Physiol. 260:R21–R26.

Sauberlich, H. E. 1952. Effect of aureomycin and penicillin upon the vitamin requirements of the rat. J. Nutr. 46:99–108.

Schaafsma, G., and R. Visser. 1980. Nutritional interrelationships between calcium, phosphorus and lactose in rats. J. Nutr. 110:1101–1111.

Schachter, D., D. V. Kimberg, and H. Schenker. 1961. Active transport of calcium by intestine: Action and bio-assay of vitamin D. Am. J. Physiol. 200:1263–1271.

Schaeffer, M., and M. J. Kretsch. 1987. Quantitative assessment of motor and sensory function in vitamin B_6 deficient rats. Nutr. Res. 7:851–864.

Schaeffer, M. C. 1987. Attenuation of acoustic and tactile startle responses of vitamin B_6 deficient rats. Physiol. Behav. 40:473–478.

Schaeffer, M. C., D. A. Sampson, J. H. Skala, D. W. Gietzen, and R. E. Grier. 1989. Evaluation of vitamin B_6 status and function of rats fed excess pyridoxine. J. Nutr. 119:1392–1398.

Schaeffer, M. C., E. F. Cochary, and J. A. Sadowski. 1990. Subtle abnormalities of gait detected early in vitamin B_6 deficiency in aged and weanling rats with hind leg gait analysis. J. Am. Coll. Nutr. 9:120–127.

Schemmel, R., O. Mickelsen, and K. Motowi. 1972. Conversion of dietary to body energy in rats as affected by strain, sex and ration. J. Nutr. 102:1187–1197.

Schenker, S., R. M. Qualls, R. Butcher, and D. W. McCandless. 1969. Regional renal adenosine triphophate metabolism in thiamine deficiency. J. Nutr. 99:168–176.

Schneeman, B. O., and D. Gallaher. 1980. Changes in small intestinal digestive enzyme activity and bile acids with dietary cellulose in rats. J. Nutr. 110:584–590.

Schnegg, A., and M. Kirchgessner. 1975a. Zur Essentialität von Nickel für das tierische Wachstum. Z. Tierphysiol. Tierernaehr. Futtermittelkde 36:63–74.

Schnegg, A., and M. Kirchgessner. 1975b. Veränderungen des Hämoglobingehaltes, der Erythrozytenzahl und des Hämatokrits bei Nickelmangel. Nutr. Metabol. 19:268–278.

Schnegg, A., and M. Kirchgessner. 1976a. Zur Absorption und Verfügbarkeit von Eisen bei Nickel-Mangel. Int. Z. Vitaminforsch. 46:96–99.

Schnegg, A., and M. Kirchgessner. 1976b. Zur Toxizität von Alimentär Verabreichtem Nickel. Landwirtsch. Forsch. 29:177–185.

Schnegg, A., and M. Kirchgessner. 1978. Ni deficiency and its effect on metabolism. Pp. 236–243 in Trace Element Metabolism in Man and Animals—3, M. Kirchgessner, ed. Freising-Weihenstephan, Germany: University of Munich.

Schoenmakers, A. C. M., J. Ritskes-Hoitinga, A. G. Lemmens, and A. C. Beynen. 1989. Influence of dietary phosphorus restriction on calcium and phosphorus metabolism in rats. Int. J. Vitam. Nutr. Res. 59:200–206.

Schrader, C. A., C. Prickett, and W. D. Salmon. 1937. Symptomatology and pathology of potassium and magnesium deficiencies in the rat. J. Nutr. 14:85–109.

Schreier, C. J., and R. J. Emerick. 1986. Diet calcium carbonate, phosphorus and acidifying and alkalizing salts as factors influencing silica urolithiasis in rats fed tetraethylorthosilicate. J. Nutr. 116:823–830.

Schroeder, H. A. 1966. Chromium deficiency in rats: A syndrome simulating diabetes mellitus with retarded growth. J. Nutr. 88:439–445.

Schroeder, H. A., W. H. Vinton, Jr., and J. J. Balassa. 1963. Effects of chromium, cadmium and lead on the growth and survival of rats. J. Nutr. 80:48–54.

Schulz, A. R. 1991. Interpretation of nutrient-response relationships in rats. J. Nutr. 121:1834–1843.

Schumacher, M. F., M. A. Williams, and R. L. Lyman. 1965. Effect of high intakes of thiamine, riboflavin, and pyridoxine on reproduction in rats and vitamin requirements of the offspring. J. Nutr. 86:343–349.

Schwarz, K. 1951. Production of dietary necrotic liver degeneration using American Torula yeast. Proc. Soc. Exp. Biol. Med. 77:818–823.

Schwarz, K., and C. M. Foltz. 1957. Selenium as an integral part of factor 3 against dietary necrotic liver degeneration. J. Am. Chem. Soc. 79:3292–3293.

Schwarz, K., and W. Mertz. 1959. Chromium (III) and the glucose tolerance factor. Arch. Biochim. Biophys. 85:292–295.

Schwarz, K., and D. B. Milne. 1971. Growth effects of vanadium in the rat. Science 174:426–428.

Schwarz, K., and D. B. Milne. 1972. Growth-promoting effects of silicon in rats. Science 238:333–334.

Schweigert, B. S., and B. T. Guthneck. 1953. Utilization of amino acids from foods by the rat. I. Methods of testing for lysine. J. Nutr. 49:277–287.

Schweigert, B. S., and B. T. Guthneck. 1954. Utilization of amino acids from foods by the rat. II. Methionine. J. Nutr. 54:333–343.

Scott, E. M., and I. V. Griffith. 1957. A comparative study of thiamine-sparing agents in the rat. J. Nutr. 61:421–436.

Scott, M. L. 1978. Vitamin E. Pp. 133–210 in Handbook of Lipid Research, Vol. 2, The Fat-Soluble Vitamins, H. F. DeLuca, ed. New York: Plenum Press.

Seaborn, C. D., E. D. Mitchell, and B. J. Stoecker. 1992. Vanadium and ascorbate effects on hepatic 3-hydroxy-3-methylglutaryl coenzyme A reductase, cholesterol, and tissue minerals in guinea pigs fed low chromium diets. Magnes. Trace Elem. 10:327–338.

Sealey, J. E., I. Clark, M. B. Bull, and J. H. Laragh. 1970. Potassium balance and the control of renin secretion. J. Clin. Invest. 49:2119–2127.

Seshadri, S., L. Das, and A. Patil. 1982. Effect of varying the level of lead on ascorbic acid status in rats fed low and high protein diets. Arogya 8:34–39.

Shah, B. G., and B. Belonje. 1991. Different dietary calcium levels required to prevent nephrocalcinosis in male and female rats. Nutr. Res. 11:385–390.

Shah, B. G., B. Belonje, and E. A. Nera. 1980. Reduction of nephrocalcinosis in female rats by additional magnesium and by fluoride. Nutr. Rep. Int. 22:957–963.

Shah, B. G., K. D. Trick, and B. Belonje. 1986. Factors affecting nephrocalcinosis in male and female rats fed AIN-76 salt mixture. Nutr. Res. 6:559–570.

Shane, B., and E. L. R. Stokstad. 1985. Vitamin B_{12}-folate relationships. Annu. Rev. Nutr. 5:115–141.

Shanghai, Y. D. X. 1991. Effects of riboflavin deficiency on iron nutritional status in rats. Shanghai Yike Dazue Xuebao 18:9–13.

Shastri, N. V., S. G. Nayudu, and M. C. Nath. 1968. Effect of high fat and high fat-high protein diets on biosynthesis of niacin from tryptophan in rats. J. Vitaminol. 14:198–202.

Shaw, S., V. Herbert, N. Colman, and E. Jayatilleke. 1990. Effect of ethanol-generated free radicals on gastric intrinsic factor and glutathione. Alcohol 7:153–157.

Sheehan, P. M., B. A. Clevidence, L. K. Reynolds, F. W. Thye, and S. J. Ritchey. 1981. Carcass nitrogen as a predictor of protein requirement for mature female rats. J. Nutr. 111:1224–1230.

Shelling, D. H., and D. E. Asher. 1932. Calcium and phosphorus studies. IV. The relation of calcium and phosphorus of the diet to the toxicity of viosterol. Bull. Johns Hopkins 50:318–343.

Shepard, T. H., B. Mackler, and C. A. Finch. 1980. Reproductive studies in the iron-deficient rat. Teratology 22:329–334.

Sherman, H. 1954. Pyridoxine and related compounds. Pp. 265–276 in The Vitamins, Vol. III, W. H. Sebrell, Jr., and R. S. Harris, eds. New York: Academic Press.

Shibata, K. 1986. Nutritional factors affecting the activity of liver nicotinamide methyltransferase and urinary excretion of N-1-methylnicotinamide in rats. Agr. Biol. Chem. 50:1489–1494.

Shibuya, M., and M. Okada. 1986. Effect of pyridoxine-deficiency on the turnover of aspartate aminotransferase isozymes in rat liver. J. Biochem. 99:939–944.

Shibuya, M., F. Hisaoka, and M. Okada. 1990. Effects of pregnancy on vitamin B_6-dependent enzymes and B_6 content in tissues of rats fed diets containing two levels of pyridoxine-hydrochloride. Nippon Eiyo Shokuryo Gakkaishi 43:189–195.

Shigematsu, Y., Y. Kikawa, M. Sudo, and Y. Itokawa. 1989. Branched-chain α-ketoacids and related acids in thiamin-deprived rats. J. Nutr. Sci. Vitaminol. 35:163–70.

Shirley, B. 1982. Detrimental effects of increased riboflavin intake by pregnant and lactating rats. IRCS Med. Sci. 10:632–633.

Shurson, G. C., P. K. Ku, G. L. Waxler, M. T. Yokoysma, and E. R. Miller. 1990. Physiological relationships between microbiological status and dietary copper levels in the pig. J. Anim. Sci. 68:1061–1071.

Sibbald, I. R., R. T. Berg, and J. P. Bowland. 1956. Digestible energy in relation to food intake and nitrogen retention in the weanling rat. J. Nutr. 59:385–392.

Sibbald, I. R., J. P. Bowland, A. R. Robblee, and R. T. Berg. 1957. Apparent digestible energy and nitrogen in the food of the weanling rat: Influence of food consumption, nitrogen retention and carcass composition. J. Nutr. 61:71–85.

Sinclair, H. M. 1952. Essential fatty acids and their relation to pyridoxine. Biochem. Soc. Symp. 9:80–99.

Skala, J. H., M. C. Schaeffer, D. A. Sampson, and D. Gretz. 1989. Effects of various levels of pyridoxine on erythrocyte aminotransferase activities in the rat. Nutr. Res. 9:195–204.

Small, D. M. 1991. The effects of glyceride structure on absorption and metabolism. Annu. Rev. Nutr. 11:413–434.

Smith, A. M., and M. F Picciano. 1986. Evidence for increased selenium requirement for the rat during pregnancy and lactation. J. Nutr. 116:1068–1079.

Smith, A. M., and M. F. Picciano. 1987. Relative bioavailability of seleno-compounds in the lactating rat. J. Nutr. 117:725–731.

Smith, B. S. W., and A. C. Field. 1963. Effect of age on magnesium deficiency in rats. Br. J. Nutr. 17:591–600.

Smith, E. B., and B. C. Johnson. 1967. Studies of amino acid requirements of adult rats. Br. J. Nutr. 21:17–27.

Smith, J. C., Jr., E. G. McDaniel, F. F. Fan, and J. A. Halsted. 1973. Zinc: A trace element essential in vitamin A metabolism. Science 181:954–955.

Smith, J. E. 1990. Preparation of vitamin A-deficient rats and mice. Methods Enzymol. 190:229–236.

Smith, J. T. 1973. An optimal level of inorganic sulfate for the diet of a rat. J. Nutr. 103:1008–1011.

Smith, J. E., and R. Borchers. 1972. Environmental temperature and the utilization of β-carotene by the rat. J. Nutr. 102:1017–1024.

Smolin, L. A., and N. J. Benevenga. 1984. Factors affecting the accumulation of homocyst(e)ine in rats deficient in vitamin B_6. J. Nutr. 114:103–111.

Sowers, J. E., W. L. Stockland, and R. J. Meade. 1972. L-Methionine and L-cystine requirements of the growing rat. J. Anim. Sci. 35:782–788.

Spoerl, R., and M. Kirchgessner. 1975a. Changes in the copper status and the ceruloplasmin activity of mother and suckling rats in response to the copper supplied at graded levels. Z. Tierphysiol. Tierernaehr. Futtermittelke 35:113–127.

Spoerl, R., and M. Kirchgessner. 1975b. Effect of copper deficiency during rearing and gravidity on reproduction. Z. Tierphysiol. Tierernaehr. Futtermittelkd 35:321–328.

Sprecher, H. 1991. Metabolism of dietary fatty acids. Pp. 12–20 in Health Effects of Dietary Fatty Acids, G. J. Nelson, ed. Champaign, Ill.: American Oil Chemists Society.

Steenbock, H., and D. C. Herting. 1955. Vitamin D and growth. J. Nutr. 57:449–468.

Stenflo, J. 1976. Vitamin K-dependent carboxylation of blood coagulation proteins. Trends Biochem. Sci. 1:256–258.

Stewart, C. N., D. B. Coursin, and H. N. Bhagavan. 1975. Avoidance behavior in vitamin B_6-deficient rats. J. Nutr. 105:1363–1370.

Stockland, W. L., Y. F. Lai, R. J. Meade, J. E. Sowers, and G. Oeskmer. 1971. L-Phenylalanine and L-tyrosine requirements of the growing rat. J. Nutr. 101:177–184.

Stockland, W. L., R. J. Meade, D. F. Wass, and J. E. Sowers. 1973. Influence of levels of methionine and cystine on the total sulfur amino acid requirement of the growing rat. J. Anim. Sci. 36:526–530.

Stokstad, E. L., and C. P. Nair. 1988. Effect of hypothyroidism on methylmalonate excretion and hepatic vitamin B_{12} levels in rats. J. Nutr. 118:1495–1501.

Strasia, C. A. 1971. Vanadium essentiality and toxicity in the laboratory rat. Ph.D. dissertation. Purdue University, Lafayette, Indiana.

Strause, L., P. Saltman, and J. Glowacki. 1987. The effect of deficiencies of manganese and copper on osteoinduction and on resorption of bone particles in rats. Cal. Tiss. Int. 41:145–150.

Stucki, W. P., and A. E. Harper. 1962. Effects of altering the ratio of indispensable to dispensable amino acids in diets for rats. J. Nutr. 78:278–287.

Sun, S., R. McKee, J. S. Fisler, and M. E. Swendseid. 1986. Muscle creatine content in rats given repeated large doses of nicotinamide: Effects of dietary methionine, choline, carnitine, and other supplements. J. Nutr. 116:2409–2414.

Sunde, R. A. 1990. Molecular biology of selenoproteins. Annu. Rev. Nutr. 10:451–474.

Sunde, R. A., S. L. Weiss, K. M. Thompson, and J. K. Evenson. 1992. Dietary selenium regulation of glutathione peroxidase mRNA— Implications for the selenium requirement. FASEB J. 6(Part 1):A1365.

Suttie, J. W. 1991. Vitamin K. Pp. 145–194 in Handbook of Vitamins, 2nd Ed., L. J. Machlin, ed. New York: Marcel Dekker.

Suzuki, T., and A. Yoshida. 1979. Effectiveness of dietary iron and ascorbic acid in the prevention and cure of moderately long-term toxicity in rats. J. Nutr. 109:1974–1978.

Swarup, G., S. Cohen, and D. L. Garbers. 1982. Inhibition of membrane phosphotyrosyl-protein phosphatase activity by vanadate. Biochem. Biophys. Res. Commun. 107:1104–1109.

Swift, R. W., and A. Black. 1949. Fats in relation to caloric efficiency. J. Am. Oil Chem. Soc. 26:171–176.

Swift, R. W., and R. M. Forbes. 1939. The heat production of the fasting rat in relation to the environmental temperature. J. Nutr. 18:307–318.

Tagliaferro, A. R., and D. A. Levitsky. 1982. Spillage behavior and thiamin deficiency in the rat. Physiol. Behav. 28:933–937.

Takahashi, O., and K. Hiraga. 1979. Preventive effects of phylloquinone on hemorrhagic death induced by butylated hydroxytoluene in male rats. J. Nutr. 109:453–457.

Takahashi, Y. I., J. E. Smith, M. Winick, and D. S. Goodman. 1975. Vitamin A deficiency and fetal growth and development in the rat. J. Nutr. 105:1299–1310.

Tamburro, C., O. Frank, A. D. Thomson, M. F. Sorrell, and H. Baker. 1971. Interactions of folate, nicotinate, and riboflavin deficiencies in rats. Nutr. Rep. Int. 4:185–189.

Taniguchi, M. 1980. Effects of riboflavin deficiency on lipid peroxidation of rat liver microsomes. J. Nutr. Sci. 26:401–413.

Taylor, S., and E. Poulson. 1956. Long-term iodine deficiency in the rat. J. Endocrinol. 13:439–444.

Taylor, S. A., R. E. Shrader, K. G. Koski, and F. J. Zeman. 1983. Maternal and embryonic response to a "carbohydrate-free" diet fed to rats. J. Nutr. 113:253–267.

Taylor, T. G. 1980. Availability of phosphorus in animal feeds. Pp. 23–33 in Recent Advances in Animal Nutrition-1979, W. Haresign and D. Lewis, eds. Boston: Butterworths.

Thenen, S. W. 1989. Megadose effects of vitamin C on vitamin B_{12} status in the rat. J. Nutr. 119:1107–1114.

Thenen, S. W., and E. L. R. Stokstad. 1973. Effect of methionine on specific folate coenzyme pools in vitamin B_{12} deficient and supplemented rats. J. Nutr. 103:363–370.

Thomas, M. R., and A. Kirksey. 1976a. Postnatal patterns of brain lipids in progeny of vitamin B_6-deficient rats before and after pyridoxine supplementation. J. Nutr. 106:1404–1414.

Thomas, M. R., and A. Kirksey. 1976b. Postnatal patterns of fatty acids in brain of progeny from vitamin B_6-deficient rats before and after pyridoxine supplementation. J. Nutr. 106:1415–1420.

Thompson, S. G., and E. G. McGeer. 1985. GABA-transaminase and glutamic acid decarboxylase changes in the brain of rats treated with pyrithiamine. Neurochem. Res. 10:1653–1660.

Thomsen, L. L., C. Tasman-Jones, and C. Maher. 1983. Effects of dietary fat and gel-forming substances on rat jejunal disaccharidase levels. Digestion 26:124–130.

Thomson, A. D., O. Frank, B. De Angelis, and H. Baker. 1972. Thiamin depletion induced by dietary folate deficiency in rats. Nutr. Rep. Int. 6:107–110.

Tinoco, J. 1982. Dietary requirements and functions of α-linolenic acid in animals. Prog. Lipid Res. 21:1–45.

Tinoco, J., M. A. Williams, I. Hincenbergs, and R. L. Lyman. 1971. Evidence for nonessentiality of linolenic acid in the diet of the rat. J. Nutr. 101:937–946.

Tinsley, I. J., J. R. Harr, J. G. Bone, P. H. Weswig, and R. S. Yamamoto. 1967. Selenium toxicity in rats. I. Growth and longevity. Pp. 141–152 in Symposium: Selenium in Biomedicine, O. H. Muth, ed. Westport, Conn.: AVI Publishing.

Tobian, L. 1991. Salt and hypertension: Lessons from animal models that relate to human hypertension. Hypertension 17:I52–I58.

Tobin, B. W., and J. L. Beard. 1990. Interactions of iron deficiency and exercise training relative to tissue norepinephrine turnover, triiodothyronine production and metabolic rate in rats. J. Nutr. 120:900–908.

Tokunaga, T., T. Oku, and N. Hosoya. 1989. Utilization and excretion of a new sweetener, fructooligosaccharide (neosugar), in rats. J. Nutr. 119:553–559.

Tomlinson, D. R., G. B. Willars, and J. P. Robinson. 1986. Prevention of defects of axonal transport in experimental diabetes by aldose reductase inhibitors. Drugs 32(Suppl. 2):15–18.

Track, N. S., M. M. Cannon, A. Flenniken, S. Katamay, and E. F. A. Woods. 1982. Improved carbohydrate tolerance in fibre-fed rats: Studies of the chronic effect. Can. J. Physiol. Pharmacol. 60:769–776.

Trugnan, G., G. Thomas-Benhamou, P. Cardot, Y. Rayssiguier, and G. Bereziat. 1985. Short-term essential fatty acid deficiency in rats: Influence of dietary carbohydrates. Lipids 20:862–868.

Tsuji, M., Y. Ito, H. Fukuda, N. Terada, and H. Mori. 1989. Aromatase activity and progesterone metabolism in ovaries of scorbutic mutant rats unable to synthesize ascorbic acid. Int. J. Vitam. Nutr. Res. 59:353–359.

Tufts, E. V., and D. M. Greenberg. 1938. The biochemistry of magnesium deficiency. II. The minimum magnesium requirement for growth, gestation, and lactation, and the effect of the dietary calcium level thereon. J. Biol. Chem. 122:715–726.

Tulung, B., C. Remesy, and C. Demigne. 1987. Specific effects of guar gum or gum arabic on adaptation of cecal digestion to high fiber diets in the rat. J. Nutr. 117:1556–1561.

Turkki, P. R., and G. D. Degruccio. 1983. Riboflavin status of rats fed two levels of protein during energy deprivation and subsequent repletion. J. Nutr. 113:282–292.

Turkki, P. R., and P. G. Holtzapple. 1982. Growth and riboflavin status of rats fed different levels of protein and riboflavin. J. Nutr. 112:1940–1952.

Turkki, P. R., J. Buratto, M. Corteville, and M. P. Driscoll. 1986. Effect of riboflavin status on nitrogen retention during energy restriction and repletion in the rat. Nutr. Res. 6:993–1008.

Turkki, P. R., L. Ingerman, S. B. Kurlandsky, C. Yang, and R. S. Chung. 1989. Effect of energy restriction on riboflavin retention in normal and deficient tissues of the rat. Nutrition 5:331–337.

Turner, N. D., G. T. Schelling, P. G. Harms, L. W. Greene, and F. M. Byers. 1987. A comparison of the protein requirements for growth and reproduction in the rat. Nutr. Rep. Int. 36:73–88.

U.S. Pharmacopeia. 1985. U.S. Pharmacopeia, 21st Rev. XXI:1118–1119.

Uhland, A. M., G. G. Kwiecinski, and H. F. DeLuca. 1992. Normalization of serum calcium restores fertility in vitamin D-deficient male rats. J. Nutr. 122:1338–1344.

Ulman, E. A., and H. Fisher. 1983. Arginine utilization of young rats fed diets with simple versus complex carbohydrates. J. Nutr. 113:131–137.

Underwood, B. A. 1984. Vitamin A in animal and human nutrition. Pp. 281–391 in The Retinoids, Vol. 1, M. B. Spron, A. B. Roberts, and D. S. Goodman, eds. New York: Academic Press.

Underwood, E. J., and W. Mertz. 1987. Introduction. Pp. 1–19 in Trace Elements in Human and Animal Nutrition, 5th Ed., Vol. 1, W. Mertz, ed. San Diego: Academic Press.

Underwood, J. L., and H. F. DeLuca. 1984. Vitamin D is not directly necessary for bone growth and mineralization. Am. J. Physiol. 246:E493–E498.

Unna, K. 1939. Studies on the toxicity and pharmacology of nicotinic acid. J. Pharmacol. Exp. Therap. 65:95–103.

Unna, K. 1940. Pantothenic acid requirement of the rat. J. Nutr. 20:565–576.

Unna, K., and J. G. Greslin. 1941. Studies on the toxicity and pharmacology of pantothenic acid. J. Pharmacol. Exp. Therap. 73:85–90.

Ursini, F., M. Maiorino, and C. Gregolin. 1985. The selenoenzyme phospholipid hydroperoxide glutathione peroxidase. Biochim. Biophys. Acta 839:62–70.

Uthus, E. O., and F. H. Nielsen. 1990. Effect of vanadium, iodine and their interaction on growth blood variables, liver trace elements and thyroid status indices in rats. Magnes. Trace Elem. 9:219–226.

Vadhanavikit, S., and H. E. Ganther. 1993. Selenium requirements of rats for normal hepatic and thyroidal 5′-deiodinase (Type I) activities. J. Nutr. 123:1124–1128.

Vahouny, G. V. 1982. Dietary fibers and intestinal absorption of lipids. Pp. 203–227 in Dietary Fiber in Health and Disease, G. V. Vahouny, and D. Kritchevsky, eds. New York: Plenum Press.

Valencia, R., and E. Sacquet. 1968. La carence en vitamine B_{12} chez l'animal amicrobien (Anomalies tératog nes et signes hémagtologiques). Ann. Nutr. Aliment. 22:71–76.

Van Camp, I., J. Ritskes-Hoitinga, A. G. Lemmens, and A. C. Beynen. 1990. Diet-induced nephrocalcinosis and urinary excretion of albumin in female rats. Lab. Anim. 24:137–141.

Van den Berg, H., J. J. P. Bogaards, and E. J. Sinkeldam. 1982. Effect of different levels of vitamin B_6 in the diets of rats on the content of pyridoxamine-5′-phosphate and pyridoxal-5′-phosphate in the liver. Int. J. Vitam. Nutr. Res. 52:407–416.

Van den Berg, G. J., J. P. Van Wouwe, and A. C. Beynen. 1989. Ascorbic acid supplementation and copper status in rats. Biol. Trace Elem. Res. 23:165–172.

van Es, A. J. H. 1972. Maintenance. Pp. 1–55 in Handbuch der Tierernaehrung, Vol. 2, W. Lenkeit, and K. Breirem, eds. Hamburg/Berlin: Paul Parey.

Varela-Moreiras, G., and Selhub, J. 1992. Long-term folate deficiency alters folate content and distribution differentially in rat tissues. J. Nutr. 122:986–991.

Vaswani, K. K. 1985. Effect of neonatal thiamin and vitamin A deficiency on rat brain gangliosides. Life Sci. 37:1107–1115.

Veitch, K., J. P. Draye, F. Van Hoof, and H. S. Sherratt. 1988. Effects of riboflavin deficiency and colfibrate treatment on the five acyl-CoA dehydrogenases in rat liver mitochondria. Biochem. J. 254:477–481.

Vendeland, S. C., J. A. Butler, and P. D. Whanger. 1992. Intestinal absorption of selenite, selenate, and selenomethionine in the rat. J. Nutr. Biochem. 3:359–365.

Voris, L., and E. J. Thacker. 1942. The effects of the substitution of bicarbonate for chloride in the diet of rats on growth, energy and protein metabolism. J. Nutr. 23:365–374.

Vorontsov, N. N. 1979. Evolution of the alimentary system in myomorph rodents (translated from Russian). New Delhi: Indian National Scientific Documentation Centre.

Wade, A. E., J. S. Evans, D. Holmes, and M. T. Baker. 1983. Influence of dietary thiamin on phenobarbital induction of rat hepatic enzymes responsible for metabolizing drugs and carcinogens. Drug-Nutr. Interact. 2:117–130.

Wald, G. 1968. Molecular basis of visual excitation. Science 162:230–239.

Walker, J. J., and W. N. Garrett. 1971. Shifts in the energy metabolism of male rats during their adaptation to prolonged undernutrition and their subsequent realimentation. Pp. 193–197 in Energy Metabolism of Farm Animals, Proceedings of the Fifth Symposium of the European Association for Animal Production, Pub. 13. London: Butterworths.

Wallwork, J. C., G. J. Fosmire, and H. H. Sandstead. 1981. Effect of zinc deficiency on appetite and plasma amino acid concentrations in the rat. Br. J. Nutr. 45:127–136.

Walzem, R. L., and A. J. Clifford. 1988a. Determination of blood transketolase—An improved procedure with optimized conditions. J. Micronutr. Anal. 4:17–32.

Walzem, R. L., and A. J. Clifford. 1988b. Thiamin absorption is not compromised in folate-deficient rats. J. Nutr. 118:1343–1348.

Wang, F., L. Wang, E. H. Khairallah, and R. Schwartz. 1971. Magnesium depletion during gestation and lactation in rats. J. Nutr. 101:1201–1210.

Wang, T., K. W. Miller, Y. Y. Tu, and C. S. Yang. 1985. Effects of riboflavin deficiency on metabolism of nitrosamines by rat liver microsomes. J. Natl. Cancer Inst. 74:1291–1297.

Wang, X., D. Oberleas, M. T. Yang, and S. P. Yang. 1992. Molybdenum requirement of female rats. J. Nutr. 122:1036–1041.

Ward, G. J., and Nixon, P. F. 1990. Modulation of pterylpolyglutamate concentration and length in response to altered folate nutrition in a comprehensive range of rat tissues. J. Nutr. 120:476–484.

Warnock, L. G. 1970. A new approach to erythrocyte transketolase measurement. J. Nutr. 100:1057–1062.

Weaver, C., B. R. Martin, J. S. Ebner, and C. A. Krueger. 1987. Oxalic acid decreases calcium absorption in rats. J. Nutr. 117:1903–1906.

Webster, A. J. F., G. E. Lobley, P. J. Reedsan, and J. D. Pullar. 1980. Protein mass, protein synthesis and heat loss in the Zucker rat. Pp. 125–128 in Energy Metabolism of Farm Animals, Proceedings of the Eighth Symposium of the European Association for Animal Production, Pub. 26, L. E. Mount, ed. London: Butterworths.

Weissbach, H., and R. T. Taylor. 1970. Roles of vitamin B_{12} and folic acid methionine synthesis. Vitam. Horm. 28:415–440.

Wells, W. W., and L. E. Burton. 1978. Requirement for dietary *myo*-inositol in the lactating rat. Pp. 471–485 in Cyclitols Phosphoinositides, W. W. Wells, F. Eisenberg, Jr., eds. New York: Academic Press.

Wertz, P. W., W. Abraham, E. Cho, and D. T. Downing. 1986. Linoleate-rich *O*-acylsphingolipids of mammalian epidermis: Structures and effects of essential fatty acid deficiency. Prog. Lipid Res. 25:383–389.

Wesson, L. G., and G. O. Burr. 1931. The metabolic rate and respiratory quotients of rats on a fat-deficient diet. J. Biol. Chem. 91:525–539.

Whanger, P. D. 1973. Effects of dietary nickel on enzyme activities and mineral content in rats. Toxicol. Appl. Pharmacol. 25:323–331.

Whanger, P. D., and J. A. Butler. 1988. Effects of various dietary levels of selenium as selenite or selenomethionine on tissue selenium levels and glutathione peroxidase activity in rats. J. Nutr. 118:846–852.

Whitescarver, S. A., B. J. Holtzclaw, J. H. Downs, C. E. Ott, J. R. Sowers, and T. A. Kotchen. 1986. Effect of dietary chloride on salt-sensitive and renin-dependent hypertension. Hypertension 8:56–61.

Will, B. H., and J. W. Suttie. 1992. Comparative metabolism of phylloquinone and menaquinone-9 in rat liver. J. Nutr. 122:953–958.

Williams, D. L. 1973. Biological value of vanadium for rats, chickens, and sheep. Ph.D. dissertation. Purdue University, Lafayette, Indiana.

Williams, R. B., and C. F. Mills. 1970. The experimental production of zinc deficiency in the rat. Br. J. Nutr. 24:989–1003.

Williams, D. L., and G. H. Spray. 1973. The effects of diets containing raw soya-bean flour on the vitamin B_{12} status of rats. Br. J. Nutr. 29:57–63.

Willis, W. T., K. Gohil, G. A. Brooks, and P. R. Dallman. 1990. Iron deficiency: Improved exercise performance within 15 hours of iron treatment in rats. J. Nutr. 120:909–916.

Wilson, J. G., C. B. Roth, and J. Warkany. 1953. An analysis of the syndrome of malformations induced by maternal vitamin A deficiency: Effects of restoration of vitamin A at various times during gestation. Am. J. Anat. 92:189–217.

Windebank, A. J., P. A. Low, and M. D. Blexrud. 1985. Pyridoxine

neuropathy in rats: Specific degeneration of sensory axons. Neurology 35:1617–1622.

Witting, L. A., and M. K. Horwitt. 1964. Effect of degree of fatty acid unsaturation in tocopherol deficiency-induced creatinuria. J. Nutr. 82:19–33.

Woodard, J. C., and P. M. Newberne. 1966. Relation of vitamin B_{12} and 1-carbon metabolism to hydrocephalus in the rat. J. Nutr. 88:375–381.

Woodard, J. C., and W. S. S. Jee. 1984. Effects of dietary calcium, phosphorus and magnesium on intranephronic calculosis in rats. J. Nutr. 114:2331–2338.

Woolliscroft, J., and J. Barbosa. 1977. Analysis of chromium-induced carbohydrate intolerance in the rat. J. Nutr. 107:1702–1706.

Worcester, N. A., K. R. Bruckdorfer, T. Hallinan, A. J. Wilkins, and J. A. Mann. 1979. The influence of diet and diabetes on stearoyl Coenzyme A desaturase (EC 1.14.99.5) activity and fatty acid composition in rat tissues. Br. J. Nutr. 41:239–252.

Wostmann, B. S., P. L. Knight, L. L. Keeley, and D. F. Kan. 1963. Metabolism and function of thiamine and naphthoquinones in germfree and conventional rats. Fed. Proc. 22:120–124.

Wretlind, K. A. J., and W. C. Rose. 1950. Methionine requirement for growth and utilization of its optical isomers. J. Biol. Chem. 187:697–703.

Wu, W. H., M. Meydani, S. N. Meydani, P. M. Burklund, J. B. Blumberg, and H. N. Munro. 1990. Effect of dietary iron overload on lipid peroxidation, prostaglandin synthesis and lymphocyte proliferation in young and old rats. J. Nutr. 120:280–289.

Yamamoto, N., M. Saitoh, A. Moriuchi, M. Nomura, and H. Okuyama. 1987. Effect of dietary α-linolenate/linoleate balance on brain lipid compositions and learning ability of rats. J. Lipid Res. 28:144–151.

Yamamoto, N., A. Hashimoto, Y. Takemoto, H. Okuyama, M. Nomura, R. Kitajima, T. Togashi, and Y. Tamai. 1988. Effect of dietary α-linolenate/linoleate balance on lipid composition and learning ability of rats. II. Discrimination process, extinction process, and glycolipid compositions. J. Lipid Res. 29:1013–1021.

Yamanaka, W. K., G. W. Clemans, and M. L. Hutchinson. 1980. Essential fatty acid deficiency in humans. Prog. Lipid Res. 19:187–215.

Yang, J.-G., J. Morrison-Plummer, and R. F. Burk. 1987. Purification and quantitation of a rat plasma selenoprotein distinct from glutathione peroxidase using monoclonal antibodies. J. Biol. Chem. 262:13372–13375.

Yang, J.-G., K. E. Hill, and R. F. Burk. 1989. Dietary selenium intake controls rat plasma selenoprotein P concentration. J. Nutr. 119:1010–1012.

Yang, M. G., K. Manoharan, and A. K. Young. 1969. Influence and degradation of dietary cellulose in cecum of rats. J. Nutr. 97:260–264.

Yang, N. Y. J., and I. D. Desai. 1977. Reproductive consequences of mega vitamin E supplements in female rats. Experientia 33:1460–1461.

Yokogoshi, H., K. Hayase, and A. Yoshida. 1980. Effect of carbohydrates and starvation on nitrogen sparing action of methionine and threonine in rats. Agr. Biol. Chem. 44:2503–2506.

Yoo, J. S., H. S. Park, S. M. Ning, M. J. Lee, and C. S. Yang. 1990. Effects of thiamine deficiency on hepatic cytochromes P-450 and drug-metabolizing enzyme activities. Biochem. Pharmacol. 39:519–525.

Yoshida, A., and K. Ashida. 1969. Pattern of essential amino acid requirement for growing rats fed on a low amino acid diet. Agr. Biol. Chem. 33:43–49.

Yoshida, A., A. E. Harper, and C. A. Elvehjem. 1957. Effects of protein per calorie ratio and dietary levels of fat on calorie and protein utilization. J. Nutr. 63:555–560.

Yoshida, A., A. E. Harper, and C. A. Elvehjem. 1958. Effect of dietary level of fat and type of carbohydrate on growth and food intake. J. Nutr. 66:217–228.

Yoshida, T., T. Oowada, A. Ozaki, and T. Mizutani. 1993. Role of gastrointestinal microflora in the mineral absorption of young adult mice. Biosci. Biotech. Biochem. 57:1775–1776.

Yu, H., and J. W. Whittaker. 1989. Vanadate activation of bromoperoxidase from Corallian officinalis. Biochem. Biophys. Res. Commun. 160:87–92.

Yu, J. Y., and Y. O. Cho. 1989. The effect of vitamin B_2 and (or) vitamin B_6 deficiency on hematologic profile in rats. Korean J. Nutr. 22:167–174.

Yudkin, J. 1979. The avoidance of sucrose by thiamin-deficient rats. Int. J. Vitam. Nutr. Res. 49:127–135.

Zaki, F. G., C. Bandt, and F. W. Hoffbauer. 1963. Fatty cirrhosis in the rat. III. Liver lipid and collagen content in various stages. Arch. Pathol. 75:648–653.

Zelent, A., A. Krust, M. Petkovich, P. Kastner, and P. Chambon. 1989. Cloning of murine α and β retinoic acid receptors and a novel receptor γ predominantly expressed in skin. Nature 339:714–717.

Zeman, F. J., and E. C. Stanbrough. 1969. Effect of maternal protein deficiency on cellular development in the fetal rat. J. Nutr. 99:274–282.

Zhang, X., and A. C. Beynen. 1992. Increasing intake of soybean protein or casein, but not cod meal, reduces nephrocalcinosis in female rats. J. Nutr. 122:2218–2225.

Zidenberg-Cherr, S., C. L. Keen, B. Lönnerdal, and L. S. Hurley. 1983. Superoxide dismutase activity and lipid peroxidation: Developmental correlations affected by manganese deficiency. J. Nutr. 113:2498–2504.

Ziesenitz, S. C., G. Siebert, D. Schwengers, and R. Lemmes. 1989. Nutritional assessment in humans and rats of leucrose [D-glucopyranosyl-α(1-5)-D-fructopyranose] as a sugar substitute. J. Nutr. 119:971–978.

3 Nutrient Requirements of the Mouse

Mice (*Mus musculus*) have been used extensively as animal models for biomedical research in genetics, oncology, toxicology, and immunology as well as cell and developmental biology. The widespread use of this species can be attributed to the mouse's high fertility rate, short gestation period, small size, ease of maintenance, susceptibility or resistance to different infectious agents, and susceptibility to noninfectious or genetic diseases that afflict humans. Morse (1978) wrote a detailed history of the development of the mouse as a model for biomedical research. Estimating the quantitative nutrient requirements for mice is particularly challenging because of the large genetic variation within the species and the different criteria used to assess nutritional adequacy of diets. Research to determine nutrient requirements for reproduction, lactation, and maintenance of mice has received relatively little attention.

A complicating factor in estimating the nutrient requirements for laboratory mice is that they are reared and maintained in conventional, specific-pathogen-free, or germ-free environments where the intestinal flora is undefined, defined, or absent, respectively. Because intestinal flora populations influence nutrient requirements, it is not valid to generalize data among these environments.

GENETIC DIVERSITY

Laboratory mice used in biomedical research represent noninbred stocks and inbred, congenic, and mutant strains. The number of individual stocks and strains of mice available for use in research is estimated to be near 500. In addition, there are numerous recently developed transgenic mouse strains; the exact number is not readily available, but estimates are as high as 20,000 strains. With this amount of genetic diversity within a mammalian species the probability is high that there would be differences in nutrient requirements among the different stocks and strains. Even though a small percentage of the existing stocks and strains of laboratory mice have been used in nutritional research, discussions of individual nutrients in this chapter indicate that mouse stocks or strains differ in their requirements for various nutrients.

GROWTH AND REPRODUCTION

The growth rates published by Poiley (1972) for 38 stocks and strains of mice show an approximately twofold difference between the slowest and fastest growing mice. Growth statistics for five stocks and strains representing this range of mouse genotype are shown in Table 3-1. This difference in growth suggests a marked difference in nutrient requirements among mouse genotypes. Considering this large variation in growth rates, expected or acceptable growth rates used as standards should come only from those individuals responsible for maintaining the breeding colony that provides a given genotype.

Reproduction, too, varies among genotypes (National Institutes of Health, 1982). The reproductive performance of highly inbred strains is frequently low, particularly for strains that have been selected for a metabolic defect. The reproductive characteristics of representative inbred and outbred mouse genotypes are presented in Table 3-2. Because reproductive characteristics are specific to each mouse genotype, information in this regard should be obtained from those individuals who maintain the foundation colony of the strain or substrain of interest.

ESTIMATION OF NUTRIENT REQUIREMENTS

The nutrient requirements of mice have been defined by several different criteria including growth, reproduction, longevity, nutrient storage, enzyme activity, gross or histo-

TABLE 3-1 Average Growth of Commonly Used Strains of Laboratory Mice

| Age (weeks) | Weight, g | | | |
	Swiss-Webster (OB)	BALB/c (IB)	DBA/2 (IB)	B6C3F$_1$ (HB)
Female				
3	12	12	11	9
4	14	14	15	14
5	23	16	18	16
6	24	17	19	17
7	26	18	20	18
8	26	18	21	20
9	27			
10	29			
Male				
3	12	12	11	9
4	16	14	15	15
5	26	19	17	19
6	31	21	21	22
7	34	22	23	24
8	36	23	22	25
9	38			
10	40			

Abbreviations: OB, outbred; IB, inbred; HB, hybrid.
SOURCE: Adapted from Poiley (1972).

TABLE 3-2 Some Reproductive Characteristics of Representative Strains of Inbred and Outbred Mouse Colonies Maintained at the National Institutes of Health

Strain	Sterile Matings (%)	Mean Litter Size	Preweaning Mortality (%)
Inbred strain			
NFR/N	3.0	7.6	7.0
FVB/N	1.0	7.8	8.7
I/STN	30.0	4.9	34.0
P	13.0	2.6	27.0
Outbred strain			
N:NIH	14.9	7.1	8.1
N:GP	16.3	6.7	6.1

SOURCE: Data summarized and provided by C. T. Hansen (Veterinary Resources Program, National Center for Research Resources, National Institutes of Health, personal communication, 1993).

logical appearance of tissue lesions, and nucleic acid or protein content of tissue. The requirement for any nutrient may vary with the criteria used. Traditionally, rapid growth leading to maximum body size at maturity has been the basis for measuring dietary adequacy on the assumption that a diet promoting maximum growth would be adequate for reproduction, lactation, and maintenance. Data have been published that test the validity of this assumption for the mouse. Knapka et al. (1977) found that diets producing maximum postweaning growth did not support maximum rates of reproduction. Since the mouse achieves one-third of its total growth during the suckling period, lactation imposes a heavier nutritional burden on the dam; this may influence some nutrient requirements more than others. Dubos et al. (1968) reported that a casein-starch diet that contained 0.05 percent magnesium was adequate for growth of mice, but sudden death occurred in some lactating females when they consumed that diet. An additional 0.02 percent magnesium prevented this syndrome, indicating the need for an increase in the magnesium requirement during lactation.

These results suggest that diets which support maximal growth are not optimal for reproduction. Therefore, for the mouse diet, the meaning of the term "adequate" may need to be expanded to indicate a range of nutrient intakes between minimal and harmfully excessive; the range will

vary at different stages of the life cycle. Nutrition investigators generally focus on nutrient requirements as minimal dietary concentrations. For life-span studies with mice, however, optimum dietary concentrations of energy and nutrients may have to be established. Although many studies have been conducted on the effects of diet on longevity, there are insufficient published data to estimate the nutrient requirements for long-term maintenance of mice.

Mice maintained in germ-free, gnotobiotic, or specific-pathogen-free environments, where the kinds and number of intestinal microorganisms are altered, may require different dietary concentrations of nutrients. Luckey et al. (1974) fed mice a sterilized diet, marginal in several vitamins, and observed decreased reproduction in germ-free as compared to conventionally reared mice. Allen et al. (1991) reported that adding 20 mg menadione sodium bisulfite/kg diet to the diets of hysterectomy-derived mice maintained in a specific-pathogen-free environment arrested a spontaneous outbreak of hemothorax.

Considering the many genetic and environmental factors that influence the nutrient requirements of laboratory mice, too few controlled studies have been conducted, particularly in recent years, to identify the nutrient requirements of this species. As a result the estimates of nutrient requirements are based on

- data accumulated many years ago involving mouse strains fed dietary ingredients that are no longer available or cannot be identified,
- experimental results derived from studies that were not designed to establish nutrient requirements,
- nutrient consumption by mice fed diets producing "acceptable performance," and
- the assumption that mouse nutrient requirements are similar to those of the rat.

The estimated nutrient requirements presented in Table 3-3 provide guidelines for the adequate nutrition of mice maintained in conventional animal facilities. However, mice subjected to stress, such as drug testing or surgery, or mice maintained in a germ-free environment may have altered nutrient requirements. The values in Table 3-3 may have to be adjusted to allow a margin of safety between the actual and estimated requirements.

For most nutrients a single estimate has been made. It is recognized, however, that mice are similar to other

TABLE 3-3   Estimated Nutrient Requirements of Mice

Nutrient	Unit	Amount, per kg diet	Comment/Reference
Lipid	g	50.0	
Linoleic acid	g	6.8	
Protein (N × 6.25)			
Growth	g	180.0	Equivalent to 20% casein supplemented with 0.3% DL-methionine or 24% casein
	g	200.0	Casein (see text)
Reproduction	g	180.0	Natural ingredients
Amino acids			
Arginine	g	3.0	See text
Histidine	g	2.0	See text
Isoleucine	g	4.0	See text
Leucine	g	7.0	See text
Valine	g	5.0	See text
Threonine	g	4.0	See text
Lysine	g	4.0	See text
Methionine	g	5.0	Cystine may replace 50-66.6%
Phenylalanine	g	7.6	Tyrosine may replace 50%
Tryptophan	g	1.0	Niacin may replace 0.025% (see text)
Minerals			
Calcium	g	5.0	
Chloride	g	0.5	
Magnesium	g	0.5	0.7 g/kg for lactation (see text)
Phosphorus	g	3.0	
Potassium	g	2.0	Higher concentrations may be required for lactation (see text)
Sodium	g	0.5	
Copper	mg	6.0	8.0 mg/kg for pregnancy and lactation
Iron	mg	35.0	
Manganese	mg	10.0	
Zinc	mg	10.0	30 mg/kg for reproduction and lactation
Iodine	μg	150.0	
Molybdenum	μg	150.0	
Selenium	μg	150.0	Selenite form of Se
Vitamins			
A (retinol)[a]	mg	0.72	Santhanam et al., 1987
D (cholecalciferol)[b]	mg	0.025	Adequate; no quantitative data
E (*RRR*-α-tocopherol)[c]	mg	22.0	Yasunaga et al., 1982
K (phylloquinone)	mg	1.0	Based on the requirement for the rat; Kindberg and Suttie, 1989
Biotin (*d*-biotin)	mg	0.2	Adequate; Fenton et al., 1950
Choline (choline bitartrate)	mg	2,000.0	Adequate; insufficient data to establish requirement
Folic acid	mg	0.5	Fenton et al., 1950; Heid et al., 1992
Niacin (nicotinic acid)	mg	15.0	Based on the requirement for the rat; Hundley, 1949
Pantothenate (Ca)	mg	16.0	Morris and Lippincott, 1941, suggest 36 mg/kg for reproduction and lactation
Riboflavin	mg	7.0	Adequate; insufficient data to establish requirement
Thiamin (thiamin-HCl)	mg	5.0	
B_6 (pyridoxine-HCl)	mg	8.0	1 mg/kg for maintenance
B_{12}	μg	10.0	Adequate; insufficient data to establish requirement

NOTE: Nutrient requirements are expressed on an as-fed basis for diets containing 10% moisture and 3.8 to 4.1 kcal ME/g (16–17 kJ ME/g) and should be adjusted for diets of differing moisture and energy concentrations. Unless otherwise specified, the listed nutrient concentrations represent minimal requirements and do not include a margin of safety. Higher concentrations for many nutrients may be warranted in natural-ingredient diets.

[a]Equivalent to 2,400 IU/kg of diet.

[b]Equivalent to 1,000 IU/kg of diet.

[c]Equivalent to 32 IU/kg.

mammalian species in that optimum nutrient requirements differ for growth, reproduction, lactation, and maintenance. Unfortunately, published data are not available for estimating nutrient requirements for each stage of the life cycle.

EXAMPLES OF DIETS FOR LABORATORY MICE

Mice are omnivorous; they consume a wide variety of seeds, grains, and other plant material as well as feedstuffs of animal origin. Natural-ingredient diets for mice maintained in conventional and barrier environments are commercially available. The formulations for conventional and autoclavable natural-ingredient diets that have been used successfully for many years are presented in Table 2-3 (Knapka et al., 1974; National Institutes of Health, 1982; Knapka, 1983). Note that alterations of these formulations may be appropriate to accommodate changes in ingredient availability or nutrient composition.

The purified-diet formulations presented in Table 2-5 have been developed and evaluated by a committee of the American Institute of Nutrition (AIN) as new standard reference diets (Reeves et al., 1993b). These diets replace the widely used purified diet referred to as AIN-76A shown in Table 2-4 (American Institute of Nutrition, 1977). Chemically defined diets are another option (Pleasants et al., 1970). The type of diet used depends on production or experimental objectives.

ENERGY

Troelson and Bell (1963) found that mice consumed an average of 3.5 g diet/day during 14 days postweaning. Calvert et al. (1986) reported that male mice consumed an average of 3.75 g diet/day during the 21 days postweaning and that mice with an increased genetic potential for growth because of (1) an apparent single-gene mutation that increased postweaning weight gain, (2) a genetic selection for growth, or (3) a combination of the single-gene plus genetic selection consumed 4.81, 4.57, and 6.18 g diet/day, respectively. The metabolizable energy (ME) content was estimated to be 3.9 kcal/g diet (16.2 kJ/g diet). It was not possible to determine whether the maintenance energy requirement differed for the different strains. Canolty and Koong (1976) reported that a line of mice selected for rapid postweaning growth consumed 5.0 g diet/day [18 kcal ME/day (75 kJ ME/day)] from 21 to 42 days of age while a line of mice from the same strain, but not selected for rapid growth, ate 3.8 g diet/day (57.7 kJ ME/day). The energy requirement for maintenance was exactly the same for both lines—176 kcal ME/BW$_{kg}^{0.75}$/day (736 kJ ME/BW$_{kg}^{0.75}$/day). Bernier et al. (1986) examined the effect of growth rate on maintenance energy requirements in two lines of mice with similar genetic backgrounds with the

exception that one line had a single-gene mutation that increased body weight gain without altering body composition. They reported that the normal mice consumed 12.7 kcal ME/BW$_{kg}^{0.75}$/day (53.3 kJ ME/BW$_{kg}^{0.75}$/day) from 21 to 42 days of age and gained 17.2 g BW, while mice with the gene for rapid growth consumed 17.4 kcal ME/BW$_{kg}^{0.75}$/day (72.8 kJ ME/BW$_{kg}^{0.75}$/day) and gained 29.8 g BW. On the basis of a comparative slaughter experiment, the rapidly growing mice had a maintenance energy requirement of 155 kcal ME/BW$_{kg}^{0.75}$/day (648 kJ ME/BW$_{kg}^{0.75}$/day) as compared to 164 kcal ME/BW$_{kg}^{0.75}$/day (686 kJ ME/BW$_{kg}^{0.75}$/day) in the normal mice. These values are slightly lower than the 176 kcal ME/BW$_{kg}^{0.75}$/day (736 kJ ME/BW$_{kg}^{0.75}$/day) reported by Canolty and Koong (1976). It is likely that this difference is the result of an error of 7 percent in the ME of glucose (Canolty and Koong, 1976). The estimated maintenance energy requirement reported by Bernier et al. (1986) is similar to that reported by Webster (1983). He found that the maintenance energy requirement of adult mice maintained at 24° C was 161 kcal ME/BW$_{kg}^{0.75}$/day (673 kJ ME/BW$_{kg}^{0.75}$/day).

It is not possible to specify a maintenance energy requirement for the average growing mouse without sufficient information about the mouse and its environment. Genetics and diet can have a substantial influence on the estimated maintenance energy requirement. Lin et al. (1979) reported that when fed a high-carbohydrate diet, obese (*ob/ob*) mice had an estimated maintenance energy requirement averaging 73 kcal ME/BW$_{kg}^{0.75}$/day (305 kJ ME/BW$_{kg}^{0.75}$/day), while lean mice required 118 kcal ME/BW$_{kg}^{0.75}$/day (493 kJ ME/BW$_{kg}^{0.75}$/day). Although the genetic background of these mice is different than those discussed previously, other factors influence estimates of maintenance energy requirements. Lin et al. (1979) maintained mice at an environmental temperature of 30° C for the first 2 weeks postweaning and 26° C for the following 2 weeks. Webster (1983) reported that when room temperature was increased from 24° to 28° C, heat production in mice decreased 21 percent, thus decreasing observed maintenance energy requirement. Diet composition also can alter the maintenance energy requirement. Lin et al. (1979) reported that lean mice fed a high-carbohydrate diet had a maintenance energy requirement of 118 kcal/BW$_{kg}^{0.75}$/day (493 kJ/BW$_{kg}^{0.75}$/day); when fed a high-fat diet, their estimated maintenance energy requirement increased to 130 kcal/BW$_{kg}^{0.75}$/day (544 kJ/BW$_{kg}^{0.75}$/day)—a 10 percent increase. This effect of diet was not observed in obese mice (Lin et al., 1979). Thus, environmental temperature, diet composition, and genetic background must be considered when predicting maintenance energy requirements. Sex must also be considered. Bull et al. (1976) report that female rodents appear to have a lower maintenance energy requirement than males. Based on their observations it is estimated that the daily ME requirement for maintenance is 160 kcal/$BW_{kg}^{0.75}$/day (670 kJ ME/$BW_{kg}^{0.75}$/day) in mice that have no obvious genetic or stress-induced abnormalities or pathologies. A diet containing 3.9 kcal ME/g (16.2 kJ ME/g) should meet this requirement.

As with determining a specific energy requirement for maintenance, developing a specific energy requirement for a given growth rate or to support lactation or pregnancy is difficult at best. In the case of growth, the composition of gain must be considered. Gain as fat is an energetically more efficient process than gain as lean (see "Energy" section in Chapter 2). Further, the energetic efficiency with which a diet is used to support gain depends, in part, on the composition of the diet. It is energetically more efficient to produce body fat from dietary fat than it is to produce body fat from dietary carbohydrate. Still, there may be value in estimating the efficiency of ME used for fat energy gain and that used for lean energy gain. Kielanowski (1965) proposed that total ME intake (MEI) of growing animals could be partitioned into three portions: one proportional to body weight (representing energy requirement for maintenance), a second proportional to fat energy gain, and a third proportional to lean energy gain. Several problems have been identified relative to the use of this model. Generally, body weight is scaled using the interspecific relationship of adult basal heat production to body weight ($BW_{kg}^{0.75}$). Such scaling, however, was intended to describe a relationship among species, not within species (Webster, 1988). Bernier et al. (1987) demonstrated that the body weight exponent affects the proportions of heat production assigned to maintenance and growth. Thus it is important, when using the Kielanowski (1965) model, that the body weight exponent chosen to represent basal heat production be appropriate for the given species. Using the Kielanowski (1965) model and data generated by Bernier et al. (1986, 1987), the estimated energetic efficiencies of fat energy gain in two lines of mice would be 189 and 161 percent, obviously a biological impossibility. Use of regression techniques to estimate the efficiency of lean energy gain also leads to results that are impossible (Bernier et al., 1986, 1987; Thonney et al., 1991). A more precise estimate of energy requirements will be possible only with the development and testing of mechanistic models. Given the concerns raised above, it is likely that the average daily ME requirement of growing mice from 21 to 42 days of age will be met by their consuming 263 kcal ME/$BW_{kg}^{0.75}$ (1,100 kJ ME/$BW_{kg}^{0.75}$). During early gestation the energy requirement is likely less but may be as high as 358.5 kcal ME/$BW_{kg}^{0.75}$ (1,500 kJ ME/$BW_{kg}^{0.75}$) during the third trimester. Although specific data on energy requirement and support of lactation for mice are notably lacking, from other rodent data it is estimated that peak lactation could require a daily MEI of 311 to 430 kcal ME/$BW_{kg}^{0.75}$ (1,300 to 1,800 kJ ME/$BW_{kg}^{0.75}$) to support the dam and large litters.

LIPIDS

Lipids are required by the mouse to provide essential fatty acids (EFA). Dietary fat is another concentrated energy source and a carrier for fat-soluble vitamins. It aids absorption of fat-soluble vitamins and enhances diet acceptability.

DIETARY FAT CONCENTRATION

Bossert et al. (1950) demonstrated that weanling (Dohme and Swiss-Webster strains) mice gained weight equally well when fed diets containing 0.5 to 40 percent fat; all diets contained 0.5 percent corn oil with the remainder of the fat supplied by hydrogenated cottonseed oil. A decrease in growth was noted when the dietary fat content exceeded 40 percent. Fenton and Carr (1951) demonstrated that the effect of dietary fat concentration on weight gain of mature mice depended on the strain. Strains A and C3H had higher rates of gain when dietary fat was increased from 5 to 47 percent, while strains C57 and I showed no further increases in weight gain when the fat content of the diet was increased to more than 15 percent.

Knapka et al. (1977) found a significant strain × fat interaction in reproductive capability (number of pups born per litter, pups weaned per litter) when diets containing 4, 8, and 12 percent crude fat were fed to four different strains of mice (BALB/cAnN, C3H/HeN, C57BL/6N, and DBA/2N). For example, the mean number of pups born to and weaned by each BALB/cAnN mouse over the reproductive lifetime increased from 38 to 46 when crude fat was increased from 4 to 8 percent, while the number of pups born to and weaned by DBA/2N mice decreased from 13 to 8 with increasing concentrations of crude fat. The other two strains had responses similar to (although less dramatic than) those of the DBA/2N strain, indicating that the absolute concentration of dietary crude protein and crude fat in diets for production of these inbred mouse strains is not as important as the ratio of these nutrients. Mouse reproduction also was affected by protein × fat interactions. Knapka and co-workers (1977) suggested that optimal crude protein and crude fat concentrations should be lower than 18 percent and 10 to 11 percent, respectively.

Olson et al. (1987) described an increase in mammary tumorigenesis, reduced T-cell blastogenesis, and lowered cell-mediated immunity when fat (soybean oil) fed to C3H/OUJ female mice was increased from 5 to 20 percent of the diet. Birt et al. (1989) noted that dimethylbenzanthracene-induced skin papillomas grew more rapidly in SENCA mice fed 24.6 percent versus 5 percent dietary corn oil. Kubo et al. (1987) observed that longevity of (NZB× NZW)F_1 female mice whose feeding was restricted was greater than that of controls allowed to feed ad libitum. Of the diets fed on a restricted basis, high-fat (69.8 percent fat) diets increased longevity only two-thirds as well as low-fat (4.5 percent fat) diets. For these reasons, a diet with a fat concentration of 5 percent is recommended, similar to that suggested for the rat. This concentration is adequate but may not support maximal growth and reproduction of all mouse strains.

ESSENTIAL FATTY ACIDS (EFA)

n-6 Fatty Acids

The mouse, like the rat, requires linoleic acid to avoid classical signs of EFA deficiency; however, the precise requirement for n-6 fatty acids has not been determined. Cerecedo et al. (1952) reported that 5 mg linoleate/day alleviated clinical signs of EFA deficiency in three mouse strains (C57, DBA, C3H) that had become EFA-deficient after eating a fat-free diet for more than 50 days postweaning. The n-6 stores of the mice were unknown; it is likely that the requirement is higher in the young growing mouse.

The rate of depletion of tissue linoleate is biphasic (Tove and Smith, 1959) with the most rapid loss occurring when linoleate comprised more than 20 percent of the depot fat. Also, female and immature male mice lose linoleate more quickly than mature males during the slower, second phase of depletion. The linoleate requirement for pregnant and lactating mice is unknown, although it should increase during lactation as in the rat.

The previous recommendation (National Research Council, 1978) was 0.3 percent dietary linoleate, based on the n-6 requirement of the rat. However, the present recommendation for a standard rat diet is 0.68 percent of dietary ME as linoleate; therefore, the current recommended amount of dietary linoleic acid for mice is 0.68 percent.

n-3 fatty acids

As in the rat and other mammals, n-3 fatty acids are sequestered in certain tissues in the mouse; thus, an essential function is likely. Two studies (Rivers and Davidson, 1974; Wainwright et al., 1991) have attempted to demonstrate a need for n-3 fatty acids. No attempt is made here to indicate a specific requirement. Gross depletion of tissue 22:6(n-3) in conjunction with abnormalities of the retina have been demonstrated in other species only when fed diets that contain oils with extremely high ratios (150:1) of n-6:n-3 fatty acids (as found in safflower, sunflower, and peanut oil) (Neuringer et al., 1988). Diets containing other common oils with more moderate ratios of n-6:n-3 fatty acids will not result in a depletion of n-3 fatty acids. (See Appendix Table 1 for fatty acid composition of common dietary fats and oils.)

Signs of EFA Deficiency   EFA deficiency in the mouse was first described by White et al. (1943) and later by Decker et al. (1950) using weanling mice inbred in their respective laboratories (strain was not reported). Cercedo et al. (1952) produced EFA deficiency in three strains of weanling mice (C57, DBA, C3H), all of which were reported to have dermatitis of the thorax and extremities, scaliness of the ears, alopecia, and growth retardation. Mice fed a fat-free diet also developed lighter colored hair in the lower dorsal region. A "spectacle" eye condition was reported in two-thirds of the DBA mice but only noted occasionally in the C57 mice. The DBA mice developed the EFA-deficiency syndrome more rapidly than did the C57 mice. Berkow and Campagnoni (1983) reported reduced myelination and abnormal myelin composition in C57BL/6J female mice fed EFA-deficient diets during the rapid growth phase.

CARBOHYDRATES

Typical diets fed to mice contain high concentrations of carbohydrate, although diets containing no carbohydrate (83 percent protein) have been shown to support growth rates of 0.1 g/day from 4 to 16 weeks of age in normal mice (Leiter et al., 1983). Mice fed high-fat diets (49 percent fat, 20 percent protein, 15 percent carbohydrate) grew at rates similar to those fed high-carbohydrate diets (4 percent fat, 20 percent protein, 65 percent carbohydrate; 0.13 and 0.09 g/day, respectively; Robeson et al., 1981). Similar growth rates were observed in normal mice fed high-carbohydrate ($\geq$50 percent) diets in which glucose, fructose, sucrose, or starch was the primary carbohydrate source (Leiter et al., 1983; Seaborn and Stoecker, 1989). Diets with high concentrations of fructose or sucrose increased liver fatty acid synthesis and decreased extrahepatic fatty acid synthesis as opposed to diets high in glucose or starch (Herzberg and Rogerson, 1982).

PROTEIN AND AMINO ACIDS

PROTEIN

The requirement for protein to support maximal growth or reproduction depends on the content and digestibility of the amino acids in the diet and the growth and reproductive potential of the mice in question (Keith and Bell, 1988). Growth rates of mouse strains used in research range from 0.6 to 1.2 g/day and litter size may vary from three to seven. The estimated protein requirements are based on strains with high growth and reproductive potential under the assumption that these requirements should meet the needs of strains with lower growth and reproductive potential.

Growth

Of the studies reviewed, none used natural-ingredient diets to estimate the protein requirements of growing mice. For male mice, purified diets will support growth rates up to 1.2 g/day for the 2-week period following weaning. The AIN-76 purified diet (American Institute of Nutrition, 1977), which contains 20 percent casein supplemented with 0.3 percent DL-methionine, will support gains of 0.98 to 1.2 g/day in Carworth Farms No. 1 X Swiss male mice (Bell and John, 1981; Keith and Bell, 1989) (Table 3-4). Others (Maddy and Elvehjem, 1949; Hirakawa et al., 1984; Toyomizu et al., 1988) using either Swiss or Y strain mice have reported growth rates of 0.8 g/day, 1 g/day, and 1 to 1.1 g/day when 19, 22.7, and 27 percent casein diets, respectively, were used. Improving the dietary amino acid pattern by supplementing casein with an equivalent amount of nitrogen as a mixture of indispensable amino acids lowers the protein required for CD-1 mice that gain 1.3 g/day to 15.5 or 16.5 percent of the diet (Bell and Keith, 1992). Females grow more slowly, and lower dietary protein concentrations are needed to maximize growth. Goettsch (1960) found that a diet containing 20 percent casein would support a growth rate of 0.73 g/day in male Swiss mice but that 14 percent would support the growth rate of 0.59 g/day in female Swiss mice. Thus, diets containing 18 percent crude protein, equivalent to 20 percent casein supplemented with methionine or casein alone at 23 to 27 percent of the diet, support growth rates of more than 1 g/day in male mice.

Reproduction

Both casein-based purified and natural-ingredient diets have been used in studies of the protein requirement for reproduction and lactation. In a study limited to the first gestation/lactation, a diet containing 16.7 percent casein resulted in the lowest age at first estrus (30.5 days versus 36.7 days) and largest litter size (7.5 versus 7.1), while a diet containing 20 percent casein was needed to support lactation, as demonstrated by the weight of the dam and litter at 21 days in Swiss-STM mice (Goettsch, 1960) (Table 3-5). Natural-ingredient diets varying in composition but containing 18 to 24 percent crude protein have been used to evaluate the protein requirement for mice over 6- to 9-month periods (Bruce and Parkes, 1949; Hoag and Dickie, 1962) or over four to seven litters for five strains [BALB/cAnN, C3H/HeN, C57BL/6N, N:NIH(S), and DBA/2N] (Knapka et al., 1974, 1977). Although strain differences were observed, a natural-ingredient diet containing 18 percent crude protein supported litter sizes of six to seven and a weaning percentage of 80 to 85 percent over four to five litters (Knapka et al. 1974, 1977). Thus natural-ingredient diets containing 18 percent crude protein from a mixture of animal and plant proteins will meet the protein needs of gestating/lactating mice through several pregnancies.

TABLE 3-4   Protein Requirements for Growth for Various Strains of Mice

Strain	Sex	Growth, g/day	Source[a]	Comment	Reference
Carworth Farms No. 1 X Swiss	Male	0.98	Casein (20%) + DL-Met (0.3%)		Bell and John, 1981
ICR	Male	1.0	Casein (22.3%)		Hirakawa et al., 1984
Swiss-STM	Male, female	0.73, 0.57	Casein	14% (female), 20% (male)	Goettsch, 1960
Swiss-Webster	Male	0.82	Casein (19%)		Maddy and Elvehjem, 1949
Swiss-Webster	Male	0.93	20% casein + 0.3% L-Cys		Reeves et al., 1993
	Female	0.71	20% casein + 0.3% L-Cys		Reeves et al., 1993
Y	Male	1.0–1.1	Casein (0, 5, 10, 15, 20, 30, 40, 50, 60, 70, 80, 90, and 100%)	Multiple	Toyomizu et al., 1988
Carworth Farms No. 1 X Swiss	Male	1.2 ± 0.05 (7 reps)	Casein (20%) + DL-Met (0.3%)	20%	Keith and Bell, 1989
CD-1	Male	1.15–1.3	Casein (56% of crude protein)	15.5% crude protein + IAA and DAA	Bell and Keith, 1992
	Male	1.3	Casein (20%) + DL-Met (0.3%)		

[a]Values in parentheses are amount of ingredient expressed as percent of the diet.

TABLE 3-5 Protein Requirements for Reproduction for Various Strains of Mice

Strain	Diet	Pups in Litter, no.	Comment	Reference
Swiss-STM	Casein	7	16.7%	Goettsch, 1960
Gestation			20%	
Lactation				
C57BL/6J; DBA/2N; AKR/J	Natural-ingredient	7.5; 80% weaned	20% crude protein	Hoag and Dickie, 1962
BALB/cAnN; C3H/HeN; C57BL/6N; DBA/2N	Natural-ingredient	6; 85% weaned	18% and 24% crude protein	Knapka et al., 1977
C57BL/6N; BALB/cAnN; N:NIH	Natural-ingredient	6–8; weaned 6–8 over 5–6 litters; 70–85% weaned	24% crude protein	Knapka et al., 1974

AMINO ACIDS

The estimated requirement for a single amino acid depends on the amounts of other amino acids in the diet and the rate of growth. With the exception of D-lysine (Friedman and Gumbmann, 1981) and probably D-threonine, the L-indispensable amino acid requirement may be met, in part, by D-amino acids. The efficiency of use of individual D-amino acids depends on the activity (Konno and Yasumura, 1984) and specificity of D-amino acid oxidase (Konno et al., 1982) as well as the amount and distribution of other D-amino acids in the diet because of competition for the enzyme (Marrett and Sunde, 1965). Growth rates similar to those obtained with intact protein (0.7 to 1.0 g/day) have been obtained with L-amino acid diets in 14-day growth studies (Maddy and Elvehjem, 1949; Hirakawa et al., 1984; Reicks and Hathcock, 1989). The concentration of amino acids in these diets exceeds the estimated requirement (National Research Council, 1978) by 25 to 200 percent.

Growth

Differences in growth potential among strains have been observed (see Table 3-6). Mice of the C57BL/6 strain gained 0.44 g/day, while CD-1 mice gained 0.68 g/day (Olejer et al., 1982); and Swiss-Webster mice gained 0.4 g/day, while the Rockland strain gained 0.78 g/day (Maddy and Elvehjem, 1949). Few studies have focused on estimating amino acid requirements of mice (John and Bell, 1976; Bell and John, 1981). The requirement for L-arginine is suggested to be less than 0.1 percent for mice gaining 0.8 g/day (Bell and John, 1981) and less than 0.3 percent for mice gaining 0.9 g/day (John and Bell, 1976). Milner et al. (1975) report evidence of arginine deficiency in BFD-SCH mice that gained 0.4 g/day when fed a diet devoid of L-arginine. Bauer and Berg (1943) found that arginine could be deleted from the DL-amino acid diet of mice

gaining 0.11 g/day. A level of 0.3 percent arginine should meet the requirements of mice with a growth potential of 1 g/day. The requirement for L-histidine is 0.2 percent of the diet for mice gaining more than 1 g/day (John and Bell, 1976). Olejer et al. (1982) showed that 0.1 percent would meet the needs of C57BL/6 mice growing 0.4 g/day but that 0.2 percent was required for CD-1 mice, which grew 0.7 g/day. L-carnosine at 0.29 percent can replace 0.2 percent L-histidine (Olejer et al., 1982). Parker et al. (1985) showed that deletion of L-histidine from the amino acid mixture resulted in weight loss and that the single-test concentration of 0.33 percent L-histidine met the needs of Swiss-Webster mice gaining 0.3 g/day. The L-histidine requirement for growing mice seems to be met at 0.2 percent of the diet. The L-isoleucine requirement is 0.4 percent of diet for mice gaining 1.0 g/day (John and Bell, 1976). Diets containing 0.7 percent and 0.5 percent L-leucine and L-valine, respectively (John and Bell, 1976), meet the needs of mice growing 1 g/day; therefore, these concentrations are set as the requirements. The requirement for L-threonine is set at 0.4 percent of diet since it will support a gain of 1.1 g/day in Carworth Farms No. 1 X Swiss mice (John and Bell, 1976). The L-lysine requirement of 0.4 percent of diet for mice gaining 0.9 g/day (John and Bell, 1976) was confirmed (Bell and John, 1981). The L-methionine requirement is 0.5 percent of diet for mice gaining 1 g/day (John and Bell, 1976). L-cysteine may replace as much as one-half to two-thirds of methionine in diets of mice gaining 1 g/day, and D-cysteine does not spare L-methionine (Friedman and Gumbmann, 1984a). D-methionine may have a value as high as 60 percent that of L-methionine in mice gaining 1 g/day (Friedman and Gumbmann, 1984a). The L-phenylalanine requirement of 0.4 percent of diet is supported by the work of John and Bell (1976) and Bell and John (1981). In estimates of the requirement for L-phenylalanine, dietary L-tyrosine must be taken into account as it may replace as much as 50 percent of L-phenylalanine (Friedman and Gumbmann,

TABLE 3-6   Amino Acid Requirements for Growth for Various Strains for Mice

Strain	Sex	Growth, g/day	Amino Acid	Amount, % of diet	Reference
Carworth Farms No. 1 X Swiss	Male	0.92	Arg at 40, 60, 80, 100, 120% of rat requirement	≤0.3	John and Bell, 1976
		1.08	His added to 5% casein	0.2	
		1.00	Ile added to 5% casein	0.4	
		1.10	Leu added to 5% casein	0.7	
		0.93	Lys added to 5% casein	≤0.4	
		0.97	Met added to 5% casein	0.5	
		1.15	Phe added to 5% casein	≤0.4	
		1.13	Thr added to 5% casein	0.4	
		0.92	Trp added to 5% casein	≤0.1	
		0.93	Val added to 5% casein	0.5	
Carworth Farms No. 1 X Swiss	Male	0.75	Arg 0.1, 0.2, 0.3, 0.4, and 0.7%	<01	Bell and John, 1981[a]
		0.87	Lys 0.4, 0.5, and 1.0%	0.4	
		0.90	Trp 0.10, 0.13, and 0.17%	0.1	
		0.81	Phe 0.3, 0.55, and 0.89%	0.4	
ICR	Male	1.0	L-amino acid mixture, 16% of diet	—	Hirakawa et al., 1984
Swiss-Webster	Male, female	0.3–0.4	Mixture of 16 L- and DL-amino acids	—	Maddy and Elvehjem, 1949
Rockland	Male	0.7–0.78	Diet 39 VII	—	Maddy and Elvehjem, 1949
Swiss-Webster	Male	0.84	L-amino acids, 14.1%	—	Reicks and Hathcock, 1989
C57BL/6	Male	0.44	L-amino acids, 12.0%	—	Olejer et al., 1982
CD-1	Male	0.68	L-amino acids, 12.0%	—	Olejer et al., 1982
C3HeJ	Male	0.7	Casein (6%) + gelatin (6%) ± 0.065% D- or L-Trp ± 0.05% niacin	—	MacEwan and Carpenter, 1980
Swiss-Webster	Male	1.0–1.1	L-amino acid diet (other amino acids up to 2 × rat requirement)	—	Friedman and Gumbmann, 1984a,b
	Male	1.1	0.59% L-Met 0.29% L-Met + 0.29% L-Cys	—	

[a]Based on purified proteins.

1984b). Growth rates of 1.2 g/day in Swiss-Webster mice required 0.76 percent of L-phenylalanine or 0.38 percent L-phenylalanine + 0.38 percent L-tyrosine (Friedman and Gumbmann, 1984b). D-phenylalanine has a growth promoting value that is one-third that of L-phenylalanine. Based on the above discussion, the requirement of L-phenylalanine + L-tyrosine is set at 0.76 percent of the diet (where L-tyrosine may replace 50 percent of L-phenylalanine). The requirement for L-tryptophan of 0.1 percent of diet for mice gaining 0.9 g/day (John and Bell, 1976) was confirmed (Bell and John, 1981). MacEwan and Carpenter (1980) showed that 0.05 percent niacin reduced the L-tryptophan requirement from 0.125 percent to 0.1 percent of the diet in C3HeJ mice gaining 0.7 g/day. The requirement for L-tryptophan is retained at 0.1 percent of the diet.

Reproduction

No studies were found regarding estimated amino acid requirements for gestation and lactation. Concentrations similar to those listed for growth can be expected to meet the requirements for gestation. Higher concentrations may be required for lactation.

MINERALS

MACROMINERALS

Calcium and Phosphorus

Reports regarding the quantitative calcium and phosphorus requirements of mice have not been published; therefore, the estimated requirements for these minerals are based on the dietary concentrations that have resulted in acceptable performance in mice. Purified diets that contain 4.0 g Ca/kg diet and 3 to 12 g P/kg diet (Morris and Lippincott, 1941; Mirone and Cerecedo, 1947), 5.0 g Ca/kg diet (Wolinsky and Guggenheim, 1974), and 8.0 g Ca/kg diet and 4.0 mg P/kg diet (Bell and Hurley, 1973) have been shown to support growth and reproduction in mice. Natural-ingredient diets containing 12 g Ca/kg and 8.6 g P/kg (Knapka et al., 1974, 1977) also have been reported to support growth and reproduction in BALB/cAnN, C57BL/6N, N:NIH(S), C3H/HeN, and DBA/2N mice.

Limiting dietary phosphorus to 3.0 g P/kg appears to promote bone calcification in mice. Bell et al. (1980) found that female B6D2F$_1$ mice had higher concentrations of calcium and phosphorus in their bones when fed 6.0 g Ca/kg diet with 3.0 g P/kg diet than when fed 6.0 to 24.0 g Ca/kg diet with 12.0 g P/kg. Further work demonstrated that female B6D2F$_1$ mice grew better when fed 15 percent casein with 6.0 g Ca/kg and 3.0 g P/kg diet (basal diet) than when fed 15 percent or 30 percent protein with 6.0 g Ca/kg and 12.0 g P/kg diet (Yuen and Draper, 1983). Moreover, bone calcium concentrations were higher when mice were fed the basal diet than when they were fed the diets containing elevated concentrations of phosphorus.

Little has been written about the problem of nephrocalcinosis in mice. However, Yuen and Draper (1983) observed that calcium concentrations in the kidneys of B6D2F$_1$ mice more than doubled when dietary phosphorus was increased from 3.0 g P/kg to 12.0 g P/kg and protein was held constant at 15 percent of the diet. This suggests that excess dietary phosphorus has the same negative effects in mice as in rats; and as regards calcium, there is no evidence to indicate that the requirements for calcium are greater for mice than rats. Therefore, the 5.0 g Ca/kg diet and 3.0 g P/kg diet estimated as the requirements for the rat are also the estimated requirements for the mouse.

Signs of Calcium and Phosphorus Deficiency Wolinsky and Guggenheim (1974) and Ornoy et al. (1974) reported that Swiss mice consuming a diet containing only 0.2 g Ca/kg experienced decreased weight gain, bone ash, and serum calcium. These effects were much less marked in mice than in rats, however. Mice increased the concentration of calcium-binding protein in the duodenal mucosa and reduced skeletal growth. Decreased growth rather than osteoporosis was the more prominent sign of deficiency.

Chloride

(See "Sodium and Chloride" section.)

Magnesium

Magnesium has been shown to be a dietary essential for mice, but the optimal intake for this species has not been well established. Alcock and Shils (1974) reported that mice fed diets containing 20 mg Mg/kg diet developed signs of deficiency, but these signs did not develop when the diet contained 400 mg Mg/kg diet. Fahim et al. (1990) reported that mice fed a diet containing 111 mg Mg/kg diet grew more slowly and had lower concentrations of magnesium in their serum than mice fed a diet containing 335 mg Mg/kg diet. A purified diet containing 730 mg Mg/kg diet supported normal growth and development of mice (Bell and Hurley, 1973). Dubos et al. (1968) reported sudden death in lactating mice fed a diet containing 500 mg Mg/kg diet but not in mice fed a diet containing 700 mg Mg/kg. This parallels the finding for rats (Hurley et al., 1976)

that the magnesium requirement for lactation is higher than for growth. Natural-ingredient diets containing 1,800 and 2,600 mg Mg/kg diet provided for good growth and reproduction in three mouse strains (Knapka et al., 1974). The magnesium concentration in the widely used AIN-76 (American Institute of Nutrition, 1977) purified diet is 500 mg/kg. Since the data regarding the quantitative magnesium requirements for mice are inconsistent and not definitive, 500 mg Mg/kg diet is the estimated requirement for this species, and the requirement for lactation may be as high as 700 mg/kg, at least for some strains.

Signs of Magnesium Deficiency Alcock and Shils (1974) reported magnesium-deficient mice developed rapid and usually fatal convulsions without previous hyperirritability. Soft tissue calcification resulting from magnesium deficiency has been reported in the hereditarily diabetic KK mouse strain (Hamuro et al., 1970).

Phosphorus

(See "Calcium and Phosphorus" section.)

Potassium

Bell and Erfle (1958) found that mice (Carworth Farms No. 1) fed purified diets with potassium concentrations of 2.0 g/kg diet did not exhibit signs of potassium deficiency, such as poor growth, inanition, lusterless eyes and hair, and dry scaly tails. A natural-ingredient diet containing 8.2 g K/kg diet (Knapka et al., 1974) and a purified diet containing 8.9 g K/kg (Bell and Hurley, 1973) supported good growth and reproduction in mice. Based on these results, the estimated required potassium concentration for mice is 2.0 g K/kg diet.

Sodium and Chloride

Sodium and chloride requirements of mice have not been studied. Two natural-ingredient diets containing 3.6 and 4.9 g Na/kg diet (Knapka et al., 1974) and a purified diet containing 3.9 g Na/kg diet (Bell and Hurley, 1973) are known to support good growth and reproduction. However, the estimated requirement for sodium and chloride is 0.5 g Na/kg diet and 0.5 g Cl/kg diet, which is identical to that estimated for rats. The actual requirements may be lower for mice. Rowland and Fregley (1988) observed that adrenalectomized mice are not as dependent as adrenalectomized rats on supplemental sodium. Mice are also less likely than rats to ingest saline solutions.

TRACE MINERALS

Based on work with the rat, it seems reasonable to suggest that the mouse might have similar requirements for the trace elements. Until further work is completed with the mouse, the requirements established for the rat will suffice as estimates for the mouse. For a more in-depth discussion of the trace element requirements, see Chapter 2.

Copper

Specific studies to determine copper requirement of young growing mice have not been published. Knapka et al. (1974) reported satisfactory growth and reproduction by feeding mice a natural-ingredient diet containing 16 mg Cu/kg diet. Hurley and Bell (1974) reported adequate growth and development in young mice when they were fed a purified diet containing 4.5 mg Cu/kg diet. Mulhern and Koller (1988) showed that C57BL/6J weanling male and female mice fed an egg white-glucose-based purified diet containing 2 mg Cu/kg diet maintained values for serum ceruloplasmin activity and immune response for 8 weeks that were not significantly different from mice fed diets containing 6 mg Cu/kg diet; however, 1 mg/kg was not sufficient.

Reeves et al. (1994) used nonlinear modeling techniques to estimate the copper requirement of adult male Swiss-Webster mice fed the AIN-93M purified diet. By feeding mice a range of dietary copper, from 0.8 to 6.5 mg/kg for 12 weeks, they estimated that minimal dietary concentrations of 2.5 and 4 mg Cu/kg diet were required to maintain maximal concentrations of serum copper and serum ceruloplasmin activity, respectively. However, under other environmental and dietary conditions the copper requirement for adult male mice might be more than 4 mg/kg diet.

Based on these limited results, the estimated minimal requirement for both immature and adult mice is set at 6 mg Cu/kg diet. No information is available about the requirements for pregnancy and lactation. However, because of the similarity between the estimated requirement for copper in young rats (Johnson et al., 1993; Klevay and Saari, 1993) and adult mice (Reeves et al., 1994), the estimated requirement for pregnancy and lactation in mice is similar to that for rats; 8 mg Cu/kg diet. High dietary concentrations of zinc, cadmium, and ascorbic acid may increase the dietary requirement for copper (Davis and Mertz, 1987).

Signs of Copper Deficiency   Copper-deficient mice have low plasma copper concentrations, low plasma ceruloplasmin activity, anemia, enlarged hearts, altered catecholamine metabolism, thymus and spleen atrophy, and low hepatic cytochrome P-450 concentrations (Prohaska and Lukasewycz, 1989a,b; Gross and Prohaska, 1990; Prohaska, 1990; Phillips et al., 1991; Arce and Keen, 1992).

Signs of Copper Toxicity   Mice are relatively resistant to copper toxicosis. Pregnant mice fed diets containing 2,000 mg Cu/kg diet throughout gestation did not carry litters to term; when the high-copper diet was restricted to days 7 to 12 of gestation, the resorption frequency was higher than 50 percent and surviving fetuses were normal. The diet's toxicity to embryos was apparently caused by an indirect effect of reduced food intake rather than by a direct effect of excess copper on the fetus (Keen et al., 1982).

Iron

Sorbie and Valberg (1974) reported that iron concentrations of 25 to 100 mg Fe/kg diet supported normal growth and hematopoiesis in male C57BL/6J mice, although, liver iron storage in these animals was low compared to mice fed natural-ingredient diets containing between 220 to 240 mg Fe/kg diet. When the dietary iron concentration was increased to 120 mg Fe/kg diet, liver iron stores were similar to those obtained with the natural-ingredient diet. The 120 mg Fe/kg diet supported good reproduction for three generations. The requirement for iron is set at 35 mg/kg diet based on the concentration in the widely used purified diet, AIN-76 (American Institute of Nutrition, 1977). Higher concentrations may be necessary for reproduction. Two natural-ingredient diets known to provide good health and reproduction in three mouse strains contained between 198 and 255 mg Fe/kg diet (Knapka et al., 1974).

Signs of Iron Deficiency   Compared to controls fed a diet containing 122 mg Fe/kg diet, male CD-1 mice fed a low-iron diet (2 mg Fe/kg) for 30 days were characterized by low body weights, anemia, and suppressed T-lymphocyte-dependent functions associated with antibody production and blastogenesis (Blakley and Hamilton, 1988). Kuvibidila et al. (1990) reported similar T-cell abnormalities in female C57BL/6 mice fed low-iron diets (10 mg Fe/kg) for 40 days; in addition, these investigators noted a reduction in mature B-cell populations.

Signs of Iron Toxicity   NMR1 mice fed high-iron diets (ranging from 0.5 to 3.5 percent Fe-fumarate) for 4 weeks were characterized by iron concentration-dependent increases in liver and colon iron concentrations and tissue lipid peroxidation (Younes et al., 1990).

Manganese

Manganese concentrations of 3 mg/kg diet or less are clearly inadequate for optimal growth and development of several mouse strains, while diets containing 45 to 50 mg Mn/kg diet are adequate for all criteria tested (Hurley and

Bell, 1974; Hurley and Keen, 1987). Consumption of diets containing 5 mg Mn/kg throughout gestation and lactation resulted in maternal and weanling tissue manganese concentrations and liver manganese superoxide dismutase and arginase activities that were similar to those observed for mice fed diets containing 45 mg Mn/kg diet (C. K. Keen and S. Zidenberg-Cherr, Dept. of Nutrition, University of California, Davis, 1990, unpublished data). Given the lack of data supporting a dietary requirement of manganese in excess of 5 mg/kg diet, the estimated requirement for manganese has been reduced to 10 mg/kg diet to account for possible differences among various strains. This is lower than the 1978 NRC recommendation of 45 mg/kg diet and reflects the absence of data supporting the need for such a high concentration of manganese coupled with the possible negative effects of excess manganese on iron metabolism (Hurley and Keen, 1987).

Signs of Manganese Deficiency   A deficiency of manganese during prenatal development can result in congenital irreversible ataxia, which is characterized by lack of equilibrium and retraction of the head. The ataxia is caused by abnormal development of the otoliths (Erway et al., 1970; Hurley and Keen, 1987). Prenatal manganese deficiency can result in an increased frequency of early postnatal death, although birth weight and early postnatal body weight gain are not typically affected (Hurley and Bell, 1974). Offspring fed manganese-deficient diets into later life can show obesity and fatty livers and abnormalities in cellular ultrastructure including altered integrity of cell and mitochondrial membranes, which may be linked in part to alterations in the free radical defense system (Bell and Hurley, 1973; Zidenberg-Cherr et al., 1985).

Zinc

Using weight gain, tissue zinc concentration, and response to immunization as criteria, weanling and adult mice housed individually in wire-bottom stainless steel cages have a dietary zinc requirement on the order of 10 mg/kg diet when egg white or casein is used as the primary protein source (Luecke and Fraker, 1979; Morgan et al., 1988a,b). The requirement is higher when soybean protein is used ($\approx$20 mg/kg), presumably because of its high phytic acid content (Beach et al., 1980).

Precise requirements for pregnancy and lactation have not been established. A concentration of 5 mg Zn/kg diet has been reported to be inadequate (Beach et al., 1982, 1983), while satisfactory reproduction has been demonstrated using diets containing 30 mg Zn/kg diet or more (Bell and Hurley, 1973; Knapka et al., 1974; Beach et al., 1982, 1983; Keller and Fraker, 1986). Based on these results, the estimated dietary zinc requirement for growing and adult mice is 10 mg/kg diet and for pregnant and lactating dams is 30 mg/kg diet. The dietary requirement for zinc can be influenced by housing conditions; for example, mice maintained in galvanized cages, in cages with a solid bottom, or in groups have a lower requirement for "dietary" zinc because zinc is available from cage materials and feces.

Signs of Zinc Deficiency   An inadequate intake of zinc is characterized by marked reductions in plasma zinc, which occur within a few days (Peters et al., 1991), followed by subsequent mild-to-severe anorexia (Beach et al., 1982, 1983). Prolonged consumption of a zinc-deficient diet can result in growth retardation/failure, alopecia, atrophy of lymphoid tissue, significant impairment of multiple components of the immune system, and alterations in lipid and protein metabolism (Nishimura, 1953; Beach et al., 1980; Hambidge et al., 1986; Morgan et al., 1988a,b; Keen and Gershwin, 1990). The introduction of a zinc-deficient diet during pregnancy can result in severe embryonic and fetal pathologies including prenatal death and a high incidence of central nervous system, soft tissue, and skeletal defects and postnatal behavioral abnormalities (Golub et al., 1986; Keen and Hurley, 1989).

Signs of Zinc Toxicity   Mice are relatively resistant to zinc toxicosis; Aughey et al. (1977) reported no significant effects associated with giving 500 mg Zn/L water for up to 14 months.

ULTRA-TRACE MINERALS

Iodine, Molybdenum, and Selenium

Negative effects on the physiological or biochemical status of mammals have been shown if the diet is unsupplemented with iodine, selenium, and molybdenum. Cobalt is essential but only as a part of vitamin B_{12}. No systematic effort has been made, however, to establish the requirements of iodine, selenium, and molybdenum for the mouse.

Iodine deficiency was produced in mice by feeding them diets containing 20 μg I/kg diet for 8 weeks (Many et al., 1986). These mice experienced enlarged thyroid glands when compared to controls fed 200 μg I/kg diet. Marginal iodine deficiency was produced in mice consuming diets with 42 μg I/kg diet (Van Middlesworth, 1986). Mice were able to adapt to a low-iodine intake by maintaining normal concentrations of iodine in the thyroids. When they were challenged with a mycotoxin, however, the iodine concentration decreased. There was no effect of the toxin on thyroid iodine content in mice fed 150 μg I/kg diet.

There are a number of reports on producing selenium deficiency in mice. The amount of dietary selenium that caused a considerable reduction in the activity of liver glutathione peroxidase ranged from 10 to 16 μg/kg diet.

Control mice in these studies were fed selenium concentrations ranging from 330 to 500 μg Se/kg diet (Wendel and Otter, 1987; Otter et al., 1989; Toyoda et al., 1989; Weitzel et al., 1990; Peterson et al., 1992).

Based on these works, it seems reasonable to suggest that the minimal requirement of selenium and iodine for the mouse might be at least as much as for the rat. Data for molybdenum requirements of the mouse are even more sparse than those for selenium or iodine, and it is suggested that those values established for the rat are good estimates for the mouse (see Table 3-3).

VITAMINS

FAT-SOLUBLE VITAMINS

Vitamin A

It has been shown by Wolfe and Salter (1931) that vitamin A is required by the mouse, and the mouse has been used extensively in studies of vitamin A metabolism and of the role of vitamin A in the prevention of cancer. Little work has been done to establish the vitamin A requirement of mice, however, and depleting the mouse of its vitamin A stores is difficult (McCarthy and Cerecedo, 1952). If rapid depletion is desired, it is necessary to use the pups from a pregnant female fed a vitamin A-deficient diet from about day 10 of gestation; this will produce very low vitamin A stores in the pups (Smith, 1990). Young mice weaned from dams fed a standard diet may require up to 1 year to show overt signs of deficiency. Santhanam et al. (1987) have explored methods to slowly produce vitamin A deficiency in mice. Both BALB/c and Swiss mice eventually developed deficiency signs when fed a cereal grain-based diet calculated to contain 1,200 IU vitamin A/kg (roughly equivalent to 1.2 μmol retinol/kg diet). BALB/c mice maintained good health, showed good growth, and stored modest liver reserves of retinyl esters when fed a diet calculated to contain 2,400 IU vitamin A/kg (2.5 μmol/kg). The AIN-76 diet was formulated to contain 4,000 IU vitamin A/kg (4.2 μmol retinyl esters/kg). This diet has been adequate for normal growth and reproduction in mice (American Institute of Nutrition, 1977).

When fed a natural-ingredient diet that contained 13,371 IU/kg (roughly 14 μmol/kg), A/J mice were found to accumulate vitamin A reserves in their livers as they increased in age from 9 to 216 days old (Sundboom and Olson, 1984). However, 644-day-old A/J mice were found to have about one-half the liver vitamin A stores observed in mice 216 days old.

Based on these limited data, the vitamin A (retinol) requirement of the mouse seems to be similar to the requirement of the rat. Therefore, a dietary concentration of 2,400 IU/kg diet (2.5 μmol/kg diet; 0.72 mg/kg diet) is adequate to meet the requirements of the mouse.

Ideally, retinyl esters should be added to animal diets in stabilized gelatin beadlets, which will protect the vitamin A from oxidation. An alternative procedure is to slowly dissolve the retinyl esters in the dietary lipid, which contains an antioxidant, before the lipid is mixed into the diet. If the second procedure is used, the diet should be freshly prepared at least every other week. Dissolving the retinyl esters in a solvent and adding them directly to the other dietary constituents without the protection afforded by the dietary oils or by gelatin beadlets will result in substantial oxidative destruction of the vitamin. The storage and treatment of the diet are also very important. Zimmerman and Wostmann (1963) reported that vitamin A activity was decreased by 20 percent as a result of steam sterilization.

Signs of Vitamin A Deficiency One of the early and significant consequences of vitamin A deficiency is impairment of the functional immune system (Smith et al., 1987). If conditions are sanitary the main overt sign observed early in deficiency is a decreased rate of weight gain. As the deficiency progresses many epithelial tissues become keratinized, including those of the seminal vesicles, testes (Van Pelt and De Rooij, 1990), bladder, kidney, trachea, esophagus, salivary glands, and lungs (McCarthy and Cerecedo, 1952). Xerophthalmia of the eye occurs if the mice are exposed to unsanitary conditions or are subjected to stress.

Signs of Vitamin A Toxicity The studies of vitamin A toxicity in mice focused on the teratological aspects of the toxicity. Kochhar et al. (1988) have found that a single dose of 349 μmol/kg BW on day 10.5 of gestation produced cleft palates and limb deformities in ICR mice. A lower dose of 175 μmol/kg BW did not produce the deformities. Giroud and Martinet (1962) reported that three doses of 6.5 μmol vitamin A/kg BW on days 8, 9, and 10 of gestation caused death or resorption of 63 percent of the fetuses and malformations in others.

Vitamin D

The AIN-76 diet was formulated to contain 0.025 mg cholecalciferol/kg (0.65 μmol or 1,000 IU/kg) (American Institute of Nutrition, 1977). This amount of vitamin D is adequate and may represent a considerable excess. However, a lesser amount cannot be recommended until the more sensitive criterion of vitamin D status has been evaluated at lower intakes.

Signs of Vitamin D Deficiency The mouse is quite resistant to the development of rickets—a disease caused by vitamin D deficiency. Beard and Pomerene (1929) found that mice fed vitamin D-deficient diets developed signs of rickets within 7 to 14 days. Rickets spontaneously healed between days 20 to 27 without vitamin D supplementation, but osteoporosis was present in many of the animals after healing. Delorme et al. (1983) found that both the 10,000 and 25,000 molecular weight kidney vitamin D-dependent calcium binding proteins were reduced to about one-third the normal concentrations in vitamin D-deficient Swiss mice. In contrast, milk production and the calcium content of the milk were normal in CD-1 mice fed a vitamin D-deficient diet (Allen, 1984).

Signs of Vitamin D Toxicity   The LD_{50} of a single intraperitoneal injection of cholecalciferol for CF_1 mice was found to be 355 μmol/kg BW (Hatch and Laflamme, 1989). No toxicity was found after the injection of 104 μmol/kg (1.7×10^6 IU/kg). In contrast, only 5.5 nmol 1,25-dihydroxycholecalciferol/kg was required to produce toxicity in C57BL/6J mice (Crocker et al., 1985).

Vitamin E

Bryan and Mason (1940) observed fetal resorption in vitamin E-deficient female mice similar to that observed in rats but observed no evidence of testicular injury in vitamin E-deficient males. They reported that administration of 81 nmol all-*rac*-α-tocopherol daily for the first 10 days of gestation was adequate to maintain the first pregnancy. This corresponds to a dietary concentration of 10 IU *RRR*-α-tocopherol/kg diet (15.7 μmol/kg). Goettsch (1942) found that a single dose of vitamin E equivalent to 1.8 IU to 2.4 IU *RRR*-α-tocopherol (1.16 to 1.55 μmol) given at the start of the gestation period was adequate to maintain pregnancy in mice between 3 and 6 months old. Mice 7 to 12 months old required a larger dose equivalent to 5 IU *RRR*-α-tocopherol (7.78 μmol) to maintain pregnancy. Trostler et al. (1979) found that male C57BL/6J mice fed a diet containing the equivalent of 62 μmol *RRR*-α-tocopherol/kg diet (lower dose not used) had a growth rate equal to or greater than that observed with mice fed 124 μmol/kg diet. However, more than 124 μmol/kg diet was required to prevent the accumulation of malondialdehyde in liver and adipose tissue. Yasunaga et al. (1982) gave daily intraperitoneal injections of all-*rac*-α-tocopherol to BALB/c mice and measured their response to mitogens. The best responses were obtained in mice injected with amounts equivalent to 7.8 to 28 μmol *RRR*-α-tocopherol/kg BW. The lower dose is equivalent to 50 μmol *RRR*-α-tocopherol/kg diet. Based on these data the vitamin E requirement for mice is estimated to be 22 mg/kg or 32 IU/kg *RRR*-α-tocopherol/kg diet (50 μmol/kg diet) when lipids comprise less than 10 percent of the diet. When all-*rac*-α-tocopheryl acetate is used as the dietary source, the equivalent amount would be 32 mg/kg diet.

Signs of Vitamin E Deficiency   Pappenheimer (1942) reported muscular dystrophy and hyaline degeneration in vitamin E-deficient mice but at a lower incidence than was observed in rats. No lesions were found in the central nervous system. Davies et al. (1987) did not find an accumulation of lipofuscin in neural tissues. The only tissue to show lipofuscin accumulation in vitamin E deficiency was the liver (Csallany et al., 1977). Spermatogenesis remained active in vitamin E-deficient mice for up to 439 days (Pappenheimer, 1942).

Signs of Vitamin E Toxicity   Yasunaga et al. (1982) found that male C3H/He mice injected intraperitoneally daily with 212 μmol all-*rac*-α-tocopherol/kg BW showed a weight loss by day 7. Injections of 846 μmol/kg BW/day were lethal. α-Tocopheryl quinone, a major metabolite of α-tocopherol, has been found to interfere with the mouse's ability to metabolize vitamin K and resulted in bleeding (Woolley, 1945).

Vitamin K

Vitamin K has not been considered essential for mice reared under conventional conditions because of the substantial contribution from coprophagy. However, with the increased use of specific-pathogen-free animals for research, this is probably no longer true. Both specific-pathogen-free CF_1 mice (Fritz et al., 1968) and germ-free ICR/JCL mice (Komai et al., 1987) were reported to die quickly from hemorrhagic diathesis when fed vitamin K-free diets. Addition of 16 μmol menadione/kg to the diet prevented hemorrhaging problems in the specific-pathogen-free mice. Studies using the more sensitive criterion of vitamin K status have not been conducted with mice as they have been with rats (Kindberg and Suttie, 1989). Therefore, the estimated requirement of vitamin K for mice is 1 mg phylloquinone/kg diet (2.22 μmol/kg diet), based on the requirement of the rat.

WATER-SOLUBLE VITAMINS

Vitamin B_{12}

Intestinal bacteria in mice synthesize undetermined amounts of vitamin B_{12} that can be utilized by the host. The presence of endogenous vitamin B_{12} generally confounds attempts to determine the quantitative requirements of this vitamin for mice. Jaffé (1952), however, reported a vitamin B_{12} requirement in excess of 5 μg/kg diet for growth and between 4 and 5 μg/kg diet for reproduction and lactation. Lee et al. (1962) demonstrated that mice require vitamin B_{12} for gestation. The widespread use of the AIN-76 (American Institute of Nutrition, 1977) purified diet containing 10 μg vitamin B_{12}/kg has not resulted in any reports of vitamin B_{12} deficiency signs. This indicates that the vitamin B_{12} concentration in the AIN-76 diets is adequate for mice. In the absence of more recent and definitive data regarding the vitamin B_{12} requirements for mice, 10 μg vitamin B_{12}/kg diet is the estimated requirement for this species. However, it is noted that lower dietary concentrations may be adequate for mice with conventional intestinal flora, but higher concentrations may be required when the availability of endogenous B_{12} is limited under conditions such as antibiotic feeding, germ-free environments, or coprophagy prevention.

Signs of Vitamin B_{12} Deficiency   Young mice deficient in vitamin B_{12} show retarded growth and renal atrophy (Lee et al., 1962). Deficiency causes death both before and after birth.

Biotin

In contrast to other rodent species, mice fed purified casein-based diets appear to have a requirement for biotin that exceeds the amount obtained from coprophagy. Several investigators have observed signs of biotin deficiency or suboptimal weight gain when mice were fed biotin-deficient diets (Nielsen and Black, 1944; Fenton et al., 1950; Lakhanpal and Briggs, 1966). Fenton et al. (1950) found that 0.823 μmol biotin/kg diet was adequate for the mouse, but they did not use other concentrations. The AIN-76 diet was formulated to contain 0.2 mg biotin/kg diet (0.82 μmol biotin/kg). In the absence of more recent and definitive data regarding the dietary requirements of mice, the concentration of 0.2 mg biotin/kg diet is the estimated safe and adequate dietary concentration.

Signs of Biotin Deficiency   Watanabe and Endo (1989, 1991) observed teratogenic effects of biotin deficiency in mice fed a spray-dried egg white diet (containing avidin). The teratogenic effects were much more severe in the fast-growing ICR and C57BL/6N strains than in the slower growing A/Jax mice. The other signs of deficiency include alopecia, achromotrichia, and growth failure, as well as decreased reproduction and lactation efficiency (Nielsen and Black, 1944).

Choline

Choline was first recognized as a dietary essential for the mouse by Best et al. (1932), who observed fatty livers in choline-deficient mice. Since choline can be synthesized from methionine (see Chapter 2), and its metabolism is influenced by folic acid and vitamin B_{12}, a minimum requirement for choline is difficult to establish. Meader and Williams (1957) found that mice fed a diet containing 80 g casein/kg and 400 g lard/kg required 5 g choline chloride/

kg diet (35,800 μmol/kg) to support growth and prevent lipid accumulation in the liver. However, Williams (1960) found this level of choline to be toxic in long-term studies. Therefore, caution should be exercised in adding high concentrations of choline to the diet. The widely used AIN-76 diet was formulated to contain 2 g choline bitartrate/kg (7,900 μmol/kg). This amount provides an adequate concentration of choline for diets containing optimal concentrations of methionine. Thus 2 g choline bitartrate/kg diet (7,900 μmol/kg) is the estimated safe and adequate dietary concentration.

Signs of Choline Deficiency   Choline-deficient mice had fatty livers with modular parenchymal hyperplasia and lower conception rates with low viability of the young (Mirone, 1954; Buckley and Hartroft, 1955; Meader and Williams, 1957). In contrast to earlier descriptions of fibrosis, Rogers and MacDonald (1965) observed that C57BL mice, unlike rats, did not develop cirrhosis or fibrosis of the liver but only fatty livers. There were acute and chronic inflammations on necrosis of individual hepatic cells. Proliferation of parenchymal cells increased with fat deposition. There was increased thymidine uptake by endothelial, perivascular, and parenchymal cells. Fifty-four percent of the choline-deficient mice died during a 24-week period.

Signs of Choline Toxicity   Choline is a very toxic nutrient with a narrow margin of safety. Williams (1960) observed that a dietary concentration of 5 g choline chloride/kg diet (35,800 μmol/kg) induced weight loss in BALB/c mice after they were fed that diet for 6 months, and there were no survivors after 9 months. After 15 weeks 52 percent of the mice had myocardial lesions and by 33 weeks 100 percent of the mice had myocardial lesions. The lesions were most frequently fibrosis with limited necrosis of muscle fibers and fibroid necrosis of coronary arteries (Thomas et al., 1968).

Folates

Weir et al. (1948) documented the essentiality of folic acid in growing mice. Fenton et al. (1950) obtained satisfactory growth in mice fed defined diets containing 0.5 mg folic acid/kg diet (1.1 μmol/kg). Heid et al. (1992) found that 0.4 to 0.5 mg/kg diet (0.9 to 1.1 μmol/kg) was necessary for successful pregnancy outcome in Swiss-Webster mice. Based on these results, 0.5 mg folic acid/kg diet (1.1 μmol/kg) is the estimated requirement for mice.

Signs of Folate Deficiency   Weir et al. (1948) observed the following effects after feeding mice a folate-deficient diet for 50 days: a decrease in white cell count from 6,000 to 4,000/mm^3, disappearance of megakaryocytes and nucleated cells from the spleen and hemosiderin accumulation, and disappearance of normal cell types from bone marrow. Other signs of deficiency reported include impaired antibody response (Rothenberg et al., 1973), decreased organ growth—particularly brain and liver (Shaw et al., 1973)—and decreased fetal implantations and increased resorption (Heid et al., 1992).

Niacin

Adequate data are not available to estimate the niacin requirement of mice. Male BK albino mice have been shown to convert [^{14}C]tryptophan to N-methyl-nicotinamide, a urinary metabolite of niacin (Bender et al., 1990). The mouse may require increased niacin when tryptophan is fed at suboptimal concentrations. Based on the requirements of the rat, a dietary concentration of 15 mg nicotinic acid/kg diet (120 μmol/kg) is the estimated requirement for mice under the most adverse conditions (Hundley, 1949).

Pantothenic Acid

Sandza and Cerecedo (1941) found that subcutaneous injections of 63 nmol Ca-d-pantothenate 6 days each week would maintain an optimal growth rate in albino mice. Morris and Lippincott (1941) reported that a diet containing 21 μmol Ca-pantothenate/kg produced growth equivalent to a diet containing 168 μmol/kg in C3H mice. Fenton et al. (1950) obtained maximal growth in C57 mice with diets containing 13 μmol/kg diet, but more than 17 μmol/kg diet was required for optimal growth in the A and C3H strains of mice. Based on these limited data, 10 mg Ca-d-pantothenate/kg diet (21 μmol/kg) seems to be adequate for optimal growth in most strains of mice, but some strains may have higher requirements. Data on the requirement for pregnancy and lactation are not available, but the AIN-76 diet was formulated to contain 16 mg Ca-d-pantothenate/kg diet (33.6 μmol/kg). This diet has been shown to be adequate to support pregnancy and lactation in mice.

Signs of Pantothenic Acid Deficiency The following pantothenic acid-deficiency signs in growing mice were reported by Morris and Lippincott (1941): loss of weight; loss of hair, particularly of the ventral surface, flanks and legs; dermatosis; partial posterior paralysis; other neurological abnormalities; and achromotrichia.

Vitamin B_6 (Pyridoxine, Pyridoxal, Pyridoxamine)

According to Miller and Baumann (1945) and Morris (1947), mice grew satisfactorily when fed diets containing 1 mg pyridoxine-HCl/kg diet. Pyridoxamine and pyridoxal were found to be less active than pyridoxine. Bell et al. (1971) found 0.2 mg pyridoxine-HCl/kg diet limited growth in two strains, whereas 8.2 mg/kg supported normal growth. A comparison of reproductive performance of C57BL and I strains found that concentrations of 1 to 6 mg/kg diet resulted in fewer productive matings, smaller litters, and a lower survival rate in the I strain. Increasing dietary pyridoxine concentrations to 8 to 12 mg/kg improved the reproductive performance, but further improvement was not obtained using concentrations of 410 or 1,230 mg/kg diet; however, these concentrations did improve the survival rate of C57BL mice over a concentration of 1 to 6 mg/kg (Hoover-Plow et al., 1988). The concentrations of pyridoxal-5'-phosphate and pyridoxamine-5'-phosphate were determined in female mice fed purified diets containing 0.5, 1.0, 2.0, 3.0, 5.0, and 7.0 mg/pyridoxine-HCl/kg diet for 5 weeks. Plasma, erythrocyte, whole blood, liver, and brain pyridoxal-5'-phosphate and liver and brain pyridoxamine-5'-phosphate concentrations correlated with dietary concentrations (r = 0.81 to 0.94) and did not plateau over the entire dietary ranges of values (Furth-Walker et al., 1990). During pregnancy mice fed open-formula diets containing 8.13 mg pyridoxine-HCl/kg had increased erythrocyte and whole blood (2.9- and 1.6-fold) and decreased plasma (50 percent) pyridoxal-5'-phosphate. Liver pyridoxal-5'-phosphate, and pyridoxamine-5'-phosphate decreased 25 percent, but brain concentration remained unchanged (Furth-Walker et al., 1989). The recommended concentration of vitamin B_6 for reproduction is set at 8 mg/kg diet. The concentration of 1 mg/kg set by Miller and Baumann (1945) and by Morris (1947) seems to be adequate for maintenance and growth.

Signs of Vitamin B_6 Deficiency Vitamin B_6 deficiency signs include poor growth, hyperirritability, posterior paralysis, necrotic degeneration of the tail, and alopecia (Beck et al., 1950). Investigators (Keyhani et al., 1974) observed in B_6-deficient CF$_1$ mice a progressive hypochromic microcytic anemia with hypersideremia. It was accompanied by an increase in reticulocyte count not observed in vitamin B_6 deficiencies of other species.

Riboflavin

Based on data reported by Fenton and Cowgill (1947a,b) and Wynder and Kline (1965), mice require 4 mg riboflavin/kg diet for normal growth. However, diets resulting in normal reproduction in mouse colonies generally contain 6 or 7 mg riboflavin/kg diet (American Institute of Nutrition, 1977). In the absence of more definitive data regarding the requirements for reproduction, the estimated riboflavin requirements for this species is 7 mg/kg diet.

Signs of Riboflavin Deficiency Ariboflavinosis in the

mouse was described by Lippincott and Morris (1942). They reported the development of either atrophic or hyper-keratotic epidermis with normal sebaceous glands, myelin degeneration in the spinal cord, and corneal vascularization with ulceration. Morris and Robertson (1943) found that adult mice lost weight and young mice grew poorly and died within 9 weeks when fed diets containing 0.4 to 0.6 mg riboflavin/kg diet. Kligler et al. (1944) showed that riboflavin-deficient mice had lowered resistance to *Salmonella* infection. Hoppel and Tandler (1975) reported striking increases in the size of hepatic mitochondria and a greatly decreased capacity for ADP-stimulated respiration in riboflavin-deficient mice. In some animals, livers were yellow and the cytoplasm of the cells was engorged with small lipid droplets. In other animals, the livers were redder than normal, and their hepatocytes contained few lipid droplets. Genetically diabetic (KK) mice had a higher riboflavin requirement than Swiss albino mice based on activity coefficients of erythrocyte glutathione reductase (Reddi, 1978). Riboflavin deficiency during gestation led to brain, orofacial, limb and gastrointestinal malformations in the offspring. The degree of severity and malformation pattern varied with the strain of mice studied (Kalter, 1990).

Thiamin

Hauschildt (1942) established the minimum requirement of thiamin for normal growth of mice at 10 μg/day. This would correspond to a concentration of approximately 3 mg/kg diet. Morris and Dubnik (1947) later found the growth requirement to be 4 to 6 μg/day for mice fed a diet containing 22 percent fat.

Results of studies on the specific requirements for reproduction and lactation have not been reported, but Mirone and Cerecedo (1947) found that 20 mg/kg diet were adequate. The purified diet (American Institute of Nutrition, 1977) containing 6 mg thiamin-HCl/kg diet has been used in numerous mouse colonies resulting in normal growth and reproduction. In the absence of more definitive data the concentration of 5 mg thiamin-HCl/kg diet that was the estimated requirement in the previous issue of this report (National Research Council, 1978) is being retained.

Signs of Thiamin Deficiency   Morris (1947) and Jones et al. (1945) reported violent convulsions, especially when the animal was held a few seconds by the tail; cartwheel or circular movements; brain hemorrhages; decreased food intake; poor growth; early mortality; silvery-streaked muscle lesions; and testicular degeneration. The onset of ataxia in thiamin-deficient Swiss-Webster mice was preceded by a rapid rise in brain α-ketoglutarate (Seltzer and McDougal, 1974). Deficiency (4 to 21 days) led to increased activity in hepatic thiamin pyrophosphatase, alkaline phosphatase, and acid phosphatase (Tumanov and Trebukhina, 1983).

Exposure of thiamin-deficient mice to ethanol resulted in brain damage that was more severe than either treatment alone (Phillips, 1987).

POTENTIALLY BENEFICIAL DIETARY CONSTITUENTS

FIBER

A fiber source is routinely included to increase bulk in diets for mice, but at high concentrations it depresses performance. Dilution of diets with cellulose at concentrations of 15, 30, and 50 percent increased feed intake by 3.5, 12.0, and 26.9 percent, respectively, resulting in consumption of noncellulose constituents of 88.4, 78.4, and 63.5 percent of what mice fed the undiluted diet consumed (Dalton, 1963). Bell (1960) diluted mouse diets with wheat bran, alfalfa, beet pulp, oat hulls, wheat straw, corn cobs, or cellulose in amounts designed to dilute digestible energy (DE) to concentrations ranging from 2.2 to 3.4 kcal/g diet (9.2 to 14.2 kJ/g diet). In general, growth rate decreased when DE was less than 2.9 kcal/g diet (12.1 kJ/g diet). Fiber sources produced different effects. Mice fed 33 to 39 percent wheat straw died. Mice performed poorly when their diets consisted of 38 to 45 percent beet pulp. However, mice fed up to 68 percent wheat bran or up to 43 percent oat hulls grew at rates similar to those fed lower fiber concentrations. When corn starch was replaced by barley bran, oat bran, rice bran, or soybean fiber in amounts supplying 7 percent fiber (TDF) to diets containing 30 percent ground beef, growth of mice was not affected (Hundemer et al., 1991). Addition of 10 percent guar gum, bagasse, or wheat bran to a natural-ingredient diet did not affect feed intake or growth of mice, although guar increased and bagasse decreased liver lipogenic enzymes (Stanley and Newsholme, 1985a,b; Stanley et al., 1986).

VITAMINS

Ascorbic acid

The successful maintenance of mouse colonies fed diets devoid of ascorbic acid has confirmed the demonstration by Ball and Barnes (1941) that the mouse requires no dietary source of vitamin C. The plasma concentration of dehydroascorbate in mice fed graded concentrations of ascorbic acid (from 0 to 80 g/kg diet) increased with greater dietary concentrations; ≥10 g/kg resulted in significantly greater dehydroascorbate concentration in heart, kidney, lung, and spleen. Concentrations in eyes, were only slightly increased; and brain, adrenal gland, and leukocyte concentrations were unchanged in mice consuming diets containing 80 g/kg (Tsao et al., 1987; Tsao and Leung, 1989).

During the first 8 days of pregnancy an inverse relationship was found between ascorbic acid intake and the concentration of peroxidase in the corpus luteum, blastocyst, and endometrium (Agrawal and Laloraya, 1979). A diet of 10 g ascorbic acid/kg increased average life span by 8.6 percent, decreased body weight by 6 to 7 percent, and increased the maximal life span 2.9 percent (from 965 to 993 days) in C57BL/6J male mice (Friedman et al., 1987).

Myo-*inositol*

Although Woolley (1941) reported that *myo*-inositol would alleviate a condition characterized by hair loss, other studies (Martin, 1941; Cerecedo and Vinson, 1944; Fenton et al., 1950; Shepherd and Taylor, 1974a) did not confirm the essentiality of *myo*-inositol for growth of mice. Studies with other rodent species indicate that they require *myo*-inositol under conditions of microbial suppression and physiological stress. Shepherd and Taylor (1974b) found that *myo*-inositol enhanced intestinal lipid transport in rats fed a 31 percent fat diet. Burton and Wells (1977) observed that rats fed 0.5 percent dietary phthalysulfathiazole required *myo*-inositol to prevent fatty liver during lactation; 500 mg *myo*-inositol/kg diet was sufficient. Anderson and Holub (1976) found that either tallow or the highly unsaturated canola oil caused liver fat accumulation in *myo*-inositol-deficient rats fed succinyl sulfathiazole, whereas corn oil or soybean oil did not; 0.5 percent *myo*-inositol was protective. Unlike rats (Bondy et al., 1990), the peripheral nerves of mice fed diets containing galactose (20 percent) were not depleted of *myo*-inositol (Calcutt et al., 1990).

If gnotobiotic, germ-free, or antibiotic-treated mice are fed diets containing tallow or the highly saturated rapeseed oil, then *myo*-inositol may be required in the diets. A purified diet fed to germ-free rats and mice contained 1,000 mg *myo*-inositol/kg diet (Wostmann and Kellogg, 1967). Chemically defined diets that supported growth and limited reproduction in germ-free CF_W or C3H mice contained 238 mg *myo*-inositol/kg diet (Pleasants et al., 1970, 1973). A concentration of 500 mg/kg diet was adequate for any combination of antibiotics, lactation, and unusual fat intake in rats and gerbils and seemed to be the upper limit of the *myo*-inositol requirement. However, conventionally reared mice fed ordinary diets have not been found to require dietary *myo*-inositol since the early studies of Woolley (1941, 1942).

REFERENCES

Agrawal, P., and M. M. Laloraya. 1979. Ascorbate and peroxidase changes during pregnancy in albino rat and Swiss mouse. Am. J. Physiol. 5:E386–E390.

Alcock, N. W., and M. E. Shils. 1974. Comparison of magnesium deficiency in the rat and mouse. Proc. Soc. Exp. Biol. Med. 146:137–141.

Allen, A. M., C. T. Hansen, T. D. Moore, J. Knapka, R. D. Ediger, and P. H. Long. 1991. Hemorrhagic cardiomyopathy and hemothorax in vitamin K deficient mice. Toxicol. Pathol. 19(4-Part ll):589–596.

Allen, J. C. 1984. Effect of vitamin D deficiency on mouse mammary gland and milk. J. Nutr. 114:42–49.

American Institute of Nutrition. 1977. Report of the American Institute of Nutrition Ad Hoc Writing Committee on standards for nutritional studies. J. Nutr. 107:1340–1348.

Anderson, D. B., and B. J. Holub. 1976. The influence of dietary inositol on glyceride composition and synthesis in livers of rats fed different fats. J. Nutr. 106:529–536.

Arce, D. S., and C. L. Keen. 1992. Reversible and persistent consequences of copper deficiency in developing mice. Reprod. Toxicol. 6:211–221.

Aughey, E., L. Grant, B. L. Furman, and W. F. Dryden. 1977. The effects of oral zinc supplementation in the mouse. J. Comp. Pathol. 87:1–14.

Ball, Z. B., and R. H. Barnes. 1941. Effect of various dietary supplements on growth and lactation in the albino mouse. Proc. Soc. Exp. Biol. Med. 48:692–696.

Bauer, C. D., and C. P. Berg. 1943. The amino acids required for growth in mice and the availability of their optical isomers. J. Nutr. 26:51–63.

Beach, R. S., M. E. Gershwin, and L. S. Hurley. 1980. Growth and development in postnatally zinc-deprived mice. J. Nutr. 110:201–211.

Beach, R. S., M. E. Gershwin, and L. S. Hurley. 1982. Reversibility of developmental retardation following murine fetal zinc deprivation. J. Nutr. 112:1169–1982.

Beach, R. S., M. E. Gershwin, and L. S. Hurley. 1983. Persistent immunological consequences of gestation zinc deprivation. Am. J. Clin. Nutr. 38:579–590.

Beard, H. H., and E. Pomerene. 1929. Studies in the nutrition of the white mouse. V. The experimental production of rickets in mice. Am. J. Physiol. 89:54–57.

Beck, E. M., P. F. Fenton, and G. R. Cowgill. 1950. The nutrition of the mouse. IX. Studies on pyridoxine and thiouracil. Yale J. Biol. Med. 23:190–194.

Bell, J. M. 1960. A comparison of fibrous feedstuffs in non-ruminant rations: Effects on growth responses, digestibility, rates of passage and ingesta volume. Can. J. Anim. Sci. 40:71–82.

Bell, J. M., and J. D. Erfle. 1958. The requirement for potassium in the diet of the growing mouse. Can. J. Anim. Sci. 38:145–147.

Bell, J. M., and A.-M. John. 1981. Amino-acid requirements of growing mice: Arginine, lysine, tryptophan and phenylalanine. J. Nutr. 111:525–530.

Bell, J. M., and M. O. Keith. 1992. Effects of levels of dietary protein and digestible energy on growth, feed utilization and body composition of growing mice. Nutr. Res. 12:375–383.

Bell, L., and L. S. Hurley. 1973. Ultrastructural effects of manganese deficiency in liver, heart, kidney and pancreas of mice. Lab. Invest. 29:723–736.

Bell, R. R., C. A. Blanshard, and B. E. Haskell. 1971. Metabolism of vitamin B_6 in the I-strain mouse. II. Oxidation of pyridoxal. Arch. Biochem. Biophys. 147:602–611.

Bell, R. R., D. Y. Tzeng, and H. H. Draper. 1980. Long-term effects of calcium, phosphorus, and forced exercise on the bones of mature mice. J. Nutr. 110:1161–1168.

Bender, D. A., E. N. M. Njagi, and P. S. Danielian. 1990. Tryptophan metabolism in vitamin B_6-deficient mice. Br. J. Nutr. 63:27–36.

Berkow, S. E., and A. T. Campagnoni. 1983. Essential fatty acid deficiency: Effects of cross-fostering mice at birth on myelin levels and composition. J. Nutr. 113:582–592.

Bernier, J. F., C. C. Calvert, T. R. Famula, and R. L. Baldwin. 1986. Maintenance energy requirement and energetic efficiency of mice with a major gene for rapid postweaning gain. J. Nutr. 116:419–428.

Bernier, J. F., C. C. Calvert, T. R. Famula, and R. L. Baldwin. 1987. Energetic efficiency of protein and fat deposition in mice with a major gene for rapid postweaning gain. J. Nutr. 117:539–548.

Best, C. H., M. E. Huntsman, and O. M. Solandt. 1932. A preliminary report on the effect of choline on fat deposition in species other than the white rat. Trans. R. Soc. Can. 26:175–176.

Birt, D. F., J. C. Pelling, G. Tibbels, and L. Schweickert. 1989. Acceleration of papilloma growth in mice fed high fat diets during promotion of two-stage skin carcinogenesis. Nutr. Cancer 12:161–168.

Blakley, B. R., and D. L. Hamilton. 1988. The effect of iron deficiency on the immune response in mice. Drug-Nutr. Interact. 5:249–256.

Bondy, C., B. D. Cowley, Jr., S. L. Lightman, and P. F. Kador. 1990. Feedback inhibition of aldose reductase gene expression in rat renal medulla: Galactitol accumulation reduces enzyme mRNA levels and depletes cellular inositol content. J. Clin. Invest. 86:1103–1108.

Bossert, D. K., W. J. Paul, and R. H. Barnes. 1950. The influence of diet composition on vitamin B_{12} activity in mice. J. Nutr. 40:595–604.

Bruce, H. M., and A. S. Parkes. 1949. Feeding and breeding of laboratory animals. IX. A complete cubed diet for mice and rats. J. Hyg. 47:202–208.

Bryan, W. L., and K. E. Mason. 1940. Vitamin E deficiency in the mouse. Am. J. Physiol. 131:263–267.

Buckley, G. F., and W. S. Hartroft. 1955. Pathology of choline deficiency in the mouse. Arch. Pathol. 59:185–197.

Bull, L. S., H. F. Tyrrell, and J. T. Reid. 1976. Energy utilization by growing male and female sheep and rats, by comparative slaughter and respiration techniques. Pp. 137–140 in Proceedings of the Seventh Symposium on Energy Metabolism, European Association of Animal Production (EAPP) Pub. No. 19. London: Butterworths.

Burton, L. E., and W. W. Wells. 1977. Characterization of the lactation-dependent fatty liver in myo-inositol deficient rats. J. Nutr. 107:1871–1883.

Calcutt, N. A., D. R. Tomlinson, and S. Biswas. 1990. Coexistence of nerve conduction deficit with increased Na^+-K^+-ATPase activity in galactose-fed mice: Implications for polyol pathway and diabetic neuropathy. Diabetes 39:663–666.

Calvert, C. C., T. R. Famula, J. F. Bernier, N. Khalaf, and G. E. Bradford. 1986. Efficiency of growth in mice with a major gene for rapid postweaning gain. J. Anim. Sci. 62:77–85.

Canolty, N. L., and L. J. Koong. 1976. Utilization of energy for maintenance and for fat and lean gains by mice selected for rapid postweaning growth rate. J. Nutr. 106:1202–1208.

Cerecedo, L. R., and L. J. Vinson. 1944. Growth, reproduction and lactation in mice on highly purified diets, and the effect of folic acid on lactation. Arch. Biochem. 5:157–161.

Cercedo, L. R., F. P. Panzarella, A. B. Vasta, and E. C. DeRenzo. 1952. Studies on essential fatty acid deficiency in three strains of mice. J. Nutr. 48:41–47.

Crocker, J. F. S., S. F. Muhtadie, D. C. Hamilton, and D. E. C. Cole. 1985. The comparative toxicity of vitamin D metabolites in the weanling mouse. Toxicol. Appl. Pharmacol. 80:119–126.

Csallany, A. S., K. L. Ayaz, and L.-C. Su. 1977. Effect of dietary vitamin E and aging on tissue lipofuscin pigment concentration in mice. J. Nutr. 107:1792–1799.

Dalton, D. C. 1963. Effect of dilution of the diet with an indigestible filler on feed intake in the mouse. Nature 197:909–910.

Davies, I., Y. Davidson, and A. P. Fotheringham. 1987. The effect of vitamin E deficiency on the induction of age pigment in various tissues of the mouse. Exp. Gerontol. 22:127–137.

Davis, G. K., and W. Mertz. 1987. Copper. Pp. 301–364 in Trace Elements in Human and Animal Nutrition, W. Mertz, ed. Orlando, Fla.: Academic Press.

Decker, A. B., D. L. Fillerup, and J. F. Mead. 1950. Chronic essential fatty acid deficiency in mice. J. Nutr. 41:507–521.

Delorme, A.-C., J.-L. Danan, and H. Mathieu. 1983. Biochemical evidence for the presence of two vitamin D-dependent calcium-binding proteins in mouse kidney. J. Biol. Chem. 258:1878–1884.

Dubos, R., R. S. Schaedler, and R. Costello. 1968. Lasting biological effects of early environmental influences. J. Exp. Med. 127:783–799.

Erway, L., L. S. Hurley, and A. Fraser. 1970. Congenital ataxia and otolith defects due to manganese deficiency in mice. J. Nutr. 100:643–654.

Fahim, F. A., N. Y. S. Morcos, and A. Y. Esmat. 1990. Effects of dietary magnesium and/or manganese variables on the growth rate and metabolism of mice. Ann. Nutr. Metab. 34:183–192.

Fenton, P. F., and C. J. Carr. 1951. The nutrition of the mouse. XI. Response of four strains to diets differing in fat content. J. Nutr. 45:225–233.

Fenton, P. F., and G. R. Cowgill. 1947a. Studies on the vitamin requirements of highly inbred strains of mice: Riboflavin and pantothenic acid. Fed. Proc. 6:407 (abstr.).

Fenton, P. F., and G. R. Cowgill. 1947b. The nutrition of the mouse. 1. A difference in the riboflavin requirements of two highly inbred strains. J. Nutr. 34:273–283.

Fenton, P. F., G. R. Cowgill, M. A. Stone, and D. H. Justice. 1950. Nutrition of the mouse. VIII. Studies on pantothenic acid, biotin, inositol, and p-aminobenzoic acid. J. Nutr. 42:257–269.

Friedman, M., and M. R. Gumbmann. 1981. Bioavailability of some lysine derivatives in mice. J. Nutr. 111:1362–1369.

Friedman, M., and M. R. Gumbmann. 1984a. The utilization and safety of isomeric sulfur-containing amino acids in mice. J. Nutr. 114:2301–2310.

Friedman, M., and M. R. Gumbmann. 1984b. The nutritive value and safety of D-phenylalanine and D-tyrosine in mice. J. Nutr. 114:2089–2096.

Friedman, M., M. R. Gumbmann, and I. I. Ziderman. 1987. Nutritional value and safety in mice of proteins and their admixtures with carbohydrates and vitamin C after heating. J. Nutr. 117:508–518.

Fritz, T. E., D. V. Tolle, and R. J. Flynn. 1968. Hemorrhagic diathesis in laboratory rodents. Proc. Soc. Exp. Biol. Med. 128:228–234.

Furth-Walker, D., D. Leibman, and A. Smolen. 1989. Changes in pyridoxal phosphate and pyridoxamine phosphate in blood liver and brain in the pregnant mouse. J. Nutr. 119:750–756.

Furth-Walker, D., D. Leibman, and A. Smolen. 1990. Relationship between blood, liver and brain pyridoxal phosphate and pyridoxamine phosphate concentrations in mice. J. Nutr. 120:1338–1348.

Giroud, A., and M. Martinet. 1962. Légèreté de la dose tératogène de la vitamine A. C. R. Soc. Biol. 156:449–450.

Goettsch, M. A. 1942. Alpha-tocopherol requirement of the mouse. J. Nutr. 23:513–523.

Goettsch, M. A. 1960. Comparative protein requirement of the rat and mouse for growth, reproduction and lactation using casein diets. J. Nutr. 70:307–312.

Golub, M. S., C. L. Keen, M. E. Gershwin, and V. K. Vijayan. 1986. Growth, development, and brain zinc levels in mice marginally or severely deprived of zinc during postembryonic brain development. Nutr. Behav. 3:169–180.

Gross, A. M., and J. R. Prohaska. 1990. Copper-deficient mice have higher cardiac norepinephrine turnover. J. Nutr. 120:88–96.

Hambidge, K. M., C. E. Casey, and N. F. Krebs. 1986. Zinc. Pp. 1–137 in Trace Elements in Human and Animal Nutrition, W. Mertz, ed. Orlando, Fla.: Academic Press.

Hamuro, Y., A. Shino, and Z. Suzuoki. 1970. Acute induction of soft tissue calcification with transient hyperphosphatemia in the KK mouse by modification in dietary contents of calcium, phosphorus and magnesium. J. Nutr. 100:404–412.

Hatch, R. C., and D. P. Laflamme. 1989. Acute intraperitoneal chole-calciferol (vitamin D$_3$) toxicosis in mice: Its nature and treatment with diverse substances. Vet. Hum. Toxicol. 31:105–112.

Hauschildt, J. D. 1942. Thiamin requirement of albino mice. Proc. Soc. Exp. Biol. Med. 49:145–147.

Heid, M. K., N. D. Bills, S. H. Hinrichs, and A. J. Clifford. 1992. Folate deficiency alone does not produce neural tube defects in mice. J. Nutr. 122:888–894.

Herzberg, G. R., and M. Rogerson. 1982. Interaction of the level of dietary fat and type of carbohydrate in the regulation of hepatic lipogenesis in the mouse. Can. J. Physiol. Pharmacol. 60:912–919.

Hirakawa, D. A., L. M. Olson, and D. H. Baker. 1984. Comparative utilization of a crystalline amino acid diet and a methionine-fortified casein diet by young rats and mice. Nutr. Res. 4:891–895.

Hoag, W. G., and M. M. Dickie. 1962. Studies of the effect of various dietary protein and fat levels on inbred laboratory mice. Proc. Anim. Care Panel 12:7–10.

Hoover-Plow, J., P. Elliott, and B. Moynier. 1988. Reproductive performance in C57BL and I strain mice. Lab. Anim. Sci. 38:595–602.

Hoppel, C. L., and B. Tandler. 1975. Riboflavin and mouse hepatic cell structure and function: Mitochondrial oxidative metabolism in severe deficiency states. J. Nutr. 105:562–570.

Hundemer, J. K., S. P. Nabar, B. J. Shriver, and L. P. Forman. 1991. Dietary fiber sources lower blood cholesterol in C57BL/6 mice. J. Nutr. 121:1360–1365.

Hundley, J. M. 1949. Influence of fructose and other carbohydrates on the niacin requirement of the rat. J. Biol. Chem. 181:1–9.

Hurley, L. S., and L. T. Bell. 1974. Genetic influence on response to dietary manganese deficiency. J. Nutr. 104:133–137.

Hurley, L. S., and C. L. Keen. 1987. Manganese. Pp. 185–223 in Trace Elements in Human and Animal Nutrition, W. Mertz, ed. Orlando, Fla.: Academic Press.

Hurley, L. S., G. Cosens, and L. L. Theriault. 1976. Teratogenic effects of magnesium deficiency in rats. J. Nutr. 106:1254–1260.

Jaffé, W. J. 1952. Influence of cobalt on reproduction of mice and rats. Science 115:265–267.

John, A. M., and J. M. Bell. 1976. Amino acid requirements of the growing mouse. J. Nutr. 106:1361–1367.

Johnson, W. T., S. N. Dufault, and A. C. Thomas. 1993. Platelet cytochrome *c* oxidase activity is an indicator of copper status in rats. Nutr. Res. 13:1153–1162.

Jones, J. H., C. Foster, F. Dorman, and G. L. Hunter. 1945. Effects on the albino mouse of feeding diets very deficient in each of several vitamin B factors (thiamine, riboflavin, pryidoxine, pantothenic acid). J. Nutr. 29:127–136.

Kalter, H. 1990. Analysis of the syndrome of congenital malformations induced in genetically defined mice by acute riboflavin deficiency. Terato. Carcino. Mutagen. 10:385–397.

Keen, C. L., and L. S. Hurley. 1989. Zinc and reproduction: Effects of deficiency on foetal and postnatal development. Pp. 183–220 in Zinc in Human Biology, C. F. Mills, ed. New York: Springer-Verlag.

Keen, C. L., and M. E. Gershwin. 1990. Zinc deficiency and immune function. Annu. Rev. Nutr. 10:415–431.

Keen, C. L., B. Lönnerdal, and L. S. Hurley. 1982. Teratogenic effects of copper deficiency and excess. Pp. 109–122 in Inflammatory Diseases and Copper, J. R. J. Sorenson, ed. New Jersey: Humana Press.

Keith, M. O., and J. M. Bell. 1988. Digestibility of nitrogen and amino acids in selected protein sources fed to mice. J. Nutr. 118:561–568.

Keith, M. O., and J. M. Bell. 1989. The utilization of nitrogen for growth in mice fed blends of purified proteins. Proc. Soc. Exp. Biol. Med. 190:246–253.

Keller, P. R., and P. J. Fraker. 1986. Gestational zinc requirement of the A/J mouse: Effects of a marginal zinc deficiency on in utero B-cell development. Nutr. Res. 6:41–50.

Keyhani, M., D. Giuliani, and B. S. Morse. 1974. Erythropoiesis in pyridoxine-deficient mice. Proc. Soc. Exp. Biol. Med. 146:114–119.

Kielanowski, J. 1965. Estimates of the energy costs of protein deposition in growing animals. Pp. 13–20 in Proceedings of the Fifth Symposium on Energy Metabolism, European Association of Animal Production (EAPP) Pub. No. 11. London: Butterworths.

Kindberg, C. G., and J. W. Suttie. 1989. Effect of various intakes of phylloquinone on signs of vitamin K deficiency and serum and liver phylloquinone concentrations in the rat. J. Nutr. 119:175–180.

Klevay, L. M., and J. T. Saari. 1993. Comparative responses of rats to different copper intakes and modes of supplementation. Proc. Soc. Exp. Biol. Med. 203:214–220.

Kliger, I. J., K. Guggenheim, and E. Buechler. 1944. Relation of riboflavin deficiency to spontaneous epidemics of *Salmonella* in mice. Proc. Soc. Exp. Biol. Med. 57:132–133.

Knapka, J. J. 1983. Nutrition. Pp. 51–67 in The Mouse in Biomedical Research, Vol. 3, H. L. Foster, J. D. Small, and J. G. Fox, eds. New York: Academic Press.

Knapka, J. J., K. P. Smith, and F. J. Judge. 1974. Effect of open and closed formula rations on the performance of three strains of laboratory mice. Lab. Anim. Sci. 24:480–487.

Knapka, J. J., K. P. Smith, and F. J. Judge. 1977. Effect of crude fat and crude protein on reproduction and weanling growth in four strains of inbred mice. J. Nutr. 107:61–69.

Kochhar, D. M., J. D. Penner, and M. A. Satre. 1988. Derivation of retinoic acid and metabolites from a teratogenic dose of retinol (vitamin A) in mice. Toxicol. Appl. Pharmacol. 96:429–441.

Komai, M., H. Shirakawa, and S. Kimura. 1987. Newly developed model for vitamin K deficiency in germfree mice. Int. J. Vit. Nutr. Res. 58:55–59.

Konno, R., S. Uchiyama, and Y. Yasumura. 1982. Intraspecies and interspecies variations in the substrate specificity of D-amino acid oxidase. Comp. Biochem. Physiol. 71B:735–738.

Konno, R., and Y. Yasumura. 1984. Involvement of D-amino-acid oxidase in D-amino acid utilization in the mouse. J. Nutr. 114:1617–1621.

Kubo, C., B. C. Johnson, A. Gajjar, and R. A. Good. 1987. Crucial dietary factors in maximizing life span and longevity in autoimmune-prone mice. J. Nutr. 117:1129–1135.

Kuvibidila, S., M. Dardenne, W. Savino, and F. Lepault. 1990. Influence of iron-deficiency anemia on selected thymus functions in mice: Thymulin biological activity, T-cell subsets and thymocyte proliferation. Am. J. Clin. Nutr. 51:228–232.

Lakhanpal, R. K., and G. M. Briggs. 1966. Biotin deficiency in mice and depigmentation: Effect of dietary carbohydrates. Proc. Soc. Exp. Biol. Med. 121:472–475.

Lee, Y. C. P., M. B. Visscher, and J. T. King. 1962. Role of manganese and vitamin E deficiency in mouse paralysis. Am. J. Physiol. 203:1103–1108.

Leiter, E. H., D. L. Coleman, D. K. Ingram, and M. A. Reynolds. 1983. Influence of dietary carbohydrate on the induction of diabetes in C57BL/KsJ-*db/db* diabetes mice. J. Nutr. 113:184–195.

Lin, P., D. R. Ramos, J. G. Vander Tuig, and G. A. Leveille. 1979. Maintenance energy requirements, energy retention and heat production of young obese (*ob/ob*) and lean mice fed a high-fat or a high-carbohydrate diet. J. Nutr. 109:1143–1153.

Lippincott, S. W., and H. P. Morris. 1942. Pathologic changes associated with riboflavin deficiency in the mouse. J. Natl. Cancer Inst. 2:601–610.

Luckey, T. D., M. H. Bengson, and H. Kaplan. 1974. Effect of bioisolation and the intestinal flora of mice upon evaluation of an Apollo diet. Aerosp. Med. 45:509–518.

Luecke, R. W., and P. J. Fraker. 1979. The effect of varying zinc levels on growth and antibody mediated response in two strains of mice. J. Nutr. 109:1373–1376.

MacEwan, K. L., and K. J. Carpenter. 1980. The nutritional value of supplementary D tryptophan for growing mice. Nutr. Rep. Int. 21:279–284.

Maddy, K. H., and C. A. Elvehjem. 1949. Studies of growth of mice fed rations containing free amino acids. J. Biol. Chem. 177:577–590.

Many, M.-C., J.-F. Denef, S. Hamude, C. Cornette, S. Haumont, and C. Beckers. 1986. Effects of iodide and thyroxine on iodine-deficient mouse thyroid: A morphological and functional study. J. Endocrinol. 110:203–210.

Martin, G. J. 1941. The mouse antialopecia factor. Science 93:422–423.

Marrett, L. E., and M. L. Sunde. 1965. The effect of other D amino acids on the utilization of the isomers of methionine and its hydroxy analogue. Poult. Sci. 44:957–964.

McCarthy, P. T., and L. R. Cerecedo. 1952. Vitamin A deficiency in the mouse. J. Nutr. 46:361–376.

Meader, R. D., and W. L. Williams. 1957. Choline deficiency in the mouse. Am. J. Anat. 100:167–203.

Miller, E. C., and C. A. Baumann. 1945. Relative effects of casein and tryptophane on the health and xanthurenic acid excretion of pyridoxine-deficient mice. J. Biol. Chem. 157:551–562.

Milner, J. A., R. L. Prior, and W. J. Visek. 1975. Arginine deficiency and orotic aciduria in mammals. Proc. Soc. Exp. Biol. Med. 150:282–288.

Mirone, L. 1954. Effect of choline-deficient diets on growth, reproduction and mortality of mice. Am. J. Physiol. 179:49–52.

Mirone L., and L. R. Cerecedo. 1947. The beneficial effect of xanthopterin on lactation, and of biotin on reproduction and lactation, in mice maintained on highly purified diets. Arch. Biochem. 15:324–326.

Morgan, P. N., C. L. Keen, and B. Lönnerdal. 1988a. Effect of varying dietary zinc intake of weanling mouse pups during recovery from early undernutrition on tissue mineral concentrations, relative organ weights, hematological variables and muscle composition. J. Nutr. 118:699–711.

Morgan, P. N., C. L. Keen, C. C. Calvert, and B. Lönnerdal. 1988b. Effect of varying dietary zinc intake of weanling mouse pups during recovery from early undernutrition on growth, body composition and composition of gain. J. Nutr. 118:690–698.

Morris, H. P. 1947. Vitamin requirements of the mouse. Vitam. Horm. 5:175–195.

Morris, H. P., and S. W. Lippincott. 1941. The effect of pantothenic acid on growth and maintenance of life in mice of the C₃H strain. J. Natl. Cancer Inst. 2:29–38.

Morris, H. P., and W. V. B. Robertson. 1943. Growth rate and number of spontaneous carcinomas and riboflavin concentration of liver, muscle and tumor of C₃H mice as influenced by dietary riboflavin. J. Natl. Cancer Inst. 3:479–489.

Morris, H. P., and C. S. Dubnik. 1947. Thiamine deficiency and thiamine requirements of C₃H mice. J. Natl. Cancer Inst. 8:127–137.

Morse, H. C., III. 1978. The laboratory mouse—A historical perspective. Pp. 1–16 in The Mouse in Biomedical Research, Vol. 1, H. L. Foster, J. D. Small, and J. G. Fox, eds. New York: Academic Press.

Mulhern, S. A., and L. D. Koller. 1988. Severe or marginal copper deficiency results in a graded reduction in immune status in mice. J. Nutr. 118:1041–1047.

National Institutes of Health. 1982. NIH Rodents 1980 Catalogue. NIH No. 83-606. Washington, D.C.: U.S. Department of Health and Human Services.

National Research Council. 1978. Nutrient Requirements of Laboratory Animals, Third Revised Ed. Washington, D.C.: National Academy Press.

Neuringer, M., G. J. Anderson, and W. E. Connor. 1988. The essentiality of n-3 fatty acids for the development and function of the retina and brain. Annu. Rev. Nutr. 8:517–541.

Nielsen, E., and A. Black. 1944. Biotin and folic acid deficiency in the mouse. J. Nutr. 28:203–207.

Nishimura, H. 1953. Zinc deficiency in suckling mice deprived of colostrum. J. Nutr. 49:79–97.

Olejer, V., H. Fisher, and F. L. Margolis. 1982. The histidine requirement of 2 strains of mice with genetic differences in level of carnosinase activity. Nutr. Rep. Int. 26:879–885.

Olson, L. M., S. K. Clinton, J. I. Everitt, P. V. Johnston, and W. J. Visek. 1987. Lymphocyte activation, cell-mediated cytotoxicity and their relationship to dietary fat-enhanced mammary tumorigenesis in C3H/OUJ mice. J. Nutr. 117:955–963.

Ornoy, A., I. Wolinsky, and K. Guggenheim. 1974. Structure of long bones of rats and mice fed a low calcium diet. Calcif. Tissue Res. 15:71–76.

Otter, R., R. Reiter, and A. Wendel. 1989. Alterations in the protein-synthesis, -degradation and/or -secretion rates in hepatic subcellular fractions of selenium-deficient mice. Biochem. J. 258:535–540.

Pappenheimer, A. M. 1942. Muscular dystrophy in mice on vitamin E-deficient diets. Am. J. Pathol. 18:169–175.

Parker, C. J., Jr., T. G. Riess, and V. M. Sardesai. 1985. Essentiality of histidine in adult mice. J. Nutr. 115:824–826.

Peters, J. M., L. M. Wiley, S. Zidenberg-Cherr, and C. L. Keen. 1991. Influence of short-term maternal zinc deficiency on the in vitro development of preimplantation mouse embryos. Proc. Soc. Exp. Biol. Med. 198:561–568.

Peterson, F. J., N. F. Lindemann, P. H. Duquette, and J. L. Holtzman. 1992. Potentiation of acute acetaminophen lethality by selenium and vitamin E deficiency in mice. J. Nutr. 122:74–81.

Phillips, M., J. Camakaris, and D. M. Danks. 1991. A comparison of phenotype and copper distribution in blotchy and brindled mutant mice and in nutritionally copper deficient controls. Biol. Trace Elem. Res. 29:11–29.

Phillips, S. C. 1987. Neuro-toxic interaction in alcohol-treated, thiamin-deficient mice. Acta Neuropathol. 73:171–176.

Pleasants, J. R., B. S. Reddy, and B. S. Wostmann. 1970. Qualitative adequacy of a chemically defined diet for reproducing germfree mice. J. Nutr. 100:498–508.

Pleasants, J. R., B. S. Wostmann, and B. S. Reddy. 1973. Improved lactation in germfree mice following changes in the amino acid and fat components of a chemically defined diet. Pp. 245–250 in Germfree Research, J. B. Heneghan, ed. New York: Academic Press.

Poiley, S. M. 1972. Growth tables for 66 strains and stocks of laboratory animals. Lab. Anim. Sci. 22:759–799.

Prohaska, J. R. 1990. Development of copper deficiency in neonatal mice. J. Nutr. Biochem. 1:415–419.

Prohaska, J. R., and O. A. Lukasewycz. 1989a. Biochemical and immunological changes in mice following postweaning copper deficiency. Biol. Trace Elem. Res. 22:101–112.

Prohaska, J. R., and O. A. Lukasewycz. 1989b. Copper deficiency during perinatal development: Effects on the immune response of mice. J. Nutr. 119:922–931.

Reddi, A. S. 1978. Riboflavin nutritional status and flavoprotein enzymes in normal and genetically diabetic KK mice. Metab. Clin. Exp. 27:531–537.

Reeves, P. G., F. H. Nielsen, and G. C. Fahey, Jr. 1993. AIN-93 purified diets for laboratory rodents: Final report of the American Institute of Nutrition Ad Hoc Writing Committee on the reformation of the AIN-76A rodent diet. J. Nutr. 123:1939–1951.

Reeves, P. G., K. L. Rossow, and L. Johnson. 1994. Maintenance requirements for copper in adult male mice fed AIN-93 rodent diet. Nutr. Res. 14:1219–1226.

Reicks, M., and J. N. Hathcock. 1989. Prolonged acetaminophen ingestion in mice effects on the availability of methionine for metabolic functions. J. Nutr. 119:1042–1049.

Rivers, J. P., and B. C. Davidson. 1974. Linolenic acid deprivation in mice. Proc. Nutr. Soc. 33:48A.

Robeson, B. L., E. J. Eisen, and J. M. Leatherwood. 1981. Adipose cellularity, serum glucose, insulin and cholesterol in polygenic obese mice fed high-fat or high-carbohydrate diets. Growth 45:198–215.

Rogers, A. E., and R. A. MacDonald. 1965. Hepatic vasculature and cell proliferation in experimental cirrhosis. Lab. Invest. 14:1710–1726.

Rothenberg, S. P., M. da Costa, and F. Siy. 1973. Impaired antibody response in folate-deficient mice persisting after folate repletion. Life Sci. 12:177–184.

Rowland, N. E., and M. J. Fregley. 1988. Sodium appetite: Species and strain differences and role of renin-angiotensin-aldosterone system. Appetite 11:143–178.

Sandza, J. G., and L. R. Cerecedo. 1941. Requirement of the mouse for pantothenic acid and for a new factor of the vitamin B complex. J. Nutr. 21:609–615.

Santhanam, U., U. J. Nair, and S. V. Bhide. 1987. Development of dietary regimen to achieve long-term survival and subclinical vitamin A deficient status in mice. Ind. J. Exp. Biol. 25:164–168.

Seaborn, C. D., and B. J. Stoecker. 1989. Effects of starch, sucrose, fructose and glucose on chromium absorption and tissue concentrations in obese and lean mice. J. Nutr. 119:1444–1451.

Seltzer, J. L., and D. B. McDougal, Jr. 1974. Temporal chances of regional cocarboxylase levels in thiamin-depleted mouse brain. Am. J. Physiol. 227:714–718.

Shaw, W., R. A. Schreiber, and J. W. Zemp. 1973. Perinatal folate deficiency: Effects on the developing brain in C57BL/6J mice. Nutr. Rep. Int. 8:219–228.

Shepherd, N. D., and T. G. Taylor. 1974a. A reassessment of the status of *myo*-inositol as a vitamin. Proc. Nutr. Soc. 33:63A (abstr.).

Shepherd, N. D., and T. G. Taylor. 1974b. The lipotropic action of *myo*-inositol. Proc. Nutr. Soc. 33:64A (abstr.).

Smith, J. E. 1990. Preparation of vitamin A-deficient rats and mice. Methods Enzymol. 190:229–236.

Smith, S. M., N. S. Levy, and C. E. Hayes. 1987. Impaired immunity in vitamin A-deficient mice. J. Nutr. 117:857–865.

Sorbie, J., and L. S. Valberg. 1974. Iron balance in the mouse. Lab. Anim. Sci. 24:900–904.

Stanley, J. C., and E. A. Newsholme. 1985a. The effect of dietary bagasse on the activities of some key enzymes of carbohydrate and lipid metabolism in mouse liver. Br. J. Nutr. 54:415–420.

Stanley, J. C., and E. A. Newsholme. 1985b. The effect of dietary guar gum on the activities of some key enzymes of carbohydrate and lipid metabolism in mouse liver. Br. J. Nutr. 53:215–222.

Stanley, J. C., J. A. Lambadarios, and E. A. Newsholme. 1986. Absence of effects of dietary wheat bran on the activities of some key enzymes of carbohydrate and lipid metabolism in mouse liver and adipose tissue. Br. J. Nutr. 55:287–294.

Sundboom, J., and J. A. Olson. 1984. Effect of aging on the storage and catabolism of vitamin A in mice. Exp. Gerontol. 19:257–265.

Thomas, H. M., W. L. Williams, and B. R. Clower. 1968. Cardiac lesions in C mice: Results of choline-deficient and choline-supplemented diets. Arch. Pathol. 85:532–538.

Thonney, M. L., A. M. Arnold, D. A. Ross, S. L. Schaaf, and T. R. Rounsaville. 1991. Energetic efficiency of rats fed low or high protein diets and grown at controlled rates from 80 to 205 grams. J. Nutr. 121:1397–1406.

Tove, S. B., and F. H. Smith. 1959. Kinetics of the depletion of linoleic acid in mice. Arch. Biochem. Biophys. 85:352–365.

Toyoda, H., S. Himeno, and N. Imura. 1989. The regulation of glutathione peroxidase gene expression relevant to species difference and the effects of dietary selenium manipulation. Biochim. Biophys. Acta 1008:301–308.

Toyomizu, M., K. Hayashi, K. Yamashita, and Y. Tomita. 1988. Response surface analyses of the effects of dietary protein on feeding and growth patterns in mice from weaning to maturity. J. Nutr. 118:86–92.

Troelson, J. E., and J. M. Bell. 1963. A comparison of nutritional effects in swine and mice: Responses in feed intake, feed efficiency and carcass characteristics to similar diets. Can. J. Anim. Sci. 43:294–304.

Trostler, N. P., S. Brady, D. R. Romsos, and G. A. Leveille. 1979. Influence of dietary vitamin E on malondialdehyde levels in liver and adipose tissue and on glutathione peroxidase and reductase activities in liver and erythrocytes of lean and obese (*ob/ob*) mice. J. Nutr. 109:345–352.

Tsao, C. S., and P. Y. Leung. 1989. Effect of ascorbic acid intake on tissue dehydroascorbic acid in mice. Nutr. Res. 9:1371–1379.

Tsao, C. S., P. Y. Leung, and M. Young. 1987. Effect of dietary ascorbic acid intake on tissue vitamin C in mice. J. Nutr. 117:291–297.

Tumanov, V. N., and R. V. Trebukhina. 1983. Degradation of thiamin diphosphate in subcellular liver fractions of mice in the development of experimental nutritional vitamin B_1 deficiency. Vopr. Med. Khim. 29:99–102.

Van Middlesworth, L. 1986. T-2 mycotoxin intensifies iodine deficiency in mice fed low iodine diet. Endocrinology 118:583–586.

Van Pelt, A. M. M., and D. G. De Rooij. 1990. Synchronization of the seminiferous epithelium after vitamin A replacement in vitamin A-deficient mice. Biol. Reprod. 43:363–367.

Wainwright, P. E., Y. S. Huang, B. Bulman-Fleming, D. E. Mills, P. Redden, and D. McCutcheon. 1991. The role of n-3 essential fatty acids in brain and behavioral development: A cross-fostering study in the mouse. Lipids 26:37–45.

Watanabe, T., and A. Endo. 1989. Species and strain differences in teratogenic effects of biotin deficiency in rodents. J. Nutr. 119:255–261.

Watanabe, T., and A. Endo. 1991. Biotin deficiency per se is teratogenic in mice. J. Nutr. 121:101–104.

Webster, A. J. F. 1983. Energetics of maintenance and growth. Pp. 178–207 in Mammalian Thermogenesis, L. Girardier and M. J. Stock, eds. New York: Chapman and Hill.

Webster, A. J. F. 1988. Comparative aspects of energy exchange. Pp. 37–54 in Comparative Nutrition, K. Blaxter and I. McDonald, eds. London: Libbey.

Weir, D. R., R. W. Heinle, and A. D. Welch. 1948. Pterylglutamic acid deficiency in mice: Hematologic and histologic findings. Proc. Soc. Exp. Biol. Med. 69:211–215.

Weitzel, F., F. Ursini, and A. Wendel. 1990. Phospholipid hydroperoxide glutathione peroxidase in various mouse organs during selenium deficiency and repletion. Biochim. Biophys. Acta 1036:88–94.

Wendel, A., and R. Otter. 1987. Alterations in the intermediary metabolism of selenium-deficient mice. Biochim. Biophys. Acta 925:94–100.

White, E. A., J. R. Foy, and L. R. Cerecedo. 1943. Essential fatty acid deficiency in the mouse. Proc. Soc. Exp. Biol. Med. 54:301–302.

Williams, W. L. 1960. Hepatic liposis and myocardial damage in mice fed choline-deficient or choline-supplemented diets. Yale J. Biol. Med. 33:1–14.

Wolfe, J. M., and H. P. Salter, Jr. 1931. Vitamin A deficiency in the albino mouse. J. Nutr. 4:185–192.

Wolinsky, I., and K. Guggenheim. 1974. Effect of low calcium diet on bone and calcium metabolism in rats and mice—A differential species response. Comp. Biochem. 49A:183–195.

Woolley, D. W. 1941. Identification of the mouse antialopecia factor. J. Biol. Chem. 139:29–34.

Woolley, D. W. 1942. Synthesis of inositol in mice. J. Exp. Med. 75:227–284.

Woolley, D. W. 1945. Some biological effects produced by α-tocopherol quinone. J. Biol. Chem. 159:59–66.

Wostmann, B. S., and T. F. Kellogg. 1967. Purified starch casein diet for nutritional research with germfree rats. Lab. Anim. Care 17:589–593.

Wynder, E. L., and U. E. Kline. 1965. The possible role of riboflavin deficiency in epithelial neoplasia. I. Epithelial changes of mice in simple deficiency. Cancer 18:167–180.

Yasunaga, T., H. Kato, K. Ohgahi, T. Inamoto, and Y. Hikasa. 1982. Effect of vitamin E as an immunopotentiation agent for mice at optimal dosages and its toxicity at high dosage. J. Nutr. 112:1075–1084.

Younes, M., H. D. Trepkau, and C. P. Siegers. 1990. Enhancement by dietary iron of lipid peroxidation in mouse colon. Res. Commun. Chem. Pathol. Pharmacol. 70:349–354.

Yuen, D. E., and H. H. Draper. 1983. Long-term effects of excess protein and phosphorus on bone homeostasis in adult mice. J. Nutr. 113:1374–1380.

Zidenberg-Cherr, S., L. S. Hurley, B. Lönnerdal, and C. L. Keen. 1985. Manganese deficiency: Effects on susceptibility to ethanol toxicity in rats. J. Nutr. 115:460–467.

Zimmerman, D. R., and B. S. Wostmann. 1963. Vitamin stability in diets sterilized for germfree animals. J. Nutr. 79:318–322.

4 Nutrient Requirements of the Guinea Pig

The domestic guinea pig (*Cavia porcellus*) has been bred in captivity for at least 400 years and probably originated in Peru, Argentina, or Brazil (Weir, 1974). Many laboratory guinea pigs were bred from a strain established by Dunkin and Hartley in 1926 (Dunkin et al., 1930). [See National Institutes of Health (1982) for other strains.] Unless otherwise indicated, the strain referred to in this chapter is the outbred Hartley.

In its natural habitat this herbivorous animal consumes large quantities of vegetation (Navia and Hunt, 1976). The molar teeth are especially suited to grinding and, like other species of rodents, the guinea pig has open-rooted incisors that grow continuously throughout its life. Like the rat, mouse, and rabbit, the guinea pig is simple-stomached; but in contrast to these species, the entire stomach of the guinea pig is lined with glandular epithelium (Breazile and Brown, 1976; Navia and Hunt, 1976). The intestine allows the development of predominantly gram-positive bacterial flora, which may contribute to the nutritional requirements of the host perhaps through direct absorption of bacterial metabolites or digestion and absorption of intestinal bacteria and other materials following coprophagy. The guinea pig has a large semicircular cecum with numerous lateral pouches. This organ resembles that of the rabbit and possibly has similar digestive functions—e.g., synthesis of B vitamins and indispensable amino acids by microorganisms and recycling of intestinal contents by coprophagy (Hunt and Harrington, 1974). Few serious attempts have been made to determine the contribution of coprophagy to the nutrition of the guinea pig.

BEHAVIORAL AND NUTRITIONAL CHARACTERISTICS

In the laboratory, the guinea pig's diet is much higher in energy density and lower in fiber content than the diet of green vegetation and fruits it consumes in the wild. The guinea pig consumes many small meals throughout the day, is fastidious in choice of foods, and may resist abrupt changes in composition or form of the diet. Animals fed pelleted natural-ingredient diets often do not readily accept a powdered purified diet unless introduced gradually. Pelleting the powdery diet (Ostwald et al., 1971), moistening the food with water (O'Dell and Regan, 1963; Singh et al., 1968), and using gel diets (Navia and Lopez, 1973; Apgar and Everett, 1991b) have been successful in promoting diet acceptance. These behavioral characteristics and special nutritional requirements need to be considered when designing nutritional or metabolic studies.

Water intake is variable and food intake is largely influenced by the form and composition of the diet and the age of the animal. Liu (1988) reported a mean water intake of 21.7 mL/100 g BW/day and mean consumption of a natural-ingredient diet 3.0 Mcal/kg (12.6 MJ/kg) to be 6.9 g/100 g BW/day in 6-week-old male guinea pigs (312 ± 13 g) that were individually housed on sawdust. Water and food consumption was 7.5 mL/100 g BW/day and 4 g/100 g BW/day, respectively, in male guinea pigs weighing 698 ± 19 g and fed a nonpurified diet containing 20 percent crude protein (Tsao and Young, 1989). Adult male guinea pigs weighing 725 to 750 g consumed daily 32 g of a purified diet containing crystalline amino acids (equivalent to 160 g protein/kg diet) as the sole nitrogen source (Schiller, 1977).

The guinea pig is best known, from a nutritional standpoint, by its requirement for dietary vitamin C. This feature has made the guinea pig particularly useful in studies of collagen biosynthesis, wound healing, and bone growth. The young guinea pig seems to have a relatively high dietary requirement for arginine, folic acid, and selected minerals, although this may not prove to be true as more information on specific nutrient requirements becomes available. These characteristics and others mentioned

103

TABLE 4-1   Estimated Nutrient Requirements for Growth for Guinea Pigs

Nutrient	Unit	Amount, per kg diet	Comments	Reference
Protein (28.6 g N × 6.25)	g	180.0[a]		Shelton, 1971; Lister and McCance, 1965; Typpo et al., 1990a,b
Essential fatty acids (n-6)	g	1.33–4.0	10 g corn oil/kg diet is satisfactory	Reid et al., 1964
Fiber	g	150.0	Used cellulose and/or materials of low digestibility to supply bulk	Heinicke and Elvehjem, 1955
Amino acids[b]				
Arginine	g	12.0		Yoon, 1977
Histidine	g	3.6		Anderson and Typpo, 1977
Isoleucine	g	6.0		Ayers et al., 1987
Leucine	g	10.8		Mueller, 1978
Lysine	g	8.4		Typpo et al., 1985
Methionine	g	6.0[c]		Typpo et al., 1990b
Phenylalanine	g	10.8[d]		Chueh, 1973
Threonine	g	6.0		Horstkoetter, 1974
Tryptophan	g	1.8		Smith, 1979
Valine	g	8.4		Typpo et al., 1990b
Dispensable nitrogen	g	16.9[e]		Typpo et al., 1990b
Minerals				
Calcium	g	8.0	Requirements for calcium, phosphorus, magnesium and potassium seem to reflect interactions among them	Morris and O'Dell, 1961, 1963; O'Dell et al., 1956, 1960
Phosphorus	g	4.0		
Magnesium	g	1.0		
Potassium	g	5.0		
Chloride	g	0.5	From the estimate for rats fed purified diet	
Sodium	g	0.5		
Copper[f]	mg	6.0		Everson et al., 1967, 1968
Iron	mg	50.0	Estimate	
Manganese	mg	40.0		Everson et al., 1959
Zinc	mg	20.0		Alberts et al., 1977; Navia and Lopez, 1973
Iodine[g]	μg	150.0	Based on rat requirement	
Molybdenum	μg	150.0	Based on rat requirement	
Selenium	μg	150.0	Based on rat requirement diets	
Vitamins				
A (retinol)[h] or	mg	6.6		Gil et al., 1968.
(β-carotene)	mg	28.0	Used 40% as efficiently as preformed vitamin A	Bentley and Morgan, 1945
D (cholecalciferol)[i]	mg	0.025	Adequate; no quantitative data	
E (RRR-α-tocopherol)[j]	mg	26.7	Adequate	Hsieh and Navia, 1980
K (phylloquinone)	mg	5.0	Adequate; dietary deficiency has not been produced	
Ascorbic acid	mg	200.0		Mannering, 1949
Biotin (d-biotin)	mg	0.2	Adequate; simple dietary deficiency has not been produced	
Choline (choline bitartrate)	mg	1,800		Reid, 1955
Folic acid	mg	3.0-6.0		Mannering, 1949; Reid, 1954a; Reid et al., 1956; Woodruff et al., 1953
Niacin	mg	10.0		Reid, 1961
Pantothenic acid (Ca-d-pantothenate)	mg	20.0		Reid and Briggs, 1954
Pyridoxine	mg	2.0-3.0		Reid, 1964
Riboflavin	mg	3.0	Estimated	Slanetz, 1943
Thiamin (thiamin-HCl)	mg	2.0		Liu et al., 1967; Reid and Bieri, 1967

(Table footnotes on next page.)

TABLE 4-1   (*Continued*)

NOTE: Nutrient requirements are expressed on an as-fed basis for diets containing 10% moisture; 2.8–3.5 kcal ME/g (11.7–14.6 kJ ME/g) and should be adjusted for diets of differing moisture and energy concentrations. Unless otherwise specified, the listed nutrient concentrations represent minimal requirements and do not include a margin of safety. Higher concentrations for many nutrients may be warranted in natural-ingredient diets.

[a]See text. Growth is equivalent with 300 g casein plus 3 g L-arginine per kg or 200 g soybean protein plus 10 g DL-methionine per kg.

[b]The quantities reflect a 20% adjustment for efficiency of utilization for maximum growth. See text for discussion.

[c]Cystine may replace 40%.

[d]Tyrosine may replace 50%.

[e]Mixture of L-alanine, L-asparagine·H$_2$O, L-aspartic acid, L-glutamic acid, sodium glutamate, glycine, L-proline and L-serine. See text and Typpo et al. (1990b).

[f]Minerals measured in mg/kg correspond to ppm.

[g]Iodine, molybdenum, and selenium are measured in μg/kg, corresponding to ppb.

[h]Equivalent to 21,960 IU/kg. β-Carotene requirement measured is equivalent to 47,425 IU/kg.

[i]Equivalent to 1,000 IU/kg.

[j]Equivalent to 40 IU/kg. Higher concentrations may be required if high-fat diets are used.

above are discussed in greater detail in *The Biology of the Guinea Pig* (Wagner and Manning, 1976).

Germ-free guinea pigs have been used in the study of specific disease states. Diets for germ-free and specific-pathogen-free guinea pigs have been discussed by Wagner and Foster (1976).

GROWTH AND REPRODUCTION

The guinea pig has a mean gestation period of 68 ± 2 SE days (range 59 to 72 days) (Labhsetwar and Diamond, 1970), which may contribute to its advanced development at birth. Dams usually bear 3 to 4 (range 1 to 8) offspring weighing an average of 85 to 100 g each (Ediger, 1976; Sisk, 1976; Apgar and Everett, 1991a). Guinea pigs born weighing less than 50 g have a low probability of survival (Ediger, 1976). Newborn animals can consume semisolid and solid food immediately, although weaning occurs at about 21 days of age when body weight is approximately 250 g (Ediger, 1976). Guinea pigs normally gain as much as 5 to 7 g/day during the rapid growth period when allowed to eat commercial natural-ingredient or purified diets ad libitum (Shelton, 1971; Navia and Lopez, 1973; Jeffery and Typpo, 1982; Liu, 1988; Typpo et al., 1990b). These gains occur routinely and are greater than those obtained with some of the diets used earlier (4 g/day; Woolley and Sprince, 1945). Growth slows after 2 months and maturity is reached at about 5 months. Weight gain can continue until 12 to 15 months of age and levels off at 700 to 850 g for females and 950 to 1,200 g for males (Ediger, 1976). Mating is most often successful when females are 450 to 600 g (2.5 to 3 months old; Ediger, 1976).

ESTIMATION OF NUTRIENT REQUIREMENTS

Estimates of the energy and nutrient requirements for growth of the guinea pig are presented in this chapter. Considerable variation in requirements can occur as a consequence of several factors—the same as those affecting the nutrient requirements of the rat or mouse: developmental stage, reproductive activity, and age; gender; strain. The nutrient requirements listed in this chapter represent mean values that are thought to be representative but not necessarily sufficient in all circumstances. Further research to quantify nutrient requirements and to identify sources of variation in nutrient requirements of the guinea pig is needed.

Recommendations in this chapter for nutrient concentrations have not been increased to allow a margin of safety for variations in dietary ingredients or for differences among guinea pigs. The data on which requirements are based were reported from several different laboratories using different colony management practices. They are adequate for guinea pigs in most laboratory conditions, but particular laboratory protocols, such as maintenance of germ-free colonies or testing of experimental drugs (see Chapter 1), may alter the requirements for one or more nutrients. The data are not sufficient to differentiate between adult maintenance requirements and growth, pregnancy, or lactation requirements; hence, estimates are provided for growth only (Table 4-1). When data were insufficient to determine requirements, adequate concentrations were determined on the basis of feeding studies that produced adequate growth or on well-established concentrations that produce adequate growth in the laboratory rat. If cited papers provided nutrient intake per day but did not specify dietary concentrations, the values have been converted to dietary content by assuming a dietary intake of 20 to 25 g/ guinea pig/day for growth.

EXAMPLES OF DIETS FOR GUINEA PIGS

The composition of an open-formula diet used successfully for growth, reproduction, and longevity is shown in Table 4-2. This natural-ingredient diet is the formulation developed by the National Institutes of Health for production and research colonies of conventional guinea pigs.

Several purified diets have been used that successfully support growth in guinea pigs, although it is not always clear why. Of historical importance is the purified diet developed by Reid and Briggs (1953) for the growing guinea pig; it is still used in original or modified (Typpo et al., 1985) forms. The Typpo et al. (1985) diet and three other examples of purified diets that support satisfactory growth (Navai and Lopez, 1973; O'Dell et al., 1989) and reproduction (Apgar and Everett, 1991a,b) are given in Table 4-3. Two of these are agar gel diets—one satisfactory

TABLE 4-2 Example of a Natural-Ingredient Diet Used for Guinea Pig Breeding Colonies at the National Institutes of Health

Ingredient	Amount, g/kg
Alfalfa meal (17% protein)	350.0
Soybean meal (49% protein)	120.0
Ground whole oats	252.5
Ground whole wheat	236.0
Soybean oil	15.0
Dicalcium phosphate	5.0
Calcium carbonate	10.0
Salt	7.5
Mineral and vitamin premixes[a,b]	4.0

NOTE: Open-formula pelleted guinea pig diet NIH-34M (NIH specification NIH-11-141h, October 10, 1991). This diet has been used for a number of years. The ingredient specifications predate the present report and are not entirely in agreement with the recommendations presented herein (see Table 4-1). Amounts listed in footnotes below represent the mass of IU of the specific mineral element or vitamin rather than the added compounds.

[a]Specifications for mineral premix provided by the manufacturer (mg/kg diet): cobalt, 1.5 (as cobalt carbonate); copper, 6.6 (as copper sulfate); manganese, 39.7 (as manganese oxide); zinc, 19.8 (as zinc oxide); iodine, 1.1 (as calcium iodate).

[b]Specifications for vitamin premix provided by the manufacturer (IU/kg diet): vitamin A, 6,614 (from stabilized vitamin A palmitate or acetate); vitamin D_3, 2,200 (from d-activated animal sterols); and (mg/kg diet) vitamin E, 22 (from all-rac-α-tocopheryl acetate); vitamin K, 5 (menadione activity); thiamin, 4.4 (thiamin mononitrate); riboflavin, 3.3; niacin, 11; pantothenic acid, 11 (from Ca-d-pantothenate); choline, 529 (from choline chloride); pyridoxine, 5 (from pyridoxine-HCl); folic acid, 4.8; biotin, 2.2; ascorbic acid, 992; methionine hydroxy analogue, 500; and vitamin B_{12} (11 µg/kg diet).

TABLE 4-3 Examples of Four Satisfactory Purified Diets for Guinea Pigs

Nutrient	Amount, g/kg diet			
	Navia and Lopez, 1973	O'Dell et al., 1989	Typpo et al., 1985	Apgar and Everett, 1991a,b
Casein	300.0[a]		300.0[a]	300.0[a]
Other protein		200.0[b]		
Sucrose, granulated	431.4	488.0	50.0	
Sucrose, powdered			196.0	
Glucose monohydrate			150.0	310.0
Corn oil	40.0[c]	40.0	30.0	100.0
Fiber[d]	130.1	150.0	150.0	150.0
L-arginine			3.0	
DL-methionine	2.0	5.0[e]		
Mineral mixture	72.2[f]	85.0[g]	75.0[h]	100.0[i]
Vitamin mixture	3.3[j]	30.0[k]	42.0[l]	40.0[m]
Choline chloride	1.0	2.0	2.0	[n]
Myo-inositol			2.0	
Agar	20.0			[o]

NOTE: See each reference for special diet preparation. Nutrient requiements are expressed on an as-fed basis for diets containing 10 percent moisture. Diet used by Apgar and Everett (1991a,b) is satisfactory for pregnant guinea pigs.

[a]Vitamin-free casein. Apgar and Everett (1991a,b) treated casein with EDTA primarily to remove zinc.

[b]Isolated soybean protein or heat-treated casein egg white. Lactalbumin has been used by Hsieh and Navia (1980).

[c]Cottonseed oil instead of corn oil.

[d]Cellulose, except wood pulp (O'Dell et al., 1989) and cellophane (Typpo et al., 1985).

[e]With isolated soybean protein only.

[f]Mineral ingredients (g/kg diet): $CaHPO_4$, 8.30; $CaCO_3$, 14.50; $KC_2O_2H_3$, 27.00; KCl, 4.50; NaCl, 2.80; MgO, 5.00; $MgSO_4$, 0.50; $MgCO_3$, 1.00; $Fe_3(PO_4)$, 1.60; $MnSO_4 \cdot H_2O$, 0.80; KIO_3, 0.038; $ZnSO_4 \cdot 7H_2O$, 0.025; $CuSO_4$, 0.036; $CoCl_2 \cdot 6H_2O$, 0.03; $AlK(SO_4)_2 \cdot 12H_2O$, 0.007, NaF, 0.04.

[g]Mineral ingredients (g/kg diet): $CaHPO_4 \cdot 2H_2O$, 25.4; $CaCO_3$, 9.0; $NaHPO_4$, 6.4; $KC_2O_2H_3$, 25; NaCl, 2.6; MgO, 5.0; $MgSO_4$, 3.0; $MnSO_4 \cdot H_2O$, 0.61; Fe Citrate, 0.36; $CuSO_4$, 0.02; KIO_3, 0.017. Diet supplemented with 100 mg Zn/kg as $ZnCO_3$.

[h]Mineral ingredients (g/kg diet): $CaHPO_4$, 34.92; $CaCO_3$, 5.94; $KC_2O_2H_3$, 24.93; KCl, 7.74; NaCl, 5.76; MgO, 4.96; $MgSO_4$, 4.59; Fe citrate, 0.64; $MnSO_4 \cdot H_2O$, 0.37; KIO_3, 0.015; $ZnCO_3$, 0.13; $CuSO_4$, 0.005; $KCr(SO_4)_2 \cdot 12H_2O$, 0.010; $Na_2MoO_4 \cdot 2H_2O$, 0.0005; $NiCl_2 \cdot 6H_2O$, 0.0002; Na_2SeO_3, 0.0002.

[i]Mineral ingredients (g/kg diet): $CaHPO_4$, 7.4; $CaCO_3$, 12.9; $NaHPO_4 \cdot 7H_2O$, 28; $KC_2O_2H_3$, 24; NaCL, 2.5; KCL, 4; $MgSO_4 \cdot 7H_2O$, 4.9; MgO, 4.4; $MgCO_3$, 0.9, $FeSO_4 \cdot 7H_2O$, 2.04; $MnSO_4 \cdot H_2O$, 0.71; KIO_3, 0.034; $CuSO_4 \cdot 5H_2O$, 0.05; $CoCl_2 6H_2O$, 0.027. Zinc was supplied in drinking water at 15 mg/L (as Zn acetate).

[j]Vitamin mixture supplied (mg/kg diet): ascorbic acid, 2,000; biotin, 0.2; folic acid, 10; inositol, 1,000; niacin, 50; Ca-pantothenate, 30; pyridoxine-HCl, 10; riboflavin, 10; thiamin, 10; vitamin B_{12} (triturated with mannitol at a concentration of 0.1 percent), 30; menadione, 10; (IU/kg diet): vitamin A, 28,500; vitamin D_2, 285; all-rac-α-tocopherol, 40.

[k]Vitamin mixture in sucrose supplied (mg/kg diet): biotin, 0.2; folic acid, 6; niacin, 50; Ca-pantothenate, 30; pyridoxine-HCl, 10; riboflavin, 10; thiamin, 10; vitamin B_{12}, 0.03; menadione, 10; (IU/kg diet): retinyl acetate, 20,000; cholecalciferol, 2,800; α-tocopherol, 20. Ascorbic acid was given in 30 mg doses 6 days/week per os.

[l]Vitamin mixture supplied (mg/kg diet): ascorbic acid, 2,000; biotin, 0.5; folic acid, 10; niacinamide, 200; Ca-pantothenate, 40; pyridoxine-

(*Table footnotes continue next page.*)

TABLE 4-3 (*Continued*)

HCl, 16; riboflavin, 16; thiamin-HCl, 16; vitamin B_{12} (0.1% trituration in mannitol), 50; retinal palmitate in oil, 52 (5,200 IU); cholecalciferol, 0.04; DL-tocopherol acetate, 20; menadione, 2.

[m] Vitamin mixture supplied (mg/kg diet): ascorbic acid, 4,000; biotin, 12.6; choline, 3,100; folic acid, 12; *myo*-inositol, 4,000; niacin, 400; Ca-pantothenate, 60; pyridoxine-HCl, 13.5; riboflavin, 30; thiamin-HCl, 30; vitamin B_{12}, 0.02; menadione, 4.6; (IU/kg diet): retinyl palmitate 45,000; ergocalciferol, 4,400; α-tocopheryl acetate, 198.

[n] Provided in vitamin mixture.

[o] The diet was mixed 1:1 with a 2 percent agar solution.

for growth (Navia and Lopez, 1973) and the other for gestation (Apgar and Everett, 1991a,b). Proteins other than casein—such as soybean, egg white (O'Dell et al., 1989), and lactalbumin (Hsieh and Navia, 1980)—have been used, and the lipid source and amount vary, suggesting that the guinea pig does well on a wide range of lipid intakes (Navia and Hunt, 1976). Fiber sources vary from wood pulp to cellophane. Jeffery and Typpo (1982) developed a purified diet in which crystalline amino acids replaced protein. This diet has been used successfully for growth (Typpo et al., 1990a) and maintenance (Schiller, 1977) of adult guinea pigs. A summary of recommended nutrient allowances for growing guinea pigs is presented in Table 4-1. Requirements for specific nutrients for reproduction and longevity are unknown.

ENERGY

Research has not been conducted with the specific objective of determining the actual energy requirement of the guinea pig; applying energy values of feedstuffs designed for rats or mice to the guinea pig may not be appropriate, as it is likely that the guinea pig can utilize fibrous feedstuffs more efficiently than mice and rats. Hirsh (1973) diluted a diet to 50 percent with finely ground cellulose (a calculated reduction in calories of 40 percent) and reported that the guinea pig did not increase food intake but did, after some initial loss, maintain body weight. Henning and Hird (1970) found that the cecum of the guinea pig contained concentrations of short-chain fatty acids similar to those in the bovine rumen. In fact, the guinea pig is a hindgut fermentor, and its cecum—the primary site of fermentation—has a fermentive capacity similar to that of the colon and rectum of horses (Parra, 1978). The cecal fermentive capacity of the guinea pig is 2.5 times the colonic fermentive capacity of the rat. The ability of the guinea pig to derive energy from fibrous materials must be considered when developing experimental strategies to quantify the energy requirement of this animal.

A variety of commercial diets designed to meet the nutrient requirements of the guinea pig have been used (Argen-

zio et al., 1988; Tsao and Leung, 1988; Berger et al., 1989; Fernandez et al., 1990) for a variety of experimental purposes. A commercial natural-ingredient rabbit diet has also been fed to guinea pigs, apparently with satisfactory results (Johnston, 1989; Johnston and Huang, 1991). A variety of purified diets also have been described. In general, these diets have a cornstarch/sucrose mixture as the primary energy source; protein sources include soybean protein, casein, and egg white solids (Miller et al., 1990; Apgar and Everett, 1991b; Fernandez and McNamara, 1991; Simboli-Campbell and Jones, 1991). Unfortunately, food consumption was not reported; thus, estimation of energy requirements is impossible. However, the fact that the guinea pigs appeared to function in a normal physiological manner indicates that the diets contained the necessary energy density.

The energy densities of the diets discussed above have been estimated in terms of metabolizable energy (ME) using the physiological fuel value system and on an "as is" basis. In estimating the ME values of the diets, fiber was not included. As it is most likely that the guinea pig can convert fiber to useful energy, the estimated ME values are conservative. In all diets considered, fiber made up no more than 16 percent of the total weight of the diet. The commercial diets contained from 2.8 to 3.2 Mcal/kg diet (11.9 to 13.4 MJ/kg diet), not considering the fiber. From the data of Berger et al. (1989) and Argenzio et al. (1988) it can be estimated that the maintenance energy requirement of the 400- to 600-g guinea pig can be met by approximately 136 Kcal $ME/BW_{kg}^{0.75}$ (570 kJ), where $BW_{kg}^{0.75}$ represents metabolic body weight in kilograms. The purified diets contained from 3.1 to 3.5 Mcal ME/kg diet (13.2 to 14.6 MJ ME/kg diet), again not considering the dietary fiber. A crystalline amino acid diet containing 50 g of corn oil and 150 g of fiber per kg of diet contained 3.4 Mcal ME/kg (14.2 MJ ME/kg)(not considering the fiber) and resulted in a growth rate of 7 g/day when fed to 3- to 6-week-old male guinea pigs (Typpo et al., 1990a,b). At present natural-ingredient and purified diets should contain a minimum of 3.0 Mcal ME/kg (12.5 MJ ME/kg) and approximately 15 percent fiber. Further data are necessary to precisely evaluate the ability of the guinea pig to utilize dietary fiber and thus provide more precise energy requirements.

LIPIDS

An optimal concentration of dietary lipid has not been established for the guinea pig. Reid et al. (1964) fed 2- to 5-day-old male guinea pigs (Hartley strain) purified diets containing 0, 10, 30, 75, 150, and 250 g corn oil/kg diet for 6 weeks. They found an increase in weight gained in groups fed 0 and 10 g corn oil/kg diet, a plateau in weight for those fed 10 to 150 g corn oil/kg diet, and a slight decrease in weight of those fed 250 g corn oil/kg diet. The effects of dietary fat concentration on optimal reproduction, lactation, or longevity have not been determined.

ESSENTIAL FATTY ACIDS (EFA)

n-6 Fatty Acids

Reid et al. (1964) conducted several studies to assess the requirement for n-6 fatty acids. In one experiment, they measured the appearance of dermatitis (a sign of EFA deficiency) in young male guinea pigs. These animals were fed diets containing 0, 0.17, 0.33, 0.67, 1.33, or 4.0 g methyl linoleate and 10 g corn oil/kg. The diets were found to actually contain 0.15, 0.20, 0.24, 0.34, 0.54, 1.31, and 1.89 percent, respectively, of the calories as linoleic acid after accounting for the linoleic acid supplied by the cornstarch. Dermatitis occurred until the concentration of linoleic acid in the dietary treatment reached 1.31 percent of calories; thus, the amount of linoleic acid needed to prevent dermatitis falls between 0.54 and 1.31 percent of total calories. Guinea pigs grew normally when the concentration was only 0.24 percent. Thus, a higher concentration of linoleic acid is required to prevent dermatitis than is needed for normal growth.

Reid et al. (1964) also measured the requirement for linoleic acid by measuring the triene:tetraene ratio [20:3(n-9) to 20:4(n-6)] in erythrocytes in response to increasing concentrations of dietary linoleic acid, a method developed previously by Holman (1960) for rats. The requirement for linoleic acid was estimated as the percentage of the total calories of linoleic acid at the breakpoint of this dose-response curve. The actual dietary treatments provided 0.15, 0.46, 0.88, 1.04, 1.25, and 1.42 percent linoleic acid (supplied by safflower oil). Based on this method, the linoleic acid requirement for guinea pigs was determined to be between 0.88 and 1.04 percent of total calories.

n-3 Fatty Acids

As in other mammals, in the guinea pig n-3 fatty acids concentrate in certain tissues, including brain and testes (Tinoco, 1982), but the precise function of these fatty acids is unknown. Unlike most other mammals, however, the guinea pig has been found to have relatively lower concentrations of 22:6(n-3) and higher concentrations of 22:5(n-6) in their photoreceptor outer segment membranes. This may, in part, explain the absence of changes in the electro-retinogram (a measure of retina function) of offspring of guinea pigs fed sunflower oil [high linoleic acid 18:2(n-6), very low α-linolenic acid 18:3(n-3)] treatments for several generations (Leat et al., 1986). Neuringer et al. (1988) speculated that the shift in fatty acid composition of the retina

as a result of n-3 fatty acid-deficient diets may affect retinyl function less than in other species that maintain higher concentrations of retinal 22:6(n-3). Essentiality of n-3 fatty acids in the diet of the guinea pig has not been studied.

Signs of EFA Deficiency   Classic signs of EFA deficiency (lack of n-6 fatty acids) were reported by Reid (1954b) and Reid and Martin (1959) and include ulcers about the neck and ears, loss of hair on the ventral surface, retarded growth, dermatitis, and mortality. They described additional signs of priapism; underdevelopment of the spleen, testes, and gallbladder; and enlargement of the kidneys, liver, adrenals, and heart. Specific skin changes were found to be confined to the surface layers and, therefore, less extensive than those of the EFA-deficient rat. Fat deprivation was not found to increase water consumption, which also is in contrast to signs described in the EFA-deficient rat.

CARBOHYDRATES

Sucrose, glucose, lactose, and starch have been used as primary energy sources in purified diets for guinea pigs (Reid and Briggs, 1953; Heinicke and Elvehjem, 1955; Heinicke et al., 1955). Guinea pigs consuming diets that contain lactose as the sole carbohydrate in the diet grew at rates about one-third that of controls fed sucrose-based diets (Heinicke and Elvehjem, 1955). Few differences were observed between guinea pigs fed sucrose- and dextrin-containing diets (Booth et al., 1949; Heinicke et al., 1955). Consuming sucrose or a mixture of glucose and fructose, when added to natural-ingredient diets at concentrations equal to 20 percent of energy, led to equal growth rates (6.6 g/day; Ahrens et al., 1985).

PROTEIN AND AMINO ACIDS

PROTEIN

The guinea pigs' protein requirement for growth depends on the nitrogen source in the diet (Table 4-4). Woolley and Sprince (1945) observed that the guinea pig had an unusually high protein requirement when casein was the only nitrogen source in a purified diet. Highest weight gains were obtained when the diet contained 300 g casein/kg, but equivalent growth occurred when arginine, cystine, and glycine were added to a diet supplying 180 g casein/kg. These findings have been confirmed (Reid and Briggs, 1953) and extended (Heinicke et al., 1955, 1956; Reid, 1963; Reid and Mickelsen, 1963) to demonstrate that the most limiting amino acid in casein for the growing guinea pig is arginine. As the casein content of the diet is reduced

TABLE 4-4 Protein Requirement for Growth for Various Strains of Guinea Pigs

Strain	Sex	Initial Wt, g	Growth, g/day	Source	Amount, % of diet	Reference
Not identified	Male, female	60–90	3.9[a]	Casein (vitamin-free)	30	Woolley and Sprince, 1945
			3.9[a]		18 + 1 L-Arg, 0.1 Cysscy, 0.2 Gly	
Hartley	Male, female	95–115	6.2	Casein (vitamin-free)	30	Reid and Briggs, 1953
Mixed	Male	180	5.4	Casein (alcohol extracted)[b]	30	Heinicke et al., 1955
			6.7		30 + 0.5 L-Arg-HCl	
			6.2		25 + 0.5 L-Arg-HCl, 0.3 DL-Met	
			6.9		20 + 0.5 L-Arg-HCl, 0.3 DL-Met, 0.1 DL-Trp	
Mixed	Male	206	7.5	Casein (alcohol extracted)[b]	30	Heinicke et al., 1956
			6.4		24 + 0.5 L-Arg-HCl, 0.3 DL-Met	
		571	5.2		24 + 0.3 L-Arg-HCl	
		717	4.5		24	
Hartley	Male	95–115	6.7	Casein (vitamin-free)	35, 40	Reid, 1963
			5.9		30	
			4.2		25	
			3.2		20	
			6.7	Purified soybean protein	35	
			5.9		30	
			3.6		25	
			3.1		20	
Hartley	Male	95–115	5.8	Casein (vitamin-free)	30	Reid and Mickelsen, 1963
			7.3		30 + 0.3 L-Arg-HCl	
			3.4		20	
			6.0		20 + 1.0 L-Arg-HCl	
			5.9		20 + 1.0 L-Arg-HCl, 0.5 DL-Met	
			6.4		20 + 1.0 L-Arg-HCl, 0.5 DL-Met, 0.1 L-Trp	
			5.9	Purified soybean protein	30	
			6.9		30 + 0.5 DL-Met	
			6.3		20 + 1.0 DL-Met	
			8.8	Purina Guinea Pig Chow		
			8.7	Casein	24 + 27.4 alfalfa = 30 protein	
Frant	Male, female	100	6.7	Natural ingredient	20.5 protein (N × 6.25)	Lister and McCance, 1965
Hartley	Male, female	100–130	7.8	Purina Guinea Pig Chow		Alberts et al., 1977
			5.5	Casein (vitamin-free)	30 + 0.3 L-Arg-HCl	
			6.7	Isolated soybean protein	30 + 0.5 L-Arg-HCl DL-Met	
Hartley	Male, female	200	7.4, 6.3	Natural ingredient	18.8 protein (N × 6.25)	Shelton, 1971
Hartley	Male	225	7.0	Amino acids	17.8 protein equivalent (N × 6.25)	Typpo et al., 1990b

[a] Insufficient K and Mg in diet.

[b] Sucrose is the carbohydrate source.

below 300 g/kg, methionine becomes next most limiting followed by tryptophan. Not only does the young guinea pig have a high requirement for arginine, but the arginine in casein is reported to be only about 70 percent available (Heinicke et al., 1955). Supplementing a diet containing 300 g casein/kg (12.6 g arginine/kg) with 3 g L-arginine-HCl/kg resulted in improved growth (Reid and Mickelsen, 1963) that was equivalent to a diet containing 350 g casein/kg (6.5 to 7 g/day; Reid, 1963), although theoretically 300 g casein/kg should meet the arginine requirement. Plant pro-

teins contain generous amounts of arginine, and the herbivorous guinea pig grows well when fed diets that contain 180 to 200 g protein/kg (10.8 g arginine/kg) from plant sources (Lister and McCance, 1965; Shelton, 1971). Soybean protein has been widely used in experimental diets for guinea pigs. Such diets are adequate in arginine but limiting in methionine for maximal growth at concentrations below 300 g soybean protein/kg (Reid and Mickelsen, 1963; Reid, 1966). Supplementation of a diet containing 300 g soybean protein/kg with 5 g DL-methionine/kg re-

TABLE 4-5   Amino Acid Requirements for Growth for Male Hartley Guinea Pigs

Initial Weight, g	Growth, g/day	Amino Acid	Source	Amount, % of diet	Reference
95–115	6.5–7.0	Arginine	30% casein + 0.3% L-Arg-HCl	<1.47	Reid and Mickelsen, 1963
95–115	6.7	Methionine	20% isolated soybean protein	0.36	Reid, 1966
		Cystine	(0.2% Met, 0.12% Cysscy) + 0.1% Trp supplemented with Met and Cysscy	0.35	
95–115	5.3	Tryptophan	Purified soybean protein and crystalline amino acids equivalent to 20% purified soybean protein	0.138	Reid and Von Sallmann, 1960
	5.6		(Required to prevent cataracts and promote growth)	0.2	
221–264	5.5	Arginine	Crystalline amino acids, 22.5% protein equivalent (3.6% N x 6.25)	1.0	Yoon, 1977
200–225	7.0	Histidine	Same as above	0.3	Anderson and Typpo, 1977
226	6.9	Isoleucine	Same as above	0.5	Ayers et al., 1987
235	6.0	Leucine	Same as above	0.9	Mueller, 1978
230	5.6	Lysine	Same as above	0.7	Typpo et al., 1985
135–219	6.5–6.9	Methionine	Same as above	0.3	Typpo et al., 1990b
		Cystine		0.2	
260	5.4	Phenylalanine	Same as above	0.45	Chueh, 1973; Cho, 1971
		Tyrosine		0.45	Chueh, 1973
244–269	6.8	Threonin	Same as above	0.5	Horstkoetter, 1974
225	5.2	Tryptophan	Same as above	0.15	Smith, 1979
225	6.0–6.5	Valine	Same as above	0.7	Typpo et al., 1990b
215–250	5.0	10 indispensable amino acids	Same as above at requirement levels	6.2	Condon, 1980; Blevins, 1983
215–250	6.9	10 indispensable amino acids	Same as above at 20% more than requirement concentrations for each of the 12 indispensable amino acids	7.4	Condon, 1980; Blevins, 1983

sulted in growth close to 7 g/day (Reid and Mickelsen, 1963; Alberts et al., 1977). A diet containing 200 g soybean protein/kg requires supplementation with 10 g DL-methionine/kg to produce equivalent growth (Reid and Mickelson, 1963).

AMINO ACIDS

An early estimate of the requirement for sulfur-containing amino acids of the young guinea pig fed a diet containing 200 g soybean protein/kg supplemented with 1 g tryptophan/kg was found to be 7.1 g/kg diet, with 3.6 g as cystine and 3.5 g as methionine (Reid, 1966). Growth and liver weight were greater with supplementation of 3.75 g L-methionine/kg than with 7.5 g DL-methionine/kg or 3.75 g D-methionine/kg in the diet. Thus, D-methionine does not appear to be as active as L-methionine in the guinea pig (Reid, 1966). Young guinea pigs grew well when a diet containing 190 g heated soybean protein flour/kg diet (Hasdai et al., 1989) provided 5.7 g sulfur-containing amino acids/kg (2.7 g methionine and 3 g cystine/kg). In recent studies in which crystalline amino acid diets were used, the minimum total sulfur amino acid requirement was 5 g/kg with 3 g/kg from L-methionine and 2 g/kg from L-cystine, constituting 40 percent of methionine being

replaced by cystine (Typpo et al., 1990b). When corrected for efficiency of use, 6 g total sulfur amino acids with 3.6 g/kg from L-methionine and 2.4 g/kg from L-cystine will meet the requirement (Table 4-5).

The tryptophan requirement of the growing guinea pig was reported to be between 1.6 and 2.0 g/kg diet (Reid and Von Sallmann, 1960) to promote maximum growth and prevent cataract development. Signs of tryptophan deficiency were produced by feeding guinea pigs a diet containing 100 grams each of soybean protein and gelatin per kilogram of diet supplemented with an amino acid mixture to the approximate amino acid content of 200 g soybean protein/kg and an ample supply of niacin (200 mg/kg diet). The result of this 1.08 g L-tryptophan/kg diet was poor growth, distended abdomens, alopecia, and cataracts (Reid and Von Sallmann, 1960). Adding 0.3 g L-tryptophan/kg to the diet resulted in maximum growth, but the addition of 1 g L-tryptophan/kg was necessary to obtain complete protection from cataracts. Thus, the requirement for tryptophan to prevent eye lesions was considerably greater than the requirement for maximum growth. Using crystalline amino acid diets, the tryptophan requirement for maximum growth, nitrogen retention, and freedom from cataracts was 1.5 g/kg when niacin was between 0.06 and 0.2 g/kg diet (Smith, 1979). When corrected for efficiency of use, 1.8 g/

kg meets the requirement. Reducing the niacin content of the amino acid diet to 0.05 and 0 g/kg increased the tryptophan requirement to 2 and 3 g/kg, respectively (Smith, 1979).

More recently the development of a basal crystalline L-amino acid diet suitable for studies of individual amino acid requirements in young male guinea pigs has been reported (Jeffery and Typpo, 1982; Typpo et al., 1990b). The basal diet contained 36 g nitrogen/kg as the L form of crystalline amino acids, 3.4 Mcal ME/kg (14.2 MJ ME/kg); sucrose and glucose as the carbohydrates, corn oil, fiber in the form of cellophane; and crystalline vitamins and minerals. The original mixture of indispensable amino acids (the usual 10 plus cystine and tyrosine) was based on the amino acid composition of a number of effective diets that contained natural proteins. Several dispensable amino acids (glutamic acid, asparagine, proline, alanine, aspartic acid, glycine, serine, and sodium glutamate) were included in a mixture. The indispensable nitrogen content of the original diet was 17 g/kg, giving an indispensable-to-total nitrogen ratio of 0.47. In a series of 3-week experiments conducted with guinea pigs of the Hartley strain approximately 2 to 3 weeks old (200 to 250 g), weight gain; nitrogen retention; carcass, liver, and gastrointestinal tract weights; and, for some studies, plasma concentration of the amino acid under test were used to determine the minimum requirement for each individual amino acid. As the requirement for each amino acid was established, this quantity was incorporated into the diet used to determine the requirement for the next amino acid tested. Diets were kept isonitrogenous and isoenergetic by altering the quantity of the dispensable amino acid mixture and sugar mixture, respectively.

A dietary lysine content of 7 g/kg (8.75 g/kg L-lysine-HCl) produced maximal growth and nitrogen retention in 3- to 6-week-old male guinea pigs (Typpo et al., 1985). When corrected for efficiency of use, 8.4 g/kg diet will meet the requirement. Incorporation of up to 20 g lysine/kg in the presence of 18.5 g arginine/kg in the basal diet produced no adverse effects. O'Dell and Regan (1963) found that as little as 5 g arginine/kg added to a diet containing 20 g lysine/kg prevented growth retardation resulting from lysine-arginine antagonism.

Although 12 g phenylalanine/kg was required for maximal growth and nitrogen retention when tyrosine was excluded from the diet (Cho, 1971), in the presence of tyrosine the total minimum requirement for these two amino acids was 9 g/kg (Chueh, 1973). Requirements for L-phenylalanine and L-tyrosine were determined to be 4.5 and 4.5 g/kg diet, respectively, with tyrosine replacing up to 50 percent of the total requirement (Chueh, 1973). When corrected for efficiency of use, these values become 5.4 and 5.4 g/kg diet. Raising the phenylalanine content of the diet to 6 g/kg permitted reduction of the tyrosine content

to 3 g/kg, or 33 percent of the total phenylalanine plus tyrosine requirement.

Minimum amounts for maximal growth and nitrogen retention of young guinea pigs have been determined for several other amino acids: threonine, 5 g/kg (Horstkoetter, 1974); histidine, 3 g/kg (Anderson and Typpo, 1977); isoleucine, 5 g/kg (Ayers et al., 1987); leucine, 9 g/kg (Mueller, 1978); and valine, 7 g/kg (Typpo et al., 1990b). These values have been corrected for efficiency of use in Table 4-1. Although antagonisms among the branched chain amino acids have not been investigated, dietary leucine at 14 g/kg produced some growth inhibition when dietary isoleucine and valine concentrations were 6 and 7 g/kg, respectively.

A dietary arginine concentration of 8 g/kg resulted in maximum growth and nitrogen retention, while 9 g/kg was required for minimum orotic acid excretion in urine, and a requirement of 10 g arginine/kg diet was required to maintain plasma arginine (Yoon, 1977). When corrected for efficiency of use, 12 g arginine/kg diet will meet the requirement.

In these individual amino acid studies, the requirement was determined to be the lowest concentration of the test amino acid that supported a performance (weight gain and nitrogen retention) not significantly different from that resulting from the higher concentrations. Using this criterion results in requirement values that lie somewhere between the linear, slope-ratio, or broken-line models (Hegsted and Chang, 1965a,b; Robbins et al., 1979) and the nonlinear models (Finke et al., 1987; Gahl et al., 1991) for selecting requirements (see Chapter 2, Proteins and Amino Acids). In test diets the amino acid under study was first limiting at the lowest concentrations fed, but as the dietary concentration was raised and approached the requirement, while concentrations of the other indispensable amino acids remained constant, other indispensable amino acids could become limiting. This method of estimating the "requirements" of amino acids does not provide an accurate representation of the "diminishing return" area of the response curve and promotes an inaccurate estimate of the requirement (Gahl et al., 1991). Choosing this procedure could conceivably result in underestimating the requirement. When the determined requirement concentrations of the 10 indispensable amino acids were combined in the diet containing 36 g total nitrogen/kg, weight gain was ≈5.0 g/day (Condon, 1980; Blevins, 1983). Increasing the quantity of the 10 indispensable amino acids by 20 percent raised weight gain to near 7 g/day, suggesting that the requirement concentrations were all equally limiting and, in combination, were used with lower efficiency. Therefore, the values presented in Table 4-1 reflect an adjustment of 20 percent for each indispensable amino acid.

Maximum growth and nitrogen retention in the young growing guinea pig were obtained when the crystalline amino acid diet contained 36 g total nitrogen/kg, 11.7 g indispensable amino acid nitrogen/kg, an indispensable-to-total nitrogen ratio of 0.325, and 3.4 Mcal ME/kg (14.2 MJ ME/kg). However, the total nitrogen requirement may be lowered from 36 to 28.6 g/kg by reducing the dispensable amino acid component of the chemically defined diet without reducing growth (Typpo et al., 1990a,b).

Reproduction

Although amino acid requirements for the pregnant/lactating and nonpregnant/nonlactating adult guinea pig have not been specifically determined, natural-ingredient diets that provide 18 to 20 percent protein result in satisfactory reproduction (Lister and McCance, 1965; Shelton, 1971) and maintenance of adults (Shelton, 1971). Apgar and Everett (1991b) obtained adequate and similar weight gains and reproductive performance in pregnant guinea pigs fed either a casein-agar diet containing 300 g casein/kg without added arginine or a commercial plant protein diet containing 185 g protein/kg diet, but fewer neonates from dams fed the casein-agar diet survived. These investigators suggested that the cecal flora as a possible source of essential or unrecognized nutrients may be more critical during pregnancy than during growth. In addition, it has been suggested that the elimination of waste products by the pregnant guinea pig is less difficult when an adequate but not excessive amount of protein is provided. Pregnant guinea pigs often have a fetal mass approaching the non-pregnant weight of the female, possibly making elimination of waste products difficult.

Maintenance

The adult nonpregnant, nonlactating guinea pig's requirement for protein and amino acids may be lower than the amount required for growth. The requirement may be similar to that reported for 770 g males (Schiller, 1977). Nitrogen balance and body weight were maintained when a crystalline amino acid diet containing 11 g indispensable amino acid nitrogen/kg and 25.6 or 18.2 g total nitrogen/kg diet (equivalent to 160 or 114 g crude protein/kg diet) was fed but not when the diet contained 11 g total nitrogen/kg (equivalent to 69 g crude protein/kg).

Signs of Protein and Amino Acid Deficiency   Protein deficiency produced in growing guinea pigs fed a diet containing 30 g casein/kg diet for 3 to 4 weeks causes growth retardation, marked reduction in plasma total protein and albumin, profound alterations in the plasma amino acid profile, and mild fatty liver. These animals develop clinical symptoms similar to the Kwashiorkor syndrome, including reduced activity, mild hair loss, and extensive edema of the face and forelimbs (Enwonwu, 1973). A protein deficiency produced in growing guinea pigs consuming 20 g casein/kg diet was accompanied by marked inhibition of local and systemic immune responses to vaccination with bacillus Calmette-Guerin (Bhuyan and Ramalingaswami, 1973). Enwonwu (1973) suggests that guinea pigs are suitable models for the study of human protein-calorie malnutrition.

Reports of protein deficiency produced by lowering the protein content of diets fed to adult guinea pigs during reproduction or maintenance were not found. Reducing the intakes of both protein and energy by restricting the intake of adequate protein diets (300 g casein or 185 g plant protein/kg) to 20 to 50 percent during the last half of gestation resulted in premature delivery, reduced weight for pups, and death of most pups within the immediate postnatal period (Apgar and Everett, 1991b). Reduced litter size was reported when feed was restricted to 40 percent of normal ad libitum intake beginning at day 30 of gestation (Young and Widdowson, 1975).

MINERALS

MACROMINERALS

Calcium and Phosphorous

The calcium, phosphorus, potassium, and magnesium requirements of the guinea pig, like those of the rat, appear to reflect interactions among these elements. Morris and O'Dell (1961, 1963) and O'Dell et al. (1956, 1960) found that adequate dietary concentrations of calcium (8 to 10 g Ca/kg), phosphorus (4 to 7 g P/kg), magnesium (1 to 3 g Mg/kg), and potassium (5 to 14 g K/kg) varied as the concentrations of the other three elements varied. Van Hellemond et al. (1988) observed that guinea pigs fed purified diets containing 8.4 g Ca/kg with 7.7 g P/kg and 1.0 g Mg/kg retained more calcium than those fed the same concentrations of calcium but with less phosphorus (4.4 g P/kg) and more magnesium (1.9 g Mg/kg). Thus 8 g Ca/kg and 4 g P/kg diet will meet requirements for these minerals.

There is no obvious physiological explanation for the higher apparent requirement of guinea pigs than rats for calcium and phosphorus. Further work is needed to assess growth and bone and kidney accumulation of calcium when guinea pigs are fed concentrations of calcium and phosphorus more consistent with those recommended for rats.

Signs of Calcium and Phosphorus Deficiency   Signs of calcium deficiency have been produced in young guinea pigs fed a purified diet containing 0.28 g Ca/kg, 0.20 g P/kg, and a low concentration of vitamin D (Howe et al., 1940). Nine of 21 animals fed this diet survived for 60 days. These guinea pigs lost weight and developed rachitic lesions in ribs and long bones. Generally the younger animals developed more bone abnormalities than the older animals. The teeth of all animals developed extreme enamel hypoplasia. As there were no control animals in this study, interpretation of these data is difficult.

Prevention of soft tissue calcification caused by imbalances among dietary calcium, phosphorus, potassium, and magnesium is of more concern in the formulation of practical diets for guinea pigs than the prevention of overt deficiencies. Hogan and Regan (1946) implicated excess phosphorus as a cause of soft tissue calcification in guinea pigs. These findings were confirmed when 90 percent of the guinea pigs fed a diet containing 8 g Ca/kg and 9 g P/kg developed soft tissue mineral deposits, whereas the incidence was less than 10 percent when the diet contained only 5 g P/kg (Hogan et al., 1950). In subsequent experiments, supplemental magnesium and potassium prevented the effects of excess dietary phosphorus, including soft tissue calcification, in guinea pigs (House and Hogan, 1955). These observations seem to be consistent with those for rats (see Chapter 2, Nephrocalcinosis in Rats Fed Purified Diets.)

Chloride, Sodium, and Sulfur

No published data could be found on the requirements of chloride, sodium, or sulfur for the guinea pig. The concentrations of chloride and sodium in purified diets used for rats should be used as a first approximation of the dietary concentrations of these nutrients.

Magnesium

The magnesium requirement of guinea pigs depends on the dietary concentrations of calcium, phosphorus, and potassium. Morris and O'Dell (1963) concluded that an excess of calcium or phosphorus independently increased the minimum magnesium requirement and that the effects were additive. As dietary phosphorus increased from 8 g to 17 g/kg diet, the minimum requirement for magnesium increased from 1 g to 4 g/kg diet. Similarly, as dietary calcium increased from 9 g to 25 g/kg diet, the requirement for magnesium increased.

Interactions of magnesium with potassium and fluoride are also important. Magnesium-deficient guinea pigs not only have reduced muscle extracellular and intracellular magnesium concentration (20 percent and 80 percent of control values, respectively) but also have reduced muscle potassium and increased intracellular sodium and water concentrations (Grace and O'Dell, 1970a). Supplementing the diets of magnesium-deficient guinea pigs with potassium stimulated growth, lowered blood phosphorus concentrations, decreased calcium concentrations in muscle,

extended survival times, and decreased mortality. Guinea pigs appeared to use cations rather than ammonia to neutralize and excrete acid in the urine (O'Dell et al., 1956).

Adding 0.1 to 0.4 g fluorine/kg diet to a magnesium-deficient diet (0.4 g/kg diet) significantly improved growth, increased serum magnesium concentrations, reduced incidence of soft tissue calcification, and reduced calcium concentrations in kidney, heart, and liver (Pyke et al., 1967). When magnesium was severely limiting (0.1 g/kg diet), 0.2 g fluorine/kg diet was toxic and caused lameness and swollen feet; but adequate magnesium largely overcame the deleterious effects of excess fluorine in weanling guinea pigs (O'Dell et al., 1973).

On the basis of no new research, the magnesium requirement is 1 to 3 g/kg diet, 1 g/kg diet being the minimum requirement. However, if additional work were to demonstrate that current estimates of the requirements for calcium and phosphorus are high, the magnesium requirements would also need to be reevaluated.

Signs of Magnesium Deficiency   Clinical signs of magnesium deficiency in young guinea pigs include poor growth, hair loss, decreased activity, poor muscular coordination and stiffness of hind limbs, elevated serum phosphorus, and anemia (Maynard et al., 1958; O'Dell et al., 1960; Morris and O'Dell, 1963). Convulsions, which characterize magnesium deficiency in some species, are uncommon in guinea pigs (O'Dell et al., 1960; Grace and O'Dell, 1970a), but one study reported tetany (Thompson et al., 1964). Gross tissue changes at necropsy were enlarged pale kidneys, white foci and streaks in liver, soft tissue calcification, and incisors that were darkened, eroded, and soft (Maynard et al., 1958; O'Dell et al., 1960; Morris and O'Dell, 1961). In addition, Grace and O'Dell (1970b) concluded that magnesium deficiency probably affected appetite and/or membrane transport of nutrients.

Potassium

The potassium requirements of guinea pigs depend on the dietary concentrations of calcium and phosphorus. Mortality was 100 percent within 4 weeks when young guinea pigs were fed a purified diet (30 percent casein) that supplied excess cations but only 1 g K/kg diet. The requirement for maximal growth under these circumstances was 4 to 5 g K/kg diet supplied as potassium acetate (Grace and O'Dell, 1968). Additional dietary potassium, up to 14 g K/kg diet, has been found to be required when diets combined very high concentrations of calcium, phosphorus, and magnesium (O'Dell et al., 1956; Morris and O'Dell, 1963). With moderate dietary concentrations of calcium, phosphorus, and magnesium, the requirement for potassium is 5 g/kg diet and should be considered generous.

Signs of Potassium Deficiency   Luderitz et al. (1971) found that membrane potentials in striated muscle cells from young guinea pigs fed a potassium-deficient diet were higher than in control animals. These effects were accompanied by a significant, and apparently quantitative, increase in Na^+,K^+-ATPase activity in heart muscle cells (Erdmann et al., 1971).

TRACE MINERALS

Copper and Iron

Diets containing 6 mg Cu/kg diet have been reported to be adequate for normal growth and development of the guinea pig (Everson et al., 1967, 1968). If guinea pigs are fed diets containing less than 1 mg Cu/kg during pregnancy and early postnatal development, the offspring are characterized by growth retardation, cardiovascular defects, and severe abnormalities of the central nervous system including agenesis of cerebellar folia, cerebral edema, and delayed myelination (Everson et al., 1968). The dietary requirement for copper is increased if there are high concentrations of molybdenum (Suttle, 1974). Offspring of dams given 0.18 percent $CuSO_4$ in their drinking water from day 21 of gestation on were characterized by high liver copper concentrations and subtle evidence of liver pathology (Chesta et al., 1989).

Dietary iron requirements of the guinea pig have not been directly addressed. Based on an evaluation of previous and current diets used for guinea pigs, it is estimated that a diet containing 50 mg Fe/kg will satisfy the iron requirements for reproduction, growth, and development. Dietary iron at high concentrations (200 to 300 mg/kg) can result in significant tissue iron concentration, although overt tissue pathology has not been reported (Smith and Bidlack, 1980; Caulfield and Rivers, 1990).

Manganese

A dietary concentration of 40 mg Mn/kg diet has been shown to be adequate for normal growth and development of the guinea pig (Everson et al., 1959). Similar to the mouse and rat (Hurley and Keen, 1987), diets containing 3 mg Mn/kg or less are inadequate for the guinea pig during growth and development. Although the manganese dietary requirement for the guinea pig is probably less than 40 mg/kg, in the absence of studies evaluating dietary manganese concentrations between 3 and 40 mg/kg diet, the recommended concentration is 40 mg/kg diet for all stages of life.

Signs of Manganese Deficiency   In a series of studies by Everson (Tsai and Everson, 1967; Everson, 1968; Everson and Shrader, 1968; Everson et al., 1968; Shrader and Ever-

son, 1968) it was established that signs of prenatal and early postnatal manganese deficiency include reduced litter size, abortions or stillbirths, congenital ataxia, skeletal abnormalities, and pancreatic pathology that resulted in a diabetes-like syndrome. The pancreatic pathology and diabetic syndrome can be reversed with manganese supplementation; the ataxia, however, is irreversible (Everson and Shrader, 1968; Shrader and Everson, 1968).

Zinc

Alberts et al. (1977) reported that casein-based diets containing 12 mg Zn/kg and soybean protein-based diets containing 20 mg Zn/kg were adequate to support optimal growth rate without evidence of deficiency signs in the young guinea pig. These values are consistent with the report by Navia and Lopez (1973) that purified gel diets containing 19 mg Zn/kg diet support normal growth and development. The requirement for zinc is 20 mg/kg diet for all stages of life.

Signs of Zinc Deficiency   Guinea pigs fed diets containing less than 1.25 mg Zn/kg are characterized by low plasma zinc concentration, depressed ability to elicit a delayed-type hypersensitivity, low gamma-globulin concentrations, altered glycosaminoglycan metabolism, abnormal posture, skin lesions, anorexia, and excessive vocalization (McBean et al., 1972; Hsieh and Navia, 1980; Quarterman and Humphries, 1983; Gupta et al., 1988; Verma et al., 1988; O'Dell et al., 1989). If a zinc-deficient diet (≤ 2 mg Zn/kg) is given during pregnancy, it can result in premature delivery or abortion (Apgar and Everett, 1991b).

Iodine, Molybdenum, and Selenium

Diets unsupplemented with iodine, selenium, or molybdenum have been shown conclusively to have negative effects on the physiological or biochemical status of mammals. No systematic effort has been made to establish the requirements of iodine, molybdenum, and selenium for the guinea pig. Most of the research to establish requirements has been done with the laboratory rat. (For an indepth discussion, see Chapter 2.)

The dietary requirement for selenium of the rat and mouse has been determined by using the maximization of liver glutathione peroxidase activity (GSH-Px). This procedure may prove difficult in the guinea pig because the activity of this enzyme in the liver of the guinea pig is less than 10 percent of that in other species such as hamsters, rats, and mice. Apparently this phenomenon is not related to the amount of dietary selenium. Even if guinea pigs are fed commercial natural-ingredient diets containing more selenium (140 to 200 μg/kg) than the amount required by rats, GSH-Px activities in various tissues are very low.

Lawrence and Burk (1978) found no GSH-Px activity in livers of guinea pigs fed natural-ingredient diets, while Toyoda et al. (1989) found that liver, kidney, and heart had only 4 to 6 percent of the activity normally observed in similar tissues of mice or rats. In spite of low liver GSH-Px activity, excellent reproductive performance has been observed for many years in experimental and commercial guinea pig colonies fed commercially available natural-ingredient diets containing 140 to 330 μg Se/kg (Boyd O'Dell, University of Missouri, Columbia, MO, and Dennis Renner, Sasco Inc., Lincoln, NE, 1993, personal communications). These diets are reported to contain 400 to 1,000 μg I/kg but no added molybdenum. Although liver GSH-Px activity is lower in tissues of guinea pigs than in those of other species, liver selenium concentrations are comparable (Toyoda et al., 1989).

Although no work has been done to directly establish the iodine, molybdenum, or selenium requirements for the guinea pig, some evidence indicates that these requirements might be similar to those of rats and mice. Therefore, until more research is conducted, concentrations established for the rat can be used as a first estimate for requirement in the guinea pig. By no means, however, should this be construed to indicate that guinea pigs metabolize these elements exactly as do rats. The requirement for iodine at all stages of life is 150 μg/kg diet and that for molybdenum is 150 μg/kg diet. The requirement for selenium, as selenite, for all stages of life is 150 μg/kg diet with the exception of pregnancy and lactation, for which a dietary concentration of 400 μg Se/kg diet is suggested.

VITAMINS

FAT-SOLUBLE VITAMINS

Vitamin A

Guinea pigs apparently have a high vitamin A requirement. Bentley and Morgan (1945) reported that 21 μmol vitamin A/kg diet maintained growth and a very modest store of vitamin A in the liver. Gil et al. (1968) found that 11.5 μmol/kg diet would maintain growth, but that normal tissue histology, and storage of vitamin A in the liver were only found with diets containing 23 μmol/kg diet or more. Intermediate concentrations were not tested. Although they did not measure liver vitamin A reserves, Howell et al. (1967) found that a dose of vitamin A equivalent to 18 μmol/ kg diet was adequate to maintain vision, reproduction, and growth for 460 days. Reid and Briggs (1953) found that 18 μmol/kg diet was satisfactory for optimal growth in guinea pigs fed a purified diet.

β-Carotene is used by the guinea pig as a source of vitamin A (Chevallier and Choron, 1935, 1936; Woytkiw and Esselbaugh, 1951); however, the molar efficiency of utilization may be only 40 percent that of preformed vitamin A (Bentley and Morgan, 1945) when consumed at amounts near the requirement. The reason for this low efficiency of utilization has not been identified. In rats, an intake at this concentration would have a similar efficiency. The lower efficiency of β-carotene utilization at higher intakes is mainly caused by poor absorption from the intestine. β-Carotene is a pure hydrocarbon that is very difficult to solubilize at the higher concentrations.

Diets that contain 21,960 IU retinol/kg diet (23 μmol or 6.6 mg/kg diet) appear to maintain optimal health and a slightly positive vitamin A balance in guinea pigs. If β-carotene is used as the source of vitamin A activity, then 47,425 IU β-carotene/kg diet (53 μmol or 28 mg/kg) would be needed to maintain a slightly positive vitamin A balance.

Signs of Vitamin A Deficiency   Time before onset of deficiency signs varies widely with age, liver vitamin A concentrations, and stress conditions. Young guinea pigs may develop deficiency signs in 2 weeks, whereas older pigs may require nearly 10 weeks when fed a diet devoid of vitamin A or provitamin A. The first evidence of vitamin A deficiency is poor growth, then weight loss, followed by incrustations of eyelids and severe dermatitis resulting from bacterial infection (Bentley and Morgan, 1945). Gross pathology studies often reveal accumulation of organic debris in the bile ducts and gallbladder, clouding of the cornea, and xerophthalmia. Often animals develop pneumonia prior to death. Histologically, epithelia of various organs showed squamous metaplasia and some keratinization (Howell et al., 1967).

The primary effect of a vitamin A deficiency on the incisors of guinea pigs is mainly on odontogenic epithelium with incomplete differentiation of cells, loss of organization, and formation of defective dentin by atrophic odontoblast (Wolbach, 1954). The incisors had a distinctive appearance characterized by thickened dentin on the labial side and thin dentin on the lingual and lateral sides.

Signs of Vitamin A Toxicity   Excessive amounts of vitamin A given to guinea pigs caused degenerative changes in the cartilaginous epiphyseal plates of long bones (Wolbach, 1947), and there was increased bone resorption interfering with normal remodeling. Gil et al. (1968) reported "loss of weight" in guinea pigs fed diets containing 121 mg/kg diet (230 μmol/kg) or more of retinyl palmitate. Teratogenic effects were noted by Robens (1970) when a single oral dose (210 μmol/kg BW) given to pregnant guinea pigs during fetal organogenesis (days 14 to 20) caused soft tissue and skeletal anomalies in the offspring. The most frequent defects recorded were agnathia, synotia, malpositioning of teeth, and microstomia. Administration of the same dose between days 17 and 20 frequently produced changes in the tibias and fibulas, but fetal growth was not affected.

Vitamin D

A quantitative requirement for vitamin D has not been established for guinea pigs, but currently used natural-ingredient and purified diets contain between 20 and 180 nmol/kg diet (Reid and Briggs, 1953; Navia and Lopez, 1973; O'Dell et al., 1989). These amounts seem to promote growth at rates that were average for the colony. The requirement for growth is set at 1,000 IU vitamin D/kg diet (65 nmol/kg diet; 0.025 mg/kg diet).

Signs of Vitamin D Deficiency   Guinea pigs fed diets with a normal calcium-to-phosphorus ratio do not develop gross signs of vitamin D deficiency (Kodicek and Murray, 1943). However, Sergeev et al. (1990) observed many changes in vitamin D status of guinea pigs fed a vitamin D-deficient diet containing 6 g calcium/kg diet and 6 g phosphorus/kg diet. Serum calcium and phosphorus concentrations were reduced, serum alkaline phosphatase was increased, serum 25-hydroxycholecalciferol concentrations were extremely low, kidney 1-α-hydroxylase activity was more than twice the normal concentrations, active transport of calcium in the duodenum was decreased, and bone calcium content was about four-fifths the control concentrations. Administering 5.2 nmol (15 IU) of cholecalciferol per animal every other day prevented the development of these signs. This is about twice the amount available from typical diets. Lower amounts were not used. A deficiency of ascorbic acid also altered the animal's ability to metabolize vitamin D. Howe et al. (1940) housed guinea pigs in a darkened room and fed them a low-vitamin D purified diet with 0.28 g Ca/kg and 2 g P/kg. In addition to retarded growth, typical lesions occurred in the zone of cartilage proliferation at the epiphyseal plate of long bones and ribs. Also, incisors exhibited a high degree of enamel hypoplasia, and enamel and dentin were disorganized and irregular with poor calcification.

Signs of Vitamin D Toxicity   Guinea pigs show a response to excessive vitamin D intake. An extract of the poisonous plant *Solanum malacoxylon* caused hypercalcemia and calcification of kidney, aorta, muscles, spleen, heart, and liver (Camberos et al., 1970). This plant has been shown to contain derivatives of the active metabolites of vitamin D, which are responsible for the toxicity (Boland et al., 1987).

Vitamin E

No precise quantitative requirement for vitamin E can be given, in spite of studies with guinea pigs involving vitamin E and related nutrients. The earliest estimate of a minimum requirement for the growing guinea pig, eating a vitamin E-deficient diet containing 20 g cod liver oil/kg, is equivalent to 2.6 μmol *RRR*-α-tocopherol/day (Shimotori et al., 1940). Assuming a feed intake of 20 g/day, this is equivalent to 128 μmol/kg diet (82 IU or 55 mg *RRR*-α-tocopherol/kg diet. (See Chapter 2 for a discussion of the forms and potency of vitamin E.) Farmer et al. (1950) reported that 5.1 μmol/day was required for normal reproduction in the female guinea pig receiving cod liver oil. Assuming a feed intake of 30 g/day, this is equivalent to 170 μmol/kg diet (74 mg/kg). Several popular diets have been developed for the guinea pig that do not contain cod liver oil. Reid and Briggs (1953) originally developed their diet to contain a vitamin E activity equivalent to 34 μmol *RRR*-α-tocopherol/kg diet, but later Reid (1963) modified the diet to contain 82 μmol/kg diet. The diet described by Hsieh and Navia (1980) contained a vitamin E activity equivalent to 62 μmol *RRR*-α-tocopherol per kg diet. Based on the above, a diet containing 40 IU/kg diet (62 μmol or 26.7 mg/kg diet) should meet the needs of growing guinea pigs. There have been no reports of vitamin E deficiency with these diets.

Signs of Vitamin E Deficiency   Diet-induced muscular dystrophy was produced in the guinea pig when 5 to 20 g cod-liver oil/kg was included in the diet. The research of Shimotori et al. (1940) related vitamin E deficiency to muscular dystrophy. An average oral dose of 1.5 mg of synthetic α-tocopherol per day provided protection against sign of muscular dystrophy during a 200-day period. Pappenheimer and Goettsch (1941) confirmed the role of vitamin E for maintenance of normal muscle and extended the observations to show the need for vitamin E during pregnancy. Schottelius et al. (1959) reported that vitamin E deficiency in guinea pigs precipitated a decrease in muscle myoglobin concentration. Reduced myoglobin concentration was observed before the appearance of severe tissue lesions or increased creatine excretion. Supplementation with vitamin E reduced the magnitude of the myoglobin change. Elmadfa and Feldheim (1971) have shown that creatine phosphokinase activity in skeletal muscle of young male guinea pigs was reduced significantly by feeding them a vitamin E-deficient diet for 2 weeks. The serum creatine phosphokinase activity increased during the same period. At about 4 weeks, creatine excretion in the urine increased, and by 6 weeks erythrocyte hemolysis increased, reaching a maximum at 8 weeks. Soon thereafter, the guinea pigs became prostrate with severe body weight loss and degeneration of skeletal muscle. In males, testes atrophied and developed degenerative changes in the seminiferous tubules, with clumping or complete disappearance of spermatozoa and spermatids. Fetal malformations, resorption, and death occurred in pregnant females.

Vitamin K

The information to develop a specific recommendation for the vitamin K requirement of guinea pigs does not exist. The menadione content of some of the more frequently used diets ranges from 12 to 58 μmol/kg diet (Reid and Briggs, 1953; Navia and Lopez, 1973; O'Dell et al., 1989). These diets appear to be adequate to prevent hemorrhages, but no information is available about more sensitive indicators of vitamin K status. Based on limited information, a concentration of 5 mg phylloquinone/kg diet (11 μmol phylloquinone/kg diet) is suggested.

Signs of Vitamin K Deficiency   The drug Warfarin prevents the normal recycling of vitamin K and thereby rapidly causes a nonfunctional form of vitamin K to accumulate in the tissues. In guinea pigs treated with Warfarin, prothrombin concentrations dropped to 14 percent of control concentrations within 24 hours (Carlisle et al., 1975); and a greatly reduced amount of γ-carboxyglutamic acid in plasma proteins was noted (Stenflo and Fernlund, 1984). Thus, Warfarin and vitamin K appear to interact in guinea pigs much the same as they do in rats.

WATER-SOLUBLE VITAMINS

Ascorbic Acid

The ascorbic acid requirement of the guinea pig has been reviewed by Mannering (1949). Navia and Hunt (1976) summarized the major metabolic roles for ascorbic acid in this animal. The daily requirement of ascorbic acid varied from 0.4 to 25 mg/day according to the criterion used to evaluate adequacy. Values reported to support growth were 0.4 to 2 mg/day in 250 to 350 g guinea pigs; reproduction was supported by 2 to 5 mg/day (Mannering, 1949). Scurvy was prevented by 1.3 to 2.5 mg/day; odontoblast growth, wound healing, and bone regeneration were supported by 2 mg/day; and tissue saturation occurred at 25 to 30 mg/day. Approximately 7 mg of ascorbic acid/kg BW was adequate to maintain adrenal size and odontoblast height in male guinea pigs ranging from 110 to 840 g BW (Pfander and Mitchell, 1952). Collins and Elvehjem (1958) found 5 mg/kg BW sufficient for growth of immature guinea pigs.

The liver, kidney, plasma, lens, and aqueous humor concentration of ascorbic acid reflected the amount of ascorbic acid fed (0.8 to 60 mg/animal/day) to both young and old guinea pigs. The half-life of tissue ascorbate was reported to be the same in guinea pigs provided a maintenance dose (0.5 mg/day) or a higher dose (30 mg/100 g BW/day) (Ginter et al., 1982). A concentration of 50 mg/kg diet was marginal for survival. At 200 mg/kg diet, hepatic ascorbic acid accumulated in the liver. Adrenal and splenic concentrations increased with concentrations of 900 mg/kg diet (Degkwitz

and Boedeker, 1989). Higher tissue concentrations were achieved by mixing the vitamin in the diet (500 mg/kg) than by daily oral administration (Ginter et al., 1979). An intake of 5 mg/day provides adequate amounts of ascorbic acid for growth and reproduction; normal intake of 200 mg/kg diet (1,135 μmol/kg diet) will fulfill this need.

Ascorbic acid absorption was reported to be sodium dependent and involve a transporter molecule (Siliparandi et al., 1979). Dehydroascorbic acid was readily absorbed and reduced by dehydroascorbic acid reductase to ascorbic acid in the intestinal mucosa. Transport of endogenous ascorbate across the basolateral membrane assures the mucosal cell access to the vitamin in the absence of a dietary supply (Rose et al., 1988).

The stability of ascorbic acid in diets varies with the composition of the diet, storage temperature, and humidity. Approximately one-half of the initial ascorbic acid may be oxidized and lost 90 days after the diet has been mixed. Aqueous solutions may lose vitamin C potency rapidly.

Ascorbic acid at 0.5 mg/kg BW prevented rapid fatal scurvy and 55 percent of guinea pigs survived, but after 16 weeks these animals exhibited a marked increase in serum cholesterol, LDL-cholesterol, VLDL-cholesterol, triglycerides, and total lipids. The LDL:HDL ratio rose from 1.13 to 19.02 with cholesterol added at 3 g/kg diet (Kothari and Sharma, 1988) and stimulated the oxidation of ascorbic acid to carbon dioxide (Ginter and Zloch, 1972). Magnesium L-ascorbic acid phosphate has been shown to cure scurvy in the guinea pig (Machlin et al., 1979), but L-ascorbic acid 2-sulfate did not have antiscorbutic activity (Tsujimura, 1978).

Signs of Ascorbic Acid Deficiency   Early signs of vitamin C deficiency in guinea pigs were reduced diet intake and weight loss, followed by anemia and widespread hemorrhages. An impaired clotting mechanism, as indicated by increased prothrombin time, also contributes to hemorrhaging with vitamin C deficiency. Resting body temperatures were higher in scorbutic guinea pigs than normal animals (Green et al., 1980). Deficiency resulted in muscle damage but not neuropathy (Sillevis Smitt et al., 1991). Ascorbic acid-deficient animals were dead within 3 to 4 weeks from the causes noted above or from secondary bacterial infections, to which guinea pigs are susceptible. Ascorbate deficiency has been linked to an impairment in carnitine synthesis, increased urinary carnitine excretion, and prolonged survival time with carnitine supplementation (Jones and Hughes, 1982; Alkonyi et al., 1990).

Vitamin C deficiency resulted in decreased vitamin B_{12} absorption; increased absorption of alanine and leucine; enhanced activity of brush border sucrase, alkaline phosphatase, and leucine aminopeptidase; and higher concentrations of intestinal membrane sialic acid and total lipids (Dulloo et al., 1982). Deficiency also reduced oxalate absorption (Farooqui et al., 1983). Growth and maintenance of connective tissue in skin, fetal tissues, and repairing wounds required a supply of dietary ascorbic acid (Barnes et al., 1969a,b, 1970; Rivers et al., 1970). The characteristic hemorrhages in subcutaneous tissues, joints, skeletal muscle, and intestine of scorbutic guinea pigs result in defects in connective tissue. Ascorbic acid is essential in hydroxylase reactions for the formation of hydroxyproline and hydroxylysine in the collagen molecule (Stone and Meister, 1962; Udenfriend, 1966). General weight loss in scorbutic guinea pigs resulted in decreased proteoglycan and collagen synthesis (Chojkier et al., 1983; Spanheimer and Peterkofsky, 1985; Bird et al., 1986a,b). Impaired synthesis of collagen had many effects on the guinea pig, including enlarged costochondral junctions, disturbed epiphyseal growth centers of long bones, bone loss, altered dentin, and gingivitis. Deficiency in ascorbic acid reduced cytochrome P-450 more than 50 percent (Rikans et al., 1977). Urinary excretion of fluoride was greater in guinea pigs on low ascorbic acid and protein diets (Parker et al., 1979).

Vitamin C deficiency has been reported to result in elevated serum copper and ceruloplasmin and liver copper concentrations (Milne and Omaye, 1980). Ascorbate deficiency has been reported to reduce thymus size, lower the delayed type hypersensitivity response, and active and total rosette-forming cells against sheep red blood cells (Majumder and Rahim, 1987). During the progression of a deficiency, the percentage of B lymphocytes increased and T lymphocytes decreased (Fraser et al., 1980). Leukocyte chemotaxis was impaired in guinea pigs fed 0.5 mg/kg BW compared to 20 mg/kg BW (Johnston and Huang, 1991). Antibody response to an injected antigen occurred more rapidly and was more pronounced in deficiency (Prinz et al., 1980). Splenic cell cyclic GMP and erythrocyte ATPase were depressed by a deficiency of vitamin C (Barkagan and Gelashvili, 1970; Haddox et al., 1979).

Biotin

No quantitative requirement for biotin has been demonstrated for normal healthy guinea pigs. Reid (1954b) observed no significant change in growth of young guinea pigs fed a purified diet with biotin omitted. Feeding guinea pigs a biotin-deficient diet containing raw egg white produced weight loss, alopecia, and depigmentation of the hair (Coots et al., 1959). Based on limited information a concentration of 0.2 mg biotin/kg diet (0.82 μmol/kg diet) is suggested as the requirement for all stages of the life cycle.

Choline

The inclusion of 7.2 mmol choline as 1.8 g choline bitartrate/kg diet supported acceptable growth of young guinea pigs fed a diet containing 30 percent casein (Reid, 1955).

Although methionine does seem to have some sparing effect on the choline requirement, methionine could not be used to replace choline.

Signs of Choline Deficiency   Choline deficiency has been characterized in young guinea pigs (Reid, 1954b, 1955). When 2- to 4-week-old guinea pigs were fed a 30 percent casein diet lacking added choline, but adequate in folic acid and vitamin B_{12} (23 and 0.03 μmol/kg diet, respectively), poor growth, anemia, and muscle weakness were observed. Some adrenal and subcutaneous hemorrhages occurred, but no renal hemorrhage or marked fatty infiltration of liver were reported.

Folates

The young guinea pig appears to have a high requirement for folic acid, on the order of 3 to 6 mg/kg diet (6.8 to 13.6 μmol/kg diet) (Mannering, 1949; Woodruff et al., 1953; Reid 1954a; Reid et al., 1956). Guinea pigs practice coprophagy and may obtain folic acid from bacterial synthesis in the gastrointestinal tract. As the animal matures, less folic acid is required.

Signs of Folate Deficiency   Young guinea pigs fed a folic acid-deficient diet grew slowly initially and became weaker as diet intake declined. Anemia and leukopenia developed. Hemoglobin and hematocrit values decreased, and the bone marrow became aplastic. Fatty livers and adrenal hemorrhages were prominent at necropsy (Woodruff et al., 1953; Reid, 1954a; Reid et al., 1956).

Niacin

Guinea pigs require a dietary intake of niacin (Reid, 1954b). However, because they can produce niacin from tryptophan, the niacin requirement is influenced by quantity and quality of dietary protein, especially tryptophan content and availability. According to Reid (1961), 10 mg niacin/kg diet (81 μmol/kg diet) was adequate in a purified diet containing 30 percent casein or 20 percent casein supplemented with 1 percent L-arginine and 0.25 percent DL-methionine.

Signs of Niacin Deficiency   The most definitive reports on niacin deficiency in the guinea pig are those of Reid (1954b, 1961). When niacin was omitted from a purified diet containing 30 percent casein, deficiency signs were observed in 3 to 4 weeks. All niacin-deficient animals exhibited poor growth; small appetite; pale feet, nose, and ears; drooling; anemia; and a tendency to diarrhea. The animals also had lowered hemoglobin and hematocrit. No oral or ocular lesions and no dermatitis were observed.

Pantothenic Acid

Reid and Briggs (1954) reported that 20 mg or 42 μmol calcium pantothenate/kg diet was adequate for optimal growth. They did not indicate whether calcium *d*-pantothenate or calcium *dl*-pantothenate was used. The adult requirement has not been established. It is projected to be similar to that of young animals, as nonpregnant or pregnant adults can be depleted rather rapidly (Hurley et al., 1965).

Signs of Panthothenic Acid Deficiency   Young guinea pigs fed a purified, pantothenic acid-deficient diet developed signs of deficiency such as decreased growth rate, anorexia, weight loss, rough coat, diarrhea, weakness, and death (Reid and Briggs, 1954). Hair pigmentation was unaffected, and the adrenals were enlarged and sometimes hyperemic or hemorrhagic. Adult animals fed a pantothenic acid-deficient diet died within 10 to 41 days (Hurley et al., 1965). Many of them had adrenal and gastrointestinal hemorrhages.

Pyridoxine

Based on weight gain and general appearance of the growing guinea pig, the quantitative requirement of pyridoxine is 2 to 3 mg/kg diet (9.7 to 14.6 μmol/kg diet) (Reid, 1964).

Signs of Pyridoxine Deficiency   When fed a purified diet containing 30 percent casein with no pyridoxine added, 15 of 27 animals lived for 8 weeks (Reid, 1964). These animals grew slowly, but showed no specific signs of deficiency. Some pyridoxine may have been present in the casein used in the diet.

Riboflavin

The limited research by Slanetz (1943), Reid (1954b), and Hara (1960) is inadequate to establish a riboflavin requirement for the guinea pig. The best estimate is 3 mg riboflavin/kg diet (8 μmol riboflavin/kg diet) (Slanetz, 1943).

Signs of Riboflavin Deficiency   By feeding young guinea pigs a purified diet deficient in riboflavin, Reid (1954b) found that they exhibited poor growth; rough hair; pale feet, nose and ears; and early death (2 weeks). Later, Hara (1960) described microscopic lesions, such as corneal vas-

cularization, skin atrophy and chromatolysis, and myelin degeneration in the pons and spinal cord. Myocardial alterations included hemorrhage and edema accompanied by vacuolar degeneration and atrophy.

Thiamin

The thiamin requirement of the young guinea pig is 2 mg thiamin-HCl/kg diet (5.9 μmol/kg diet) (Liu et al., 1967; Reid and Bieri, 1967). No reports are available to support a definite quantitative requirement for gestation and lactation.

Signs of Thiamin Deficiency   Young growing guinea pigs fed a thiamin-deficient diet exhibited reduced food intake and weight loss, followed by the development of central nervous system disorders. An unsteady gait and some retraction of the head also occurred as the condition progressed. Death occurred within 4 weeks (Liu et al., 1967; Reid, 1954b; Reid and Bieri, 1967).

POTENTIALLY BENEFICIAL DIETARY CONSTITUENTS

FIBER

It has long been recognized that fiber is an important ingredient in the diet of the guinea pig. Booth et al. (1949) observed low growth rates (1.9 g/day) for guinea pigs fed synthetic diets containing no fiber; additions of pectin, agar, oat straw, cellulose, and cellophane stimulated growth to some extent, but gum arabic was found to produce the best response (growth rates more than 5 g/day). Other researchers observed that cellulose was more effective in stimulating growth than either gum arabic or cellophane when added at 150 g/kg diet (Heinicke and Elvehjem, 1955). The cecum of the guinea pig contains short-chain fatty acids in concentrations comparable to those found in the rumen (Henning and Hird, 1970), and digestion of cellulose in this organ may contribute to meeting energy requirements. Hirsh (1973) showed that dilution of the diet 1:1 with cellulose did not alter food intake or body weight of guinea pigs, supporting the use of cellulose as an energy source (see "Energy" section).

TRACE MINERALS

It has been suggested that many of the minor elements, which may be supplied in minute amounts, are essential for laboratory animals, including the guinea pig. These elements include cobalt, chromium, arsenic, boron, nickel, vanadium, silicon, tin, fluorine, lead, and cadmium. Cobalt is essential but only as a part of vitamin B_{12}. For additional comments see the discussion on this subject in Chapter 2.

VITAMINS

Myo-Inositol

There is no evidence that the guinea pig requires a dietary source of inositol. Reid (1954b) did not observe significant growth retardation when inositol was omitted from a purified diet.

Vitamin B_{12}

There is no evidence that the growing guinea pig requires a dietary source of vitamin B_{12} (Reid, 1954b). Guinea pigs may ingest a significant amount of this vitamin during coprophagy.

REFERENCES

Ahrens, R. A., S. L. Garland, H. N. Kigutha, and E. Russek. 1985. The disaccharide effect of sucrose feeding on glucuronide excretion and bile concentration of injected phenolphthalein in guinea pigs. J. Nutr. 115:288–291.

Alberts, J. C., J. A. Lang, P. A. Reyes, and G. M. Briggs. 1977. Zinc requirement of the young guinea pig. J. Nutr. 107:1517–1527.

Alkonyi, I., J. Cseko, and A. Sandor. 1990. Role of the liver in carnitine metabolism: The mechanism of development of carnitine-deficient status in guinea-pigs. J. Clin. Chem. Clin. Biochem. 28:319–321.

Anderson, H. A., and J. T. Typpo. 1977. Histidine requirement of the growing guinea pig. Fed. Proc. 36:1153 (abstr.).

Apgar, J., and G. A. Everett. 1991a. The guinea pig as a model for effects of maternal nutrition on pregnancy outcome. Nutr. Res. 11:929–939.

Apgar, J., and G. A. Everett. 1991b. Low zinc intake affects maintenance of pregnancy in guinea pigs. J. Nutr. 121:192–200.

Argenzio, R. A., J. A. Liacos, and M. J. Allison. 1988. Intestinal oxalate-degrading bacteria reduce oxalate absorption and toxicity in guinea pigs. J. Nutr. 118:787–792.

Ayers, L. S., J. T. Typpo, and G. F. Krause. 1987. Isoleucine requirement of young growing male guinea pigs. J. Nutr. 117:1098–1101.

Barkagan, T. S., and S. S. Gelashvili. 1970. Guinea pig erythrocyte ATPase activity and nutritional vitamin C factor. Uch. Zap. Gor's Gos. Univ. 111:75–78.

Barnes, M. J., B. J. Constable, and E. Kodicek. 1969a. Excretion of hydroxyproline and other amino acids in scorbutic guinea pigs. Biochim. Biophys. Acta 184:358–365.

Barnes, M. J., B. J. Constable, and E. Kodicek. 1969b. Studies in vivo on the biosynthesis of collagen and elastin in ascorbic acid-deficient guinea pigs. Biochem. J. 113:387–397.

Barnes, J. J., B. J. Constable, L. F. Morton, and E. Kocicek. 1970. Studies in vivo on the biosynthesis of collagen and elastin in ascorbic acid-deficient guinea pigs: Evidence for the formation and degradation of a partially hydroxylated collagen. Biochem. J. 119:575–585.

Bentley, L. S., and A. F. Morgan. 1945. Vitamin A and carotene in the nutrition of the guinea pig. J. Nutr. 30:159–168.

Berger, J., D. Shepard, F. Morrow, and A. Taylor. 1989. Relationship between dietary intake and tissue levels of reducing and total vitamin C in the nonscorbutic guinea pig. J. Nutr. 119:734–740.

Bhuyan, U. N., and V. Ramalingaswami. 1973. Immune responses of the protein-deficient guinea pig to BCG vaccination. Am. J. Pathol. 72:489–500.

Bird, T. A., R. G. Spanheimer, and B. Peterkofsky. 1986a. Coordinate regulation of collagen and proteoglycan synthesis in costal cartilage of scorbutic and acutely fasted, vitamin C-supplemented guinea pigs. Arch. Biochem. Biophys. 246:42–51.

Bird, T. A., N. B. Schwartz, and B. Peterkofsky. 1986b. Mechanism for the decreased biosynthesis of cartilage proteoglycan in the scorbutic guinea pig. J. Biol. Chem. 261:11166–11172.

Blevins, B. G. 1983. Amino acid requirements of guinea pigs. XII. The indispensable amino acid component at levels of total nitrogen near or above the requirement. M.S. thesis. University of Missouri, Columbia, Mo.

Boland, R. L., M. I. Skliar, and A. W. Norman. 1987. Isolation of Vitamin D_3 metabolites from *Solanum malacoxylon* leaf extracts incubated with ruminal fluid. Toxicol. Lett. 53:161–164.

Booth, A. N., C. A. Elvehjem, and E. B. Hart. 1949. The importance of bulk in the nutrition of the guinea pig. J. Nutr. 37:263–274.

Breazile, J. E., and E. M. Brown. 1976. Anatomy. Pp. 53–62 in Biology of the Guinea Pig, J. E. Wagner, and P. J. Manning, eds. New York: Academic Press.

Camberos, H. R., G. K. Davis, M. I. Djafar, and C. F. Simpson. 1970. Soft tissue calcification in guinea pigs fed the poisonous plant *Solanum malacoxylon*. Am. J. Vet. Res. 31:685–696.

Carlisle, T. L., D. V. Shah, R. Schelegel, and J. W. Suttie. 1975. Plasma abnormal prothrombin and microsomal prothrombin precursor in various species. Proc. Soc. Exp. Biol. Med. 148:140–144.

Caulfield, J. E., and J. M. Rivers. 1990. Effect of increasing storage iron on ascorbic acid metabolism in the guinea pig. Am. J. Clin. Nutr. 52:529–533.

Chesta, J., S. K. S. Srai, A. K. Burroughs, P. J. Scheuer, and O. Epstein. 1989. Copper overload in the developing guinea pig liver: A historical, histochemical and biochemical study. Liver 9:198–204.

Chevallier, A., and Y. Choron. 1935. Sur la teneur du foie en vitamin A et ses variations. C. R. Soc. Biol. 120:1223–1225.

Chevallier, A., and Y. Choron. 1936. Accumulation of vitamin A reserves in the guinea pig. C. R. Soc. Biol. 121:1015–1016.

Cho, E. S. 1971. Amino acid requirements of guinea pigs. III. The phenylalanine requirement. M.S. thesis. University of Missouri, Columbia, Mo.

Chojkier, M., R. Spanheimer, and B. Peterkofsky. 1983. Specifically decreased collagen biosynthesis in scurvy dissociated from an effect on proline hydroxylation and correlated with body weight loss: In vitro studies in guinea pig calvarial bones. J. Clin. Invest. 72:826–835.

Chueh, L. M. 1973. Amino acid requirements of guinea pigs. IV. The phenylalanine and tyrosine requirements. M.S. thesis. University of Missouri, Columbia, Mo.

Collins, M., and C. A. Elvehjem. 1958. Ascorbic acid requirements of the guinea pig, using growth and tissue ascorbic acid concentrations as criteria. J. Nutr. 64:503–511.

Condon, A. E. 1980. Amino acid requirements of guinea pigs. XII. The total essential amino acid requirement. M.S. thesis. University of Missouri, Columbia, Mo.

Coots, M. C., A. E. Harper, and C. A. Elvehjem. 1959. Production of biotin deficiency in the guinea pig. J. Nutr. 67:525–530.

Degkwitz, E., and R. H. Boedeker. 1989. Indications for adaptation to differently high vitamin C supplies in guinea pigs. 1. Development of ascorbic acid levels after altered dosing. Zeit. Ernaehrung. 28:327–337.

Dulloo, R. M., S. Majumdar, R. N. Chakravarti, and A. Mahmood. 1982. Intestinal brush border membrane structure and function effect of chronic vitamin C deficiency in guinea-pigs. Biochem. Med. 27:325–333.

Dunkin, G. W., P. Hartley, E. Lewis-Faning, and W. T. Russell. 1930. Comparative biometric study of albino and coloured guinea-pigs from the point of view of their stability for experimental use. J. Hyg. 30:311–319.

Ediger, R. D. 1976. Care and management. Pp. 5–12 in Biology of the Guinea Pig, J. E. Wagner and P. J. Manning, eds. New York: Academic Press.

Elmadfa, I., and W. Feldheim. 1971. Enzyme activity, metabolites and clinically demonstrable changes in guinea pigs in tocopherol deficiency. Int. J. Vitam. Nutr. Res. 41:490–503.

Enwonwu, C. O. 1973. Experimental protein-calorie malnutrition in the guinea pig and evaluation of the role of ascorbic acid status. Lab. Invest. 29:17–26.

Erdmann, E., H. D. Bolte, and B. Ludentz. 1971. The Na^+,K^+-ATPase activity of guinea pig heart muscle in potassium deficiency. Arch. Biochem. Biophys. 145:121–125.

Everson, G. J. 1968. Preliminary study of carbohydrates in the urine of manganese-deficient guinea pigs at birth. J. Nutr. 96:283–288.

Everson, G. J., and R. E. Shrader. 1968. Abnormal glucose tolerance in manganese-deficient guinea pigs. J. Nutr. 94:89–94.

Everson, G. J., L. S. Hurley, and J. F. Geiger. 1959. Manganese deficiency in the guinea pig. J. Nutr. 68:49–56.

Everson, G. J., H. C. Tsai, and T. Wang. 1967. Copper deficiency in the guinea pig. J. Nutr. 93:533–540.

Everson, G. J., R. E. Shrader, and T. Wang. 1968. Chemical and morphological changes in the brains of copper-deficient guinea pigs. J. Nutr. 96:115–125.

Farmer, F. A., B. C. Mutch, J. M. Bell, L. D. Woolsey, and E. W. Crampton. 1950. The vitamin E requirement of guinea pigs. J. Nutr. 42:309–318.

Farooqui, S., S. K. Thind, R. Nath, and A. Mohmood. 1983. Intestinal absorption of oxalate in scorbutic and ascorbic acid supplemented guinea pigs. Acta Vitaminol. Enzymol. 5:235–241.

Fernandez, M. L., and D. J. McNamara. 1991. Regulation of cholesterol and lipoprotein metabolism in guinea pigs mediated by dietary fat quality and quantity. J. Nutr. 121:934–943.

Fernandez, M. L., A. Trejo, and D. J. McNamara. 1990. Pectin isolated from prickly pear (*Opuntia* sp.) modifies low density lipoprotein metabolism in cholesterol fed guinea pigs. J. Nutr. 120:1283–1290.

Finke, M. D., G. R. Defoliart, and N. J. Benevenga. 1987. Use of simultaneous curve fitting and a four-parameter logistic model to evaluate the nutritional quality of protein sources at growth rates of rats from maintenance to maximum gain. J. Nutr. 117:1681–1688.

Fraser, R. C., S. Pavlovic, C. G. Kurahara, A. Murata, N. S. Peterson, K. B. Taylor, and G. A. Feigen. 1980. The effect of variations in vitamin C intake on the cellular immune response of guinea pigs. Am. J. Clin. Nutr. 33:839–847.

Gahl, M. J., M. D. Finke, T. D. Crenshaw, and N. J. Benevenga. 1991. Use of a four parameter logistic equation to evaluate the response of growing rats to ten levels of each indispensable amino acid. J. Nutr. 121:1720–1729.

Gil, A., G. M. Briggs, J. Typpo, and G. MacKinney. 1968. Vitamin A requirement of the guinea pig. J. Nutr. 96:359–362.

Ginter, E., and Z. Zloch. 1972. Raised ascorbic acid consumption in cholesterol-fed guinea pigs. Int. J. Vitam. Nutr. Res. 42:72–79.

Ginter, E., P. Bobek, and D. Vargova. 1979. Tissue levels and optimum dosage of vitamin C in guinea pigs. Nutr. Metab. 23:217–226.

Ginter, E., E. Drobna, and L. Ramacsay. 1982. Kinetics of ascorbate depletion in guinea pigs after long-term high vitamin C intake. Int. J. Vitam. Nutr. Res. 52:307–311.

Grace, N. D., and B. L. O'Dell. 1968. Potassium requirement of the weanling guinea pig. J. Nutr. 94:166–170.

Grace, N. D., and B. L. O'Dell. 1970a. Interrelationship of dietary magnesium and potassium in the guinea pig. J. Nutr. 100:37–44.

Grace, N. D., and B. L. O'Dell. 1970b. Relation of polysome structure to ribonuclease and ribonuclear inhibitor activities in livers of magnesium-deficient guinea pigs. Can. J. Biochem. 48:21–26.

Green, M. D., J. Hawkins, and S. Omaye. 1980. Effect of scurvy on reserpine induced hypothermia in the guinea pig. Life Sci. 27:111–116.

Gupta, R. P., P. C. Verma, J. R. Sadana, and R. K. P. Gupta. 1988. Studies on the pathology of experimental zinc deficiency in guinea pigs. J. Comp. Pathol. 98:405–413.

Haddox, M. K., J. H. Stephenson, M. E. Moser, D. B. Glass, J. G. White, B. Holmes-Gray, and N. D. Goldberg. 1979. Ascorbic acid modulation of splenic cell cyclic GMP metabolism. Life Sci. 24:1555–1566.

Hara, H. 1960. Pathologic study on riboflavin deficiency in guinea pigs. J. Vitaminol. 6:24–42.

Hasdai, A., Z. Nitsan, and R. Volcani. 1989. Growth, digestibility, and enzyme activities in the pancreas and intestines of guinea pigs fed on raw and heated soya-bean flour. Br. J. Nutr. 62:529–537.

Hegsted, D. M., and Y. Chang. 1965a. Protein utilization in growing rats. I. Relative growth index as a bioassay procedure. J. Nutr. 85:159–168.

Hegsted, D. M., and Y. Chang. 1965b. Protein utilization in growing rats at different levels of intake. J. Nutr. 87:19–25.

Heinicke, H. R., and C. A. Elvehjem. 1955. Effect of high levels of fat, lactose, and type of bulk in guinea pig diets. Proc. Soc. Exp. Biol. Med. 90:70–72.

Heinicke, H. R., A. E. Harper, and C. A. Elvehjem. 1955. Protein and amino acid requirements of the guinea pig. I. Effect of carbohydrate, protein level and amino acid supplementation. J. Nutr. 57:483–496.

Heinicke, H. R., A. E. Harper, and C. A. Elvehjem. 1956. Protein and amino acid requirements of the guinea pig. II. Effect of age, potassium and magnesium, and type of protein. J. Nutr. 58:269–280.

Henning, S. J., and F. J. R. Hird. 1970. Concentrations and metabolism of volatile fatty acids in the fermentative organs of two species of kangaroo and guinea pig. Br. J. Nutr. 24:145–155.

Hirsh, E. 1973. Some determinants of intake and pattern of feeding in the guinea pig. Physiol. Behav. 11:687–704.

Hogan, A. G., and W. O. Regan. 1946. Diet and calcium phosphate deposits in guinea pigs. Fed. Proc. 5:138 (abstr.).

Hogan, A. G., W. O. Regan, and W. B. House. 1950. Calcium phosphate deposits in guinea pigs and phosphorus content of the diet. J. Nutr. 41:203–213.

Holman, R. T. 1960. The ratio of trienoic:tetraenoic acids in the tissue lipids as a measure of essential fatty acid requirement. J. Nutr. 70:405–410.

Horstkoetter, R. W. 1974. Amino acid requirements of guinea pigs. V. The threonine requirement. M.S. thesis. University of Missouri, Columbia, Mo.

House, W. B., and A. G. Hogan. 1955. Injury to guinea pigs that follows a high intake of phosphates. J. Nutr. 55:507–517.

Howe, P. R., L. G. Wesson, P. E. Boyle, and S. B. Wolbach. 1940. Low calcium rickets in the guinea pig. Proc. Soc. Exp. Biol. Med. 45:298–301.

Howell, J. M., J. N. Thompson, and G. A. J. Pitt. 1967. Changes in the tissues of guinea pigs fed on a diet free from vitamin A, but containing methyl retinoate. Br. J. Nutr. 21:37–44.

Hsieh, H. S., and J. M. Navia. 1980. Zinc deficiency and bone formation in guinea pig alveolar implants. J. Nutr. 110:1581–1588.

Hunt, C. E., and D. D. Harrington. 1974. Nutrition and nutritional diseases of the rabbit. In The Biology of the Laboratory Rabbit. New York: Academic Press.

Hurley, L. S., N. E. Volkert, and J. T. Eichner. 1965. Pantothenic acid deficiency in pregnant and non-pregnant guinea pigs, with special reference to effects on the fetus. J. Nutr. 86:201–208.

Hurley, L. S., and C. L. Keen. 1987. Manganese. Pp. 185–223 in Trace Elements in Human and Animal Nutrition, W. Mertz, ed. Orlando, Fla.: Academic Press.

Jeffery, D. M., and J. T. Typpo. 1982. Crystalline amino acid diet for determining amino acid requirements of growing guinea pigs. J. Nutr. 112:1118–1125.

Johnston, C. S. 1989. Effect of single oral doses of ascorbic acid on body temperature in healthy guinea pigs. J. Nutr. 119:407–425.

Johnston, C. S., and S. Huang. 1991. Effect of ascorbic acid nutriture on blood histamine and neutrophil chemotaxis in guinea pigs. J. Nutr. 121:126–131.

Jones, E., and R. E. Hughes. 1982. Influence of oral carnitine on the body weight and survival time of avitaminotic-C guinea pigs. Nutr. Rep. Int. 25:201–204.

Kodicek, E., and P. D. F. Murray. 1943. Influence of a prolonged partial deficiency of vitamin C on the recovery of guinea pigs from injury to bone and muscles. Nature 151:395–396.

Kothari, L. K., and P. Sharma. 1988. Aggravation of cholesterol induced hyperlipidemia by chronic vitamin C deficiency: Experimental study in guinea pigs. Acta Biol. Hung. 39:49–57.

Labhsetwar, A. P., and M. Diamond. 1970. Ovarian changes in the guinea pig during various reproductive stages and steroid treatments. Biol. Reprod. 2:53–57.

Lawrence, R. A., and R. F. Burk. 1978. Species, tissue and subcellular distribution of non Se-dependent glutathione peroxidase activity. J. Nutr. 108:211–215.

Leat, W. M. F., R. Curtis, N. J. Millichamp, and R. W. Cox. 1986. Retinyl function in rats and guinea pigs reared on diets low in essential fatty acids and supplemented with linoleic or linolenic acids. Ann. Nutr. Metab. 30:166–174.

Lister, D., and R. A. McCance. 1965. The effect of two diets on the growth, reproduction and ultimate size of guinea pigs. Br. J. Nutr. 19:311–319.

Liu, C. T. 1988. Energy balance and growth rate of outbred and inbred male guinea pigs. Am. J. Vet. Res. 49:1752–1756.

Liu, K. C., J. T. Typpo, J. Y. Lu, and G. M. Briggs. 1967. Thiamine requirement of the guinea pig and the effect of salt mixtures in the diets on thiamine stability. J. Nutr. 93:480–484.

Luderitz, B., H. D. Bolte, and G. Steinbeck. 1971. Single fiber potentials and cellular cation-concentration of the heart ventricle in chronic potassium deficiency. Klin. Wochenschr. 49:369–371.

Machlin, L. J., F. Garcia, W. Kuenzig, and M. Brin. 1979. Antiscorbutic activity of ascorbic acid phosphate in the rhesus monkey and the guinea pig. Am. J. Clin. Nutr. 32:325–331.

Majumder, M. S. I., and A. T. M. Rahim. 1987. Cell-mediated immune response of scorbutic guinea pigs. Nutr. Res. 7:611–616.

Mannering, G. J. 1949. Vitamin requirements of the guinea pig. Vitam. Horm. 7:201–221.

Maynard, L. A., D. Boggs, G. Fisk, and D. Sequin. 1958. Dietary mineral interrelations as a cause of soft tissue calcification in guinea pigs. J. Nutr. 64:85–97.

McBean, L. D., J. C. Smith, and J. A. Halsted. 1972. Zinc deficiency in guinea pigs. Proc. Soc. Exp. Biol. Med. 140:1207–1209.

Miller, C. C., V. A. Ziboh, T. Wong, and M. P. Fletcher. 1990. Dietary supplementation with oils rich in (n-3) and (n-6) fatty acids influences in vivo levels of epidermal lipoxygenase products in guinea pigs. J. Nutr. 120:36–44.

Milne, D. B., and T. Omaye. 1980. Effect of vitamin C on copper and iron metabolism in the guinea pig. Int. J. Vitam. Nutr. Res. 50:301–308.

Morris, E. R., and B. L. O'Dell. 1961. Magnesium deficiency in the guinea pig. Mineral composition of tissues and distribution of acid-soluble phophorus. J. Nutr. 75:77–85.

Morris, E. R., and B. L. O'Dell. 1963. Relationship of excess calcium and phosphorus to magnesium requirement and toxicity in guinea pigs. J. Nutr. 81:175–181.

Mueller, M. J. 1978. Amino acid requirement of growing guinea pigs. IX. The leucine requirement. M.S. thesis. University of Missouri, Columbia, Mo.

National Institutes of Health. 1982. NIH Rodents 1980 Catalogue. NIH No. 83–606. Washington, D.C.: Department of Health and Human Services.

Navia, J. M., and C. E. Hunt. 1976. Nutrition, nutritional diseases and nutrition research application. Pp. 235–265 in Biology of the Guinea Pig, J. E. Wagner and P. J. Manning, eds. New York: Academic Press.

Navia, J. M., and H. Lopez. 1973. A purified gel diet for guinea pigs. Lab. Anim. Sci. 23:111–114.

Neuringer, M., G. J. Anderson, and W. E. Connor. 1988. The essentiality of n-3 fatty acids for the development and function of the retina and brain. Annu. Rev. Nutr. 8:517–541.

O'Dell, B. L., and W. O. Regan. 1963. Effect of lysine and glycine upon arginine requirement of guinea pigs. Proc. Soc. Exp. Biol. Med. 112:336–337.

O'Dell, B. L., J. M. Vandepopuliere, E. R. Morris, and A. G. Hogan. 1956. Effect of a high phosphorus diet on acid-base balance in guinea pigs. Proc. Soc. Exp. Biol. Med. 91:220–223.

O'Dell, B. L., E. R. Morris, and W. O. Hogan. 1960. Magnesium requirements of guinea pigs and rats. Effect of calcium and phosphorus and symptoms of magnesium-deficiency. J. Nutr. 70:103–110.

O'Dell, B. L., R. I. Moroni, and W. O. Hogan. 1973. Interaction of dietary fluoride and magnesium in guinea pigs. J. Nutr. 103:841–850.

O'Dell, B. L., J. K. Becker, M. P. Emery, and J. D. Browning. 1989. Production and reversal of the neuromuscular pathology and related signs of zinc deficiency in guinea pigs. J. Nutr. 119:196–201.

Ostwald, R., W. Yamanaka, and D. Irvin. 1971. Effect of dietary modifications on cholesterol-induced anemia in guinea pigs. J. Nutr. 101:699–712.

Pappenheimer, A. M., and M. Goettsch. 1941. Death of embryos in guinea pigs on diets low in vitamin E. Proc. Soc. Exp. Biol. Med. 47:268–270.

Parker, C. M., R. P. Sharma, and J. L. Shupe. 1979. The interaction of dietary vitamin C, protein and calcium with fluoride toxicity. Fluoride-Quart. Rep. 12:144–154.

Parra, R. 1978. Comparison of foregut and hindgut fermentation in herbivores. Pp. 205–229 in The ecology of arboreal folivores, G. G. Montgomery, ed. Washington, D.C.: Smithsonian Institution Press.

Pfander, W. H., and H. H. Mitchell. 1952. The ascorbic acid requirement of the guinea pig when adrenal weight and odonto-blast height are used as criteria. J. Nutr. 47:503–524.

Prinz, W., J. Bloch, G. Gilich, and G. Mitchell. 1980. A systematic study of the effect of vitamin C supplementation on the humoral immune response in ascorbate-dependent mammals. I. The antibody response to sheep red blood cells (a T-dependent antigen) in guinea pigs. Int. J. Vitam. Nutr. Res. 50:294–300.

Pyke, R. E., W. G. Hoekstra, and P. H. Phillips. 1967. Effects of fluoride on magnesium deficiency in the guinea pig. J. Nutr. 92:311–316.

Quarterman, J., and W. R. Humphries. 1983. The production of zinc deficiency in the guinea pig. J. Comp. Pathol. 93:261–270.

Reid, M. E. 1954a. Nutritional studies with the guinea pig. B-vitamins other than pantothenic acid. Proc. Soc. Exp. Biol. Med. 85:547–550.

Reid, M. E. 1954b. Production and counteraction of a fatty acid deficiency in the guinea pig. Proc. Soc. Exp. Biol. Med. 86:708–712.

Reid, M. E. 1955. Nutritional studies with the guinea pig. III. Choline. J. Nutr. 56:215–229.

Reid, M. E. 1961. Nutritional studies with the guinea pig. VII. Niacin. J. Nutr. 75:279–286.

Reid, M. E. 1963. Nutritional studies with the guinea pig. IX. Effect of dietary protein level on body weight and organ weights in young guinea pigs. J. Nutr. 80:33–38.

Reid, M. E. 1964. Nutritional studies with the guinea pig. XI. Pyridoxine. Proc. Soc. Exp. Biol. Med. 116:289–292.

Reid, M. E. 1966. Methionine and cystine requirements of the young growing guinea pig. J. Nutr. 88:379–402.

Reid, M. E., and J. G. Bieri. 1967. Nutritional studies with the guinea pig. VIII. Thiamine. Proc. Soc. Exp. Biol. Med. 126:11–13.

Reid, M. E., and G. M. Briggs. 1953. Development of a semi-synthetic diet for young guinea pigs. J. Nutr. 51:341–354.

Reid, M. E., and G. M. Briggs. 1954. Nutritional studies with the guinea pig. II. Pantothenic acid. J. Nutr. 52:507–517.

Reid, M. E., and Martin, M. G. 1959. Nutritional studies with the guinea pig. V. Effects of deficiency of fat or unsaturated fatty acids. J. Nutr. 67:611–622.

Reid, M. E., and O. Mickelsen. 1963. Nutritional studies with the guinea pig. VIII. Effect of different proteins, with and without amino acid supplements, on growth. J. Nutr. 80:25–32.

Reid, M. E., and L. Von Sallmann. 1960. Nutritional studies with the guinea pig. VI. Tryptophan (with ample dietary niacin). J. Nutr. 70:329–336.

Reid, M. E., M. G. Martin, and G. M. Briggs. 1956. Nutritional studies with the guinea pig. IV. Folic acid. J. Nutr. 59:103–119.

Reid, M. E., J. G. Bieri, P. A. Plack, and E. L. Andrews. 1964. Nutritional studies with the guinea pig. X. Determination of the linoleic acid requirement. J. Nutr. 82:401–408.

Rikans, L. E., C. R. Smith, and V. G. Zannoni. 1977. Ascorbic acid and heme synthesis in deficient guinea pig liver. Biochem. Pharmacol. 26:797–799.

Rivers, J. M., L. Krook, and A. Cormier. 1970. Biochemical and histological study of guinea pig fetal and uterine tissue in ascorbic acid deficiency. J. Nutr. 100:217–227.

Robbins, K. D., H. W. Norton, and D. H. Baker. 1979. Estimation of nutrient requirements from growth data. J. Nutr. 109:1710–1714.

Robens, J. R. 1970. Teratogenic effects of hypervitaminosis A in the hamster and the guinea pig. Toxicol. Appl. Pharmacol. 16:88–99.

Rose, R. C., J. L. Choi, and M. J. Koch. 1988. Intestinal transport and metabolism of oxidized ascorbic acid (dehydroascorbic acid). Am. J. Physiol. 254:G824–G828.

Schiller, E. L. 1977. Relationships among selected dietary components and plasma transaminase activities in adult miniature swine and guinea pigs and indices of nitrogen status in adult guinea pigs. Ph.D. dissertation. University of Missouri, Columbia, Mo.

Schottelius, B. A., D. D. Schottelius, and A. D. Bender. 1959. Effect of vitamin E on myoglobin content of guinea pig skeletal muscle. Proc. Soc. Exp. Biol. Med. 102:581–583.

Sergeev, I. N., Y. P. Arkhapchev, and V. B. Spirichev. 1990. Ascorbic acid effects of vitamin D hormone metabolism and binding in guinea pigs. J. Nutr. 120:1185–1190.

Shelton, D. C. 1971. Feeding the guinea pig. Lab. Anim. 7:84–87.

Shimotori, N., G. A. Emerson, and H. M. Evans. 1940. The prevention of nutritional muscular dystrophy in guinea pigs with vitamin E. J. Nutr. 19:547–554.

Shrader, R. E., and G. J. Everson. 1968. Pancreatic pathology in manganese-deficient guinea pigs. J. Nutr. 94:269–281.

Siliparandi, L., P. Vanni, M. Kessler, and G. Semena. 1979. Na$^+$ dependent, electroneutral L-ascorbate transport across brush border membrane vesicles from guinea pig small intestine. Biochim. Biophys. Acta 552:129–142.

Sillevis Smitt, P. A., J. M. de Jong, D. Troost, and M. A. Kuipers. 1991. Muscular changes in the guinea pig caused by chronic ascorbic acid deficiency. J. Neurol. Sci. 102:4–10.

Simboli-Campbell, M., and G. Jones. 1991. Dietary phosphate deprivation increases renal synthesis and decreases renal catabolism of 1,25-dihydroxycholecalciferol in guinea pigs. J. Nutr. 121:1635–1642.

Singh, K. D., E. R. Morris, W. O. Regan, and B. L. O'Dell. 1968. An unrecognized nutrient for the guinea pig. J. Nutr. 94:534–542.

Sisk, D. B. 1976. Physiology. Pp. 63–98 in Biology of the Guinea Pig, J. E. Wagner and P. J. Manning, eds. New York: Academic Press.

Slanetz, C. A. 1943. The adequacy of improved stock diets for laboratory animals. Am. J. Vet. Res. 4:182–189.

Smith, C. H., and W. R. Bidlack. 1980. Interrelationship of dietary ascorbic acid and iron on the tissue distribution of ascorbic acid, iron and copper in female guinea pigs. J. Nutr. 110:1398–1408.

Smith, L. F. 1979. Amino acid requirements of growing guinea pigs. X. The tryptophan requirement and interrelationship with niacin. M.S. thesis. University of Missouri, Columbia, Mo.

Spanheimer, R. G., and B. Peterkofsky. 1985. A specific decrease in collagen synthesis in acutely fasted, vitamin C-supplemented, guinea pigs. J. Biol. Chem. 260:3955–3962.

Stenflo, J., and P. Fernlund. 1984. β-Hydroxyaspartic acid in vitamin K-dependent plasma proteins from scorbutic and warfarin-treated guinea pigs. FEBS Lett. 168:287–292.

Stone, N., and A. Meister. 1962. Function of ascorbic acid in the conversion of proline to collagen hydroxyproline. Nature 194:555–557.

Suttle, N. F. 1974. Recent studies of the copper-molybdenum antagonism. Proc. Nutr. Soc. 33:299–305.

Tinoco, J. 1982. Dietary requirements and functions of α-linolenic acid in animals. Prog. Lipid Res. 21:1–45.

Thompson, D. J., J. F. Heintz, and P. H. Phillips. 1964. Effect of magnesium, fluoride, and ascorbic acid on metabolism of connective tissue. J. Nutr. 84:27–30.

Toyoda, H., S. Himens, and N. Imura. 1989. The regulation of glutathione peroxidase gene expression relevant to species difference and the effects of dietary selenium manipulation. Biochim. Biophys. Acta 1008:301–308.

Tsai, H. C. C., and G. J. Everson. 1967. Effect of manganese deficiency on the acid mucopolysaccharides in the cartilage of guinea pigs. J. Nutr. 91:447–460.

Tsao, C. S., and P. Y. Leung. 1988. Urinary ascorbic acid levels following the withdrawal of large doses of ascorbic acid in the guinea pigs. J. Nutr. 118:895–900.

Tsao, C. S., and M. Young. 1989. Effect of dietary ascorbic acid on levels of serum mineral nutrients in guinea pigs. Int. J. Vitam. Nutr. Res. 59:72–76.

Tsujimura, M. 1978. Studies on the biological activity of L-ascorbic acid 2-sulfate. Joshi Eiyo Daigaku Kiyo 9:213–252.

Typpo, J. T., H. L. Anderson, G. F. Krause, and D. T. Yu. 1985. The lysine requirement of young growing male guinea pigs. J. Nutr. 115:579–587.

Typpo, J. T., J. E. Link, G. F. Krause, and D. Baravati. 1990a. The total nitrogen requirement of young, growing, male guinea pigs. FASEB J. 4:A804 (abstr.).

Typpo, J. T., D. J. Curtis, L. S. Ayers, S. C. Mokros, J. E. Link, and G. F. Krause. 1990b. Amino acid requirements of guinea pigs using chemically defined diets. Amino Acids 2:1132–1140.

Udenfriend, S. 1966. Formation of hydroxyproline in collagen. Science 152:1335–1340.

Van Hellemond, M. J., A. G. Lemmens, and A. C. Beynen. 1988. Dietary phosphorus and calcium excretion in guinea pigs. Nutr. Rep. Intl. 37:909–912.

Verma, P. C., R. P. Gupta, J. R. Sadana, and R. K. P. Gupta. 1988. Effect of experimental zinc deficiency and repletion on some immunological variables in guinea pigs. Br. J. Nutr. 59:149–154.

Wagner, J. E., and H. L. Foster. 1976. Germfree and specific pathogen-free. Pp. 21–30 in Biology of the Guinea Pig, J. E. Wagner and P. J. Manning, eds. New York: Academic Press.

Wagner, J. E., and P. J. Manning. 1976. The Biology of the Guinea Pig. New York: Academic Press.

Weir, B. J. 1974. Notes on the origin of the domestic guinea pig. In The Biology of Hystricomorph Rodents, I. W. Rowlands and B. J. Weir, eds. New York: Academic Press.

Wolbach, S. B. 1947. Vitamin-A deficiency and excess in relation to skeletal growth. J. Bone Joint Surg. 29:171–192.

Wolbach, S. B. 1954. Effects of vitamin A deficiency and hypervitaminois A in animals. Pp. 106–137 in The Vitamins, Vol. 1, W. H. Sebrell, Jr., and R. S. Harris, eds. New York: Academic Press.

Woodruff, C. W., S. L. Clark, and E. B. Bridgeforth. 1953. Folic acid deficiency in the guinea pig. J. Nutr. 51:23–34.

Woolley, D. W., and H. Sprince. 1945. The nature of some new dietary factors required by guinea pigs. J. Biol. Chem. 157:447–453.

Woytkiw, L., and N. C. Esselbaugh. 1951. Vitamin A and carotene absorption in the guinea pig. J. Nutr. 43:451–458.

Yoon, S. H. 1977. Amino Acid Requirements of Guinea Pigs. VII. The arginine requirement. M.S. thesis. University of Missouri, Columbia, Mo.

Young, M., and E. M. Widdowson. 1975. Influence of diet deficiency in energy, or in protein on conceptive weight, and the placental transfer of a non-metabolizable amino acid in the guinea pig. Biol. Neonate 27:184–191.

5   Nutrient Requirements of the Hamster

Taxonomically hamsters are classified as a subfamily, Cricetinae, with 7 genera and 18 species in the family Muridae (Musser and Carleton, 1993). They are distributed throughout the Palearctic zone of Eurasia (Anderson and Jones, 1984). Original habitats of laboratory hamsters included clay deserts, shrub-covered plains, forested steppes, and/or cultivated fields.

Golden hamsters, *Mesocricetus auratus*, collected from a burrow 8 feet deep in a wheat field near Aleppo, Syria, were established as a colony of laboratory animals at the Microbiological Institute of Jerusalem in 1930. These animals were used to complete research on kala-azar delayed by the failure of Chinese hamsters to breed in captivity (Adler and Theodor, 1931). Adler took breeding pairs to Paris and London to establish colonies at research institutions there (Bruce and Hindle, 1934; Adler, 1948; Murphy, 1985). Breeding stock were distributed to investigators in India, Egypt, and the United States (Doull and Megrall, 1939; Poiley, 1950). Golden hamsters is the hamster species most frequently used in research (Hoffman et al., 1968; Siegel, 1985; Van Hoosier and McPherson, 1987), but very little is known about their nutritional requirements.

In the past 20 years, 7 additional hamster species have been used as laboratory animals. Animals identified as strain MHH:EPH are maintained in Hannover, Germany (Reznik et al., 1978; Mohr and Ernst, 1987). Mouse-like Chinese striped hamsters, *Cricetulus barabensis*, are used in research in cytogenetics, diabetes, and toxicology (Calland et al., 1986; Diani and Gerritsen, 1987). The large guinea-pig-like European hamster, *Cricetus cricetus,* formerly considered a pest in agricultural areas, is now a model for research in carcinogenesis. Dwarf hamsters, *Phodopus campbelli* and *P. sungorus,* of southern and western Siberia, are used in research in cytogenetics, carcinogenesis, diabetes mellitus, obesity, photoperiod changes, and social behavior (Pogosianz and Sokova, 1967; Hoffmann, 1973; Daly, 1975; Gamperl et al., 1978; Hoffmann, 1978;

Steinlechner et al., 1983; Wade and Bartness, 1984; Pond et al., 1987; Ruf et al., 1991). Turkish hamsters, *Mesocricetus brandti*, are used in hibernation, taxonomy, and cytogenetics research (Lyman and O'Brien, 1977; Lyman et al., 1981, 1983; Todd et al., 1972). Colonies of the Romanian hamster, *Mesocricetus neutoni*, were established in Bucharest and used for research in cytogenetics and taxonomy (Hamar and Schutowa, 1966; Murphy, 1977; Popescu and DiPaolo, 1980). The Armenian, or migratory hamster, *Cricetulus migratorius*, is used on a limited basis in cytogenetics and oncology research (Lavappa and Yerganian, 1970; Cantrell and Padovan, 1987) (see Table 5-1).

BIOLOGICAL AND BEHAVIORAL CHARACTERISTICS

Unlike simple-stomached rats, mice, and guinea pigs, hamsters, like voles, have a stomach that consists of two distinct compartments: a keratinized, nonglandular forestomach (cardiac) separated from a glandular region (pyloric) by sphincter-like muscular marginal folds (Reznik et al., 1978) that control movement of ingesta from esophagus to duodenum. An embryological study has shown that the forestomach is gastric in origin and not an esophageal derivative (Vorontsov, 1979). The structure and function of the forestomach is similar to the rumen of herbivores (Takahashi and Tamate, 1976; Borer, 1985). Ingesta enter the forestomach from the esophagus and pass into the glandular stomach in 10 to 60 minutes (Ehle and Warner, 1978). Kunstyr (1974) noted that the concentration of microorganisms is higher in the forestomach than in the glandular region. Sakaguchi et al. (1981) demonstrated that the forestomach aids in the utilization of dietary urea.

The hamster cecum is a J-shaped structure with numerous lateral sacculations and has more volume than the stomach (Krueger and Rieschel, 1950; Magalhaes, 1968).

TABLE 5-1   Names, Characteristics, and History of Laboratory Hamsters

Genus and Species	Common Name or Origin	Size	Adult Weight, g	Origin of Collection	Where Used	Introduction Date
Cricetus cricetus	Common hamster	Large	337–500[a]	Germany	Germany	1971[b]
Mesocricetus auratus	Golden or Syrian	Medium	120–180[c]	Aleppo, Syria	Palestine	1930[d]
Mesocricetus brandti	Brandt's or Turkish	Medium	137–258[e]	Turkey, Asia Minor	United States	1965[e]
Mesocricetus newtoni	Newton's or Romanian	Medium	120[f]	E. Romania	Romania	1964[f]
Cricetulus migratorius	Migratory or Armenian	Small	40–80[g]	Armenia	United States, E. Europe	1963[h]
Cricetulus barabensis	Striped-back or Chinese	Small	28–40[i]	Wild, collected in Northeast China	Eurasia, United States	1919[j]
Phodopus campbelli[k]	Campbell's or Siberian	Dwarf	30–50[g]	Tuva, S. Siberia	Eurasia, N. America	1965[l]
Phodopus sungorus[k]	Dzungarian or Siberian	Variable; depends on season	25–45[g]	Omsk, W. Siberia	Eurasia, N. America	1968[m]

[a] Reznik et al. (1978).
[b] Reznik-Schüller et al. (1974).
[c] von Frisch (1990).
[d] Adler and Theodor (1931).
[e] Lyman and O'Brien (1977).
[f] Murphy (1977).
[g] Cantrell and Padovan (1987).
[h] Yerganian (1977).
[i] Yerganian (1958).
[j] Hsieh (1919).
[k] Wilson and Reeder (1993).
[l] Pogosianz and Sokova (1967).
[m] Hoffmann (1978).

Hamsters, like guinea pigs and gerbils, eat regularly, at 2-hour intervals throughout the day (Anderson and Shettleworth, 1977; Borer et al., 1979). Wild hamsters gather and store grains and other food in underground burrows to ensure a constant source of food (Borer et al., 1979; Micheli and Malsbury, 1982; Carleton and Musser, 1984). Hamsters are adapted to running and digging and are active primarily during twilight and during the night.

The golden hamster has a gestation period of 15 to 18 days. Members of the *Mesocricetus* species are solitary animals that live in separate burrows with one or two chambers and entrances and exits; males and females meet only for breeding (Murphy, 1985).

REPRODUCTION AND DEVELOPMENT

Developmental and reproductive indices for three species are given in Tables 5-2 and 5-3, respectively. Young hamsters weigh 2 to 4 g at birth (see Table 5-4; Poiley, 1972). Average litter size is 11 (Slater, 1972), ranging from 2 to 16 (Anderson and Shina, 1972). Newborn hamsters are fetal in appearance—hairless, eyes and ears closed, and legs underdeveloped (Balk and Slater, 1987). Incisors are erupted at birth and young animals begin to eat solid food within 7 to 10 days (Balk and Slater, 1987). Hamsters weigh 40 g when weaned at 21 days (Poiley, 1972). Male hamsters are sexually mature at 42 days old, but females can breed as early as 28 to 30 days old (Selle, 1945; Balk and Slater, 1987). Litters with the greatest average number of pups are obtained from females 8 to 10 weeks old and males 10 to 12 weeks old (Robens, 1968; Balk and Slater, 1987).

EXAMPLES OF PURIFIED AND NATURAL-INGREDIENT DIETS

Two examples of purified diets and one of a natural-ingredient diet are presented in Tables 5-5 A-C, 5-6 A-C, and 5-7 A-C. These diets supported growth that was equivalent to the highest rates reported in our review of the literature. The two purified diets supported growth rates of 1.6 to 2.0 g/day. The natural-ingredient diet was selected from three that supported a growth rate of 1.9 g/day; of the three, it was intermediate in complexity.

WATER AND ENERGY

Male and female golden hamsters consume, on average, 8.5 mL water/100 g BW/day; males consumed 5 mL water/100 g BW/day, while females consumed 14 mL water/100 g BW/day (Fitts and St. Dennis, 1981). Thompson (1971) recorded Chinese hamsters intake of water to be 11.4 mL/100 g BW/day for males and 12.9 mL/100 g BW/day for females. Water intake for golden hamsters was found to be 4.5 mL/100 g BW/day in males and 13.6 mL/100 g BW/day for females.

Little definitive work has been done on the energy requirement of the hamster, and few research studies include data on energy utilization. When fed a cereal-based diet containing 14.95 percent neutral detergent fiber (NDF) and 5.6 kcal gross energy (GE)/g diet (23.4 kJ/g diet), hamsters digested 45.2 percent of NDF and 81.5 percent of GE. Hamsters fed a 75 percent alfalfa meal diet that contained 40.6 percent NDF and 4.05 kcal GE/g diet (16.9 kJ/

TABLE 5-2 Developmental Indices for Golden, Chinese, and Siberian Hamsters

Variable	Unit	Amount	Reference
		Golden[a]	
Birth weight	g	2–3	Biven et al., 1987
Incisors present	Day	1	Balk and Slater, 1987
Ears open	Day	4–5	Balk and Slater, 1987
Eyes open	Day	14–16	Balk and Slater, 1987
Solid food eaten	Day	7–10	Balk and Slater, 1987
Weaning age	Day	21	
Weaning weight	g	35–40	Harkness and Wagner, 1983
Mature weights		95–150 female; 85–130	
	g	male	Harkness and Wagner, 1983
Life span	Year	1–3	Biven et al., 1987
		Chinese[b]	
Birth weight	g	1.5–2.5	Moore, 1965
Incisors present	Day	1	Moore, 1965
Ears open	Day	10–14	Moore, 1965
Eyes open	Day	10–14	Moore, 1965
Solid food eaten	Day	12	Smith, 1957
Weaning age	Day	21	Calland et al., 1986
Weaning weights	g	15–17 female; 16–17 male	Avery, 1968
Mature weights	g	25 female; 35 male	Smith, 1957
		Siberian[c]	
Birth weight	g	1.8	Pogosianz and Sokova, 1967
Incisors erupt	Day	0	Pogosianz and Sokova, 1967
Ears open	Day	3–4	Pogosianz and Sokova, 1967
Eyes open	Day	10	Pogosianz and Sokova, 1967
Solid food eaten	Day	10	Pogosianz and Sokova, 1967
Weaning age	Day	16–20	Pogosianz and Sokova, 1967
Weaning weight	g	23	Pogosianz and Sokova, 1967
Mature weight	g	30 female; 40–50 male	Cantrell and Padovan, 1987
Life span	Year	1–2	Cantrell and Padovan, 1987

[a] Golden hamster (*Mesocricetus auratus*).

[b] Chinese hamster (*Cricetulus barabensis*) previously (*C. griseus*).

[c] Campbell's, Djungarian, or Siberian hamster (*Phodopus campbelli*) (Pogosianz and Sokova, 1967); Siberian Dwarf, Djungarian, or Dzungrian hamster (*Phodopus sungorus*) (Cantrell and Padovan, 1987).

g diet) digested the NDF and GE to the extent of 33.4 percent and 50.2 percent, respectively (Ehle and Warner, 1978). Arrington et al. (1966) reported that hamsters fed purified diets containing 12 and 16 percent casein had a total GE intake of 27 to 29 kcal/day (113 to 121 kJ/day) and gained 40 to 100 g over a 42-day period. Smaller hamsters (45 g) consumed 58 kcal/100 g BW/day (243 kJ/ 100 g BW/day), while larger hamsters (90 g) consumed 28 kcal/100 g BW/day (117 kJ/100 g BW/day). For a summary of energy balance in golden hamsters, see Borer (1985).

LIPIDS

An optimal concentration of dietary lipid has not been established for the hamster, although they seem to thrive on diets containing 4 to 20 percent fat (w/w). Knapka and

Judge (1974) fed weanling (21 days old) male and female golden hamsters natural-ingredient pelleted diets containing 3.1, 5.0, 7.3, or 9.2 percent crude fat for 35 days. The feed:gain ratio decreased with increasing concentrations of dietary lipid. A decrease in feed intake was not associated with increased concentration of energy in the diet. Mortality was 1.0 percent when the diet contained 3.1 and 5.0 percent lipid. Higher mortality, but greater weight gain, occurred with the higher fat diets. The authors concluded that the lipid requirement for maximal growth of hamsters is slightly higher than 5 percent, but maximal growth should not be the only criterion used to determine the optimal concentration of lipid supplementation. Hamsters were maintained for a year or longer on starch gel diets (e.g., Table 5-6A) that contained up to 20 percent fat, with no mortality attributed to the fat load. However, feeding hamsters this diet for maintenance at 10 to 12 g/day [about 25 kcal ME/100 g BW/day (105 kJ ME/100 g BW/day)]

TABLE 5-3   Reproductive Indices for Golden, Chinese, and Siberian Hamsters

Variable	Unit	Amount	Reference
		Golden [a]	
Breeding age	Week	8–10, female;	Balk and Slater, 1987
		10–12, male	Balk and Slater, 1987
minimum, female	Day	28	Selle, 1945
Estrous cycle	Day	4	Balk and Slater, 1987
Gestation	Day	15.5	Balk and Slater, 1987
Litter size		4–16	Slater, 1972
average		11	
Litters/lifetime		4–6	Balk and Slater, 1987
Reproductive life	Month	10	Balk and Slater, 1987
		Chinese [b]	
Breeding age	Week	8–12, female;	Calland et al., 1986
	Week	32–48, male	Calland et al., 1986
minimum, female	Day	41	Moore, 1965
Estrous cycle	Day	4	Moore, 1965
Gestation	Day	19–21	Avery, 1968
Litter size		1–11	Calland et al., 1986
average		6.1	Calland et al., 1986
Litters/lifetime		5	Parkening, 1982
Reproductive life	Month	16 ± 0.5	Parkening, 1982
		Siberian [c]	
Breeding age	Day	35–40 male 16 hours/light/day	Cantrell and Padovan, 1987
	Day	150 male 8 hours/light/day	Cantrell and Padovan, 1987
Estrous cycle	Day	4	Iakovenko, 1974
Gestation	Day	18	Daly, 1975
Litter size		1–9	Pogosianz and Sokova, 1967
average		4–6	Pogosianz and Sokova, 1967
Litters/lifetime		12	Pogosianz and Sokova, 1967
Reproductive life	Month	12	Pogosianz and Sokova, 1967

[a] Golden hamster (*Mesocricetus auratus*).

[b] Chinese hamster (*Cricetulus griseus* and/or *C. barabensis*).

[c] Siberian Dwarf, Djungarian, or Dzungrian hamster (*Phodopus sungorus*) (Pogosianz and Sokova, 1967; Cantrell and Padovan, 1987); Campbell's, Djungarian, or Siberian hamster (*Phodopus campbelli*) (Pogosianz and Sokova, 1967; Iakovenko, 1974).

tends to lessen obesity and the hypertriglyceridemia associated with allowing the animals free access to food (Hayes et al., 1993; K. C. Hayes, Brandeis University, personal communication, 1994)

Signs of Lipid Deficiency   A review by Holman (1968) reported that weanling hamsters fed a fat-free diet showed a slow rate of growth, had pale kidneys, and developed ulcers at the mucocutaneous junction of the anus.

ESSENTIAL FATTY ACIDS

n-6 Fatty Acids

The requirement for n-6 fatty acids has not been determined for hamsters, but a deficiency has been demonstrated. Christensen and Dam (1952) found that feeding weanling hamsters a fat-free diet resulted in loss of hair,

scaly skin, and development of a profuse secretion of cerumen (ear-wax)—a light-yellow, cholesterol-containing material. Signs of n-6 deficiency may be decreased by feeding hamsters a diet with 10 percent lard or a dietary supplement of 28 mg linoleic acid/day.

n-3 Fatty Acids

No studies are available on the distribution of n-3 fatty acids in hamster tissues or on the development of n-3 fatty acid deficiency. Three studies (Cunnane et al., 1985, 1986, 1987) were conducted in which ethanol was fed to hamsters with subsequent increases in n-9 fatty acids and decreases in the n-6 and n-3 fatty acids in liver triglycerides and phospholipids. Dietary treatments supplying a range of n-6:n-3 ratios were, predictably, found to influence tissue fatty acid composition in ethanol-fed and control hamsters.

TABLE 5-4 Growth of Golden Hamster Outbred Cr:RGH (SYR)

Sex	Age, days	Weight, g Average	Range
Male	1	2.9	2.0–4.0
Male	7	6.9	5.0–13.9
Male	14	18.3	11.5–33.2
Male	21	40.0	29.2–51.0
Male	28	48.6	32.0–70.8
Male	42	86.1	68.5–99.9
Male	56	91.5	85.6–97.5
Male	70	99.4	91.8–107.1
Male	84	103.9	98.9–109.2
Male	112	121.9	112.7–131.3
Male	140	131.8	116.6–145.6
Male	168	140.5	127.8–142.4
Female	1	3.0	2.0–3.8
Female	7	7.8	5.5–13.7
Female	14	17.4	11.0–31.6
Female	21	40.3	29.5–50.0
Female	28	44.1	31.0–68.5
Female	42	93.0	86.5–102.2
Female	56	94.5	85.3–104.1
Female	70	103.2	92.0–114.5
Female	84	114.9	103.3–126.5
Female	112	135.9	125.3–146.5
Female	140	149.6	147.0–159.8
Female	168	157.8	149.6–166.5

NOTE: In 10 of the 12 age groups, the females are larger than the males.
SOURCE: Poiley (1972).

TABLE 5-5A Rutten and de Groot Purified Diet for Hamsters

Ingredient	Amount, g/kg diet
Casein	200.0[a]
Wheat starch	635.0[b]
Corn oil	50.0[c]
Cellulose	50.0[d]
Mineral mix	35.0[e]
Vitamin mix	10.0[f]
CaHPO$_4$	15.0
DL-Methionine	3.0
Choline bitartrate	2.0

[a] Acid-precipitated, containing: protein 89.1% (N × 6.38); moisture 8.9%; ash 4.67%; pH of a 10% aqueous suspension, 4.5.

[b] Ten percent of native wheat starch was replaced by pregelatinized wheat starch to provide pellets of suitable quality for feeding hamsters.

[c] No antioxidants were added.

[d] Dicacel, highly purified and bleached fibrous powder, consisting of 87–90% pure α-cellulose; average length of fibers ≈44 μm; water, 4%; ash, 0.12–0.15%; and lignin, 0.04%.

[e] See Table 5-5B. Rutten and de Groot mineral mix based on AIN-76A.

[f] See Table 5-5C. Rutten and de Groot vitamin mix based on AIN-76A.

SOURCE: Rutten and de Groot (1992).

TABLE 5-5B Rutten and de Groot Mineral Mix

Compound	Formula	Amount, g/kg Mix	Diet[a]
Salt	NaCl	110.0	1.51
Salt	NaCl		2.34
Potassium citrate	K$_3$C$_6$H$_5$O·H$_2$O	394.0	5.29
Potassium sulfate	K$_2$SO$_4$	51.8	0.81
Potassium sulfate	K$_2$SO$_4$		0.33
Magnesium oxide	MgO	28.4	0.63
Manganese carbonate	MnCO$_3$·H$_2$O	3.5	0.051
Ferric citrate	FeC$_6$H$_5$O$_7$·5H$_2$O	24.0	0.173
Zinc carbonate	ZnO·2CO$_3$·4H$_2$O	1.6	0.031
Cupric carbonate, basic	CuCO$_3$(OH)$_2$·H$_2$O	0.3	0.004
Potassium iodate	KIO$_3$	0.08	0.002
Sodium selenite	Na$_2$SeO$_3$·5H$_2$O	0.01	0.0001
Chromic potassium sulfate	CrK(SO$_4$)·12H$_2$O	0.55	0.0025
Sodium fluoride	NaF	0.063	0.001
Cobaltous chloride	CoCl$_2$·6H$_2$O	0.127	0.002
Sucrose powder		385.57	

[a] Amount of element (boldface element in the formula) provided when 35 g of mix is added per kg diet.
SOURCE: Rutten and de Groot (1992).

CARBOHYDRATES

In hamsters fed diets containing 65 percent lactose or fructose, a mortality rate of 22 percent was observed, but only a 6 percent mortality rate was observed for hamsters fed 71 percent glucose, and 3 percent mortality with 62 percent sucrose (Gustafson et al., 1955). Salley and Bryson (1957) reduced mortality by decreasing the sugar to 54 percent and substituting either cornstarch or cellulose.

TABLE 5-5C Rutten and de Groot Vitamin Mix

Compound	Amount, per kg Mix	Diet[a]
Retinyl palmitate/acetate	400,000.0 IU	4,000.0 IU
Cholecalciferol	248,000.0 IU	2,480.0 IU
All-rac-α-tocopheryl acetate	5,000.0 IU	50.0 IU
Menadione Na-bisulfite	0.4 g	0.004 g
Thiamin-HCl	2.0 g	0.020 g
Riboflavin	1.5 g	0.015 g
Pyridoxine-HCl	0.7 g	0.007 g
Nicotinic acid	9.0 g	0.090 g
Ca-d(+)-pantothenate	4.0 g	0.040 g
Folic acid	0.2 g	0.002 g
d(+)-Biotin	0.06 g	0.0006 g
B$_{12}$	0.005 g	0.00005 g
Myo-inositol	10.0 g	0.1 g
Sucrose powder to 1 kg		

[a] Amount of vitamin provided as IU or g when 10 g of mix is added per kg of diet.
SOURCE: Rutten and de Groot (1992).

TABLE 5-6A   Hayes Purified Diet for Hamsters

Ingredient	Amount, g/kg diet
Casein	200.0
Glucose	170.0
Cornstarch (or rice flour)	366.0
Corn oil	50.0
Cellulose	100.0
Wheat bran	50.0
Mineral mix[a]	46.0
Vitamin mix[b]	12.0
Choline dihydrogen citrate[c]	6.0

NOTE: Diets were fed as gel blocks, prepared by withholding from the formulation 60 g/kg of either rice flour or cornstarch and premixing it with 800 mL of water slowly heated to simmering to form a slurry that was then added to the remaining ingredients while mixing.

[a] Hayes Mineral Mix (Hayes et al., 1989). See Table 5-6B for composition of mix.

[b] Hayes-Cathcart Vitamin Mix. Mix modified based on K.C. Hayes, Brandeis University, personal communication (1993). See Table 5-6C for contents of mix.

[c] Used in place of choline chloride for increased stability.

SOURCE: Hayes et al. (1989).

TABLE 5-6B   Hayes Mineral Mix

Compound	Formula	Amount, g/kg Mix	Diet[a]
Calcium carbonate	$CaCO_3$	290.4849	5.35
Calcium phosphate	$CaHPO_4 \cdot 2H_2O$	72.5970	0.78
Calcium phosphate	$CaHPO_4 \cdot 2H_2O$		0.60
Potassium phosphate	K_2HPO_4	314.2049	6.49
Potassium phosphate	K_2HPO_4		2.57
Magnesium sulfate	$MgSO_4 \cdot 7H_2O$	98.732	0.45
Magnesium sulfate	$MgSO_4 \cdot 7H_2O$		0.59
Sodium chloride	$NaCl$	162.3664	2.94
Sodium chloride	$NaCl$		4.53
Magnesium oxide	MgO	32.0395	0.87
Ferric citrate	$FeC_6H_6O_7 \cdot 5H_2O$	27.0000	0.205
Potassium iodide	KI	0.0774	0.0008
Potassium iodide	KI		0.0027
Manganese sulfate	$MnSO_4H_2O$	1.2211	0.0183
Manganese sulfate	$MnSO_4H_2O$		0.0106
Zinc chloride	$ZnCl_2$	0.9149	0.0202
Cupric sulfate	$CuSO_4 \cdot 5H_2O$	0.2901	0.0034
Chromic acetate	$Cr(C_2H_3O_2)_3$	0.0443	0.00046
Sodium selenite	Na_2SeO_3	0.0043	0.00009
Sodium fluoride	NaF	0.0232	0.00046

[a] Amount of element (boldface in formula) provided when 46 g mix is added per kg diet.

SOURCE: Hayes et al. (1989), modified to correct published errors per K. C. Hayes, Brandeis University, personal communication, 1993. For correct version, see Hayes et al. (1993).

TABLE 5-6C   Hayes-Cathcart Vitamin Mix

Compound	Amount g/kg mix	mg/kg diet[a]
Retinyl palmitate (500,000 IU/g)	1.5	18.0 (9,000 IU)
Cholecalciferol (400,000 IU/g)	0.1	1.2 (480 IU)
All-*rac*-α-tocopheryl acetate (500 IU/g)	15.0 (7,500 IU)	180.0 (90 IU)
Menadione	0.2	2.4
Myo-inositol	5.0	60.0
Niacin	3.0	36.0
Ca-pantothenate	1.6	19.0
Folic acid	0.200	2.4
Riboflavin	0.700	8.4
Thiamin	0.600	7.2
Pyridoxine-HCl	0.700	8.4
Biotin	0.020	0.24
Cyanocobalamin	0.001	0.012
Choline dihydrogen citrate	0.000	6,000.0
Dextrin	971.379	

[a] When vitamin mix is added at 12 g/kg diet.

SOURCE: Hayes et al. (1989).

TABLE 5-7A   Natural-Ingredient Diet for Hamsters

Ingredients	Amount, g/kg diet
Alfalfa meal dehydrated (17% protein)	200.0
Corn, yellow dent ground grain	529.3
Soybean seed meal solv-extd (44% protein)	220.0
Dry beet molasses	7.0
Soybean oil	12.5
Salt	5.0
Dibasic calcium phosphate	18.0
Ground limestone	5.0
Trace mineral mix[a]	0.5
Vitamin mix[b]	0.8
Choline dihydrogen citrate[c]	1.4
DL-Methionine	0.5

[a] Trace mineral mix formulated by E. A. Ulman provides minerals (g/kg): manganese (MnO_2) 300; ferrous iron ($FeSO_4 \cdot 7H_2O$) 570; zinc (ZnO) 97; copper ($CuSo_4$),4; iodine (KIO_3), 4.7; cobalt ($CoCl_2 \cdot 6H_2O$), 3. For use, 0.5 g of mix is added per kg diet.

[b] Vitamin mix designed by E. A. Ulman for use with cereal-based diets provides vitamins (g/kg) when added at 0.8 g/kg diet: vitamin A palmitate, 500,000 IU/g, 2.2; vitamin D, 250,000 IU/g, 4.7; vitamin E acetate, 500 IU/g, 52; menadione sodium bisulfite (62.5% menadione), 2.9; niacin, 19.5; Ca-pantothenate, 11.7; thiamin-HCl, 13.0; pyridoxin-HCl, 2.2; riboflavin, 4.4; folic acid, 2.9; biotin (1% in dextrose), 9.1; cyanocobalamin (1% in mannitol), 2.6; sucrose, 872.8 to give a total 1,000 g of mix.

[c] More stable than choline chloride.

SOURCE: Birt and Conrad (1981).

TABLE 5-7B Trace Mineral Mix

Compound	Formula	Mix	Diet[a]
		Amount, g/kg	
Manganese dioxide	MnO_2	300.0	0.095
Ferrous sulfate	$FeSO_4 \cdot 7H_2O$	570.0	0.060
Zinc oxide	ZnO	97.0	0.039
Cupric sulfate	$CuSO_4$	24.0	0.0048
Potassium iodate	KIO_3	4.7	0.0014
Cobaltous chloride	$CoCl_2 \cdot 6H_2O$	4.3	0.0005

[a] Amount of element (boldface in formula) provided when 0.5 g mix is added per kg diet.

SOURCE: Formulation by E. A. Ulman based on Birt and Conrad (1981).

With purified, fiber-free diets containing 64 percent carbohydrate, cornstarch was superior to glucose or sucrose in supporting survival (Ershoff, 1956). Rice starch supported higher growth rates than lactose (Dam and Christensen, 1961). Rogers et al. (1974) obtained satisfactory growth in a long-term study when animals were fed a gel diet containing 40 percent cornstarch and 21.9 percent sucrose. Hayes et al. (1989) observed that "wet tail" could be prevented by inclusion of rice flour, fiber, or lactose in gel diets. The implication is that diarrhea and "wet tail," commonly encountered in hamsters fed purified diets, results from an insufficient amount of complex carbohydrates (fiber, starch) reaching the large bowel flora.

PROTEIN AND AMINO ACIDS

Compared to a ruminant, fermentative digestion in the hamster is not sufficient to alter the pattern of dietary amino acids enough to improve growth of hamsters fed proteins such as wheat gluten (Banta et al., 1975). It seems that hamsters can make limited use of urea as a source of

TABLE 5-7C Vitamin Mix

Compound	g/kg mix	per kg diet[a]
	Amount	
Vitamin A palmitate 500,000 IU/kg	2.2	880.0 IU
Vitamin D_2 50,000 IU/g	4.7	188.0 IU
Vitamin E acetate 500 IU/kg	52.0	20.8 IU
Menadione sodium bisulfite 62.5% menadione	2.9	1.5 mg
Nicotinic acid	19.5	15.6 mg
Ca-pantothenate	11.7	9.4 mg
Thiamin-HCl	13.0	10.4 mg
Pyridoxine-HCl	2.2	1.8 mg
Riboflavin	4.4	3.5 mg
Folic acid	2.9	2.3 mg
Biotin (1% in dextrose)	9.1	0.07 mg
Cyanocobalamin (1% in mannitol)	2.6	0.02 mg
Sucrose powder	872.8	

[a] Value when mix is added at 0.8 g/kg diet.

SOURCE: Formulation by E. A. Ulman based on Birt and Conrad (1981). This vitamin mix is designed to be used with cereal-based diets.

dietary nitrogen (Matsumoto, 1955; Sakaguchi et al., 1981). However, fermentative digestion seems to suppress the anticipated response to supplementation of amino acids expected to improve growth (Arrington et al., 1966; Banta et al., 1975) and to decrease the toxicity of high dietary concentrations of L-phenylalanine (Horowitz and Waisman, 1966).

GROWTH

No studies to determine requirements for single amino acids or mixtures of amino acids were found. Studies that focused on protein requirements used variations in natural-ingredient diets or, in a few cases, used a single protein product at a series of dietary concentrations (Table 5-8). Protein requirements for growth reported here were ob-

TABLE 5-8 Protein Requirements

Strain	Growth, g/day	Protein Source	Amounts, %	Reference
Syrian Ufnl:(SYR)[a]	1.8–2.0	Corn, soybean, casein	13.7	Arrington et al., 1979
Golden	1.8–1.9	Multiple natural ingredients	20	Banta et al., 1975
Syrian		Wayne Lab Blox	24	Birt and Conrad, 1981
		Teklad	22	Birt and Conrad, 1981
		Corn-soybean	18	Birt and Conrad, 1981
	1.8–1.9	Corn-soybean-alfalfa	18	Birt and Conrad, 1981
		Multiple natural ingredients	20	
Syrian Lak:LVG (Syr)[a]	1.2	Multiple natural ingredients	18	Feldman et al., 1982
Syrian	1.4	Casein	18	Arrington et al., 1966
Syrian	1.4	Casein	18	Horowitz and Waisman, 1966

NOTE: In all studies, both males and females were used.

[a] Only groups identified as strains.

tained from studies that used both male and female 3- to 4-week-old hamsters weighing approximately 40 g. Growth rate varied from 1 to 2 g/day in experiments lasting 3 to 5 weeks in most cases. Arrington et al. (1979) found that diets containing 13.7 percent crude protein from mixtures of corn, soybean, and casein would support gains of 1.8 to 2 g/day in both male and female golden Syrian hamsters. Banta et al. (1975) reported growth rates of 1.8 to 1.9 g/day during a 6-week period using a natural-ingredient diet containing 20 percent crude protein. Birt and Conrad (1981) compared two commercial natural-ingredient diets with three formulated diets of increasing complexity and found that diets that contained 18 to 22 percent crude protein would support gains of 1.8 to 1.9 g/day in both male and female hamsters over a period of 6 weeks. However, Feldman et al. (1982) obtained maximum gains of only 1.2 g/day even when the natural-ingredient diets contained up to 24 percent crude protein. A natural-ingredient diet containing 18 percent crude protein should support growth rates approaching 2 g/day in weanling hamsters. Addition of semipurified proteins to natural-ingredient diets may reduce the amount of crude protein required to support the expected rates of growth.

In studies using semipurified sources of protein, Arrington et al. (1966) found that diets containing 18 percent casein resulted in growth of 1.4 g/day over 6 weeks. Additional experiments with diets containing 16 percent casein or soy-protein isolate supported gains of 1.4 to 1.5 g/day in experiments lasting 5 weeks (Arrington et al., 1966). In another study using casein at 9, 18, or 25 percent of the diet, Horowitz and Waisman (1966) found that maximum gain (1.4 g/day) was obtained with 18 percent casein.

Variation in the age and size of hamsters used in the experiments reviewed make it difficult to determine whether lower rates of growth obtained when purified proteins were used is the result of the hamster or the diet. Addition of free amino acids to the diet to improve the dietary amino acid pattern and, hence, growth have not been successful. A form of encapsulated amino acid may be beneficial.

REPRODUCTION

Two studies focused on diet and reproduction in hamsters (Birt et al., 1982; Birt and Conrad, 1981). In one study (Birt and Conrad, 1981), the effects of five natural-ingredient diets containing from 18 to 24 percent crude protein (see, for example, Table 5-7A) were compared over three breeding cycles; reproductive performance of hamsters fed corn-soybean or corn-soybean-alfalfa meal diets containing 18 percent crude protein was equal to or exceeded that of hamsters fed either of two commercial natural-ingredient diets containing 22 to 24 percent crude protein. In the other study (Birt et al., 1982), which used

lactalbumin at 20 and 40 percent of the diet as the sole source of protein, reproductive efficiency (pups weaned per mating) was 20 to 40 percent that of identical females fed the commercial diet.

Although no recommendation is made for the amount of purified protein required to support reproduction, a natural-ingredient diet containing 18 percent crude protein is thought to meet the amino acid needs for reproduction in hamsters.

MINERALS

Given their widespread use as experimental animals, there is a remarkable paucity of information about the mineral requirements of hamsters.

MACROMINERALS

Calcium and Phosphorus

Normal bone formation occurred in hamsters fed diets containing 6.0 g Ca/kg and 3.5 g P/kg. In the absence of vitamin D, rickets was produced in hamsters fed 4.7 g Ca/kg and 2.0 g P/kg diet (Jones, 1945). Old female hamsters fed diets containing 4.0 g P/kg and 3.0, 5.0, or 7.0 g Ca/kg were in positive calcium balance only at the two higher calcium intakes. Young animals (52 days old) retained calcium when fed 3.0, 5.0, and 7.0 g Ca/kg diet (Kane and McCay, 1947). Stralfors (1961) obtained a 54 percent decrease in the incidence of dental caries in hamsters when the calcium content of the diet was increased from 4.0 to 6.0 g Ca/kg.

Sodium and Chloride

Rowland and Fregly (1988) reported that hamsters, unlike rats, were reluctant to ingest NaCl either spontaneously or after treatment with several natriogenic stimuli that were effective in rats. Furthermore, they noted that variations in intake of NaCl solutions made hamsters extremely refractory to either decreases or increases in functional mineralocorticoid activity.

TRACE MINERALS

No studies were located that specifically addressed the dietary requirements of the hamster for iodine, molybdenum, and selenium or for iron.

Iodine, Molybdenum, and Selenium

Iodine, molybdenum, and selenium are trace elements essential for normal growth in laboratory animals. Birt et al.

(1986) determined that male and female golden hamsters fed a diet containing 30 percent torula yeast as the protein source and 0.1 mg Se (as sodium selenite)/kg diet supported adequate growth, and 5 mg Se/kg is excessive. The iodine requirement may be met by 0.15 mg I/kg diet, and the molybdenum requirement by 0.10 mg Mo/kg diet. The selenium requirement may be met by 0.15 mg Se/kg diet for maintenance, 0.20 for growth and aging, and 0.40 for pregnancy and lactation.

Signs of Iodine Deficiency　Hamsters fed iodine-deficient diets (10 to 25 μg/kg) for several months developed enlarged thyroids when compared to controls fed adequate iodine (7.6 mg/kg) (Follis, 1959, 1962).

Iron

Signs of Iron Deficiency　Chandler et al. (1988) reported that mild iron deficiency can be induced in adult males by feeding them a diet containing 10 mg Fe/kg for several weeks. Carpenter (1982) reported that feeding females a low-iron diet (3 mg Fe/kg) during pregnancy resulted in low maternal weight gain and a high frequency of prenatal mortality compared to controls.

VITAMINS

FAT-SOLUBLE VITAMINS

Vitamin A

The vitamin A requirement of golden hamsters seems to be only slightly greater than that of the rat. Hamsters fed a purified diet containing 2 mg retinyl palmitate/kg diet (3.8 μmol/kg) grew as well as animals fed a commercial natural-ingredient diet (Rogers et al., 1974). The hamsters had normal serum vitamin A concentrations and a very modest accumulation of vitamin A in the livers. Based on these studies the minimum amount of retinol that will maintain a slightly positive vitamin A balance is approximately 1.1 mg/kg diet (3.8 μmol/kg diet).

Signs of Vitamin A Deficiency　Omission of vitamin A from a 24 percent-casein diet resulted in deficiency signs in 6 to 7 weeks. Vitamin A-deficient animals developed abnormally and had coarse and sparse hair, xerophthalmia, and keratinized stratified tracheal lining (Salley and Bryson, 1957). Stomach ulcers formed in adult male hamsters fed a vitamin A-deficient diet for 7 months (Harada et al., 1982).

Signs of Vitamin A Toxicity　Hamsters fed a diet containing 400,000 IU vitamin A/kg (419 μmol/kg) developed

liver pathology and died within 42 to 91 days. Animals fed 100,000 IU/kg diet (105 μmol/kg) and lower concentrations (4,000 and 600 IU/kg) showed no toxic effects (Beems et al., 1987).

Vitamin D

Overt signs of rickets did not appear within a 5-week period when hamsters were fed vitamin D-deficient diets that contained calcium and phosphorus in a ratio of 2:1 and calcium was included at 6 g/kg diet. Rickets may be induced in hamsters in the absence of vitamin D and when dietary calcium is 4 g/kg and phosphorus is 0.2 g/kg (Jones, 1945). No published reports on vitamin D deficiency or toxicity were found.

Vitamin E

In spite of several studies on vitamin E deficiency in hamsters and their use as an animal to bioassay compounds for vitamin E activity, data are not available to provide a good estimate of the vitamin E requirement of hamsters. Bieri and Evarts (1974) found that a diet containing 2.1 μmol/kg *RRR*-α-tocopheryl acetate/kg was adequate to prevent testicular degeneration in the rat. However, plasma creatine phosphokinase (CPK) concentrations were slightly higher at this concentration of intake. With higher dietary concentrations (6.3 μmol/kg), plasma CPK concentrations were normal (Bieri, 1972). Unfortunately the more sensitive criterion of vitamin E adequacy, such as the in vitro red blood cell hemolysis assay, has not been investigated in hamsters. In the rat 6.3 μmol *RRR*-α-tocopherol/kg diet may be adequate to prevent overt signs of vitamin E deficiency, but this concentration is quite likely not adequate for optimal performance. Therefore, 42 μmol *RRR*-α-tocopherol/kg diet (27 IU/kg), which is required by the rat, is probably a more realistic value.

A few breeding colonies have reported a higher-than-expected incidence of spontaneous hemorrhagic necrosis, a fatal disease that affects the central nervous system of fetal hamsters. Keeler and Young (1979) found that a single intraperitoneal injection of 100 μmol of vitamin E on day 7 of gestation protects fetuses from this disease. The problem may arise from improper storage of diets, which leads to destruction of vitamin E.

Signs of Vitamin E Deficiency　The absence of vitamin E in their diet causes hamsters to develop testicular degeneration. Feeding a hamster 21 μmol *RRR*-α-tocopheryl acetate/day restores normal weight and testicular histology. In contrast, rats are unable to reverse vitamin E-induced testicular degeneration (Mason and Mauer, 1975). Vitamin E-deficient hamsters show decreased growth and muscular dystrophy, which can be alleviated by administering high concentrations of vitamin E (West and Mason, 1958). Weanling hamsters fed a vitamin E-deficient diet developed muscular degeneration and died within 2 weeks. Improvement occurred within 30 hours after a single dose of 1 mg α-tocopherol (Houchin, 1942).

Vitamin K

No studies to ascertain the hamster's requirement for vitamin K could be found; however, Rogers et al. (1974) reported that a diet containing 4 mg menadione/kg (23 μmol/kg diet) was adequate for growth—the hamsters presumably received a considerable amount of vitamin K activity from coprophagy. No bleeding problems were reported. The Tolworth HS (Welsh) Warfarin-resistant strain of rats requires 1.77 μmol phylloquinone/kg BW/day (Greaves and Ayers, 1973). The vitamin K requirement of the hamster may be similar. Based on the requirement of the Tolworth HS (Welsh) Warfarin-resistant rat, 25 μmol phylloquinone/kg diet (11 mg/kg) should be a safe and adequate intake for hamsters.

Signs of Vitamin K Deficiency   Adult male hamsters fed a vitamin K-deficient diet and housed in coprophagy-preventive cages showed a drop in prothrombin concentrations to 11 percent of control concentrations within 11 days. Treatment with chloro-K (2-chloro-3-phytyl-1,4-naphthoquinone), a vitamin K antagonist, at 1 to 5 mg chloro-K/kg BW decreased plasma prothrombin to 17 to 20 percent of control values. Hamsters are Warfarin resistant and require a large amount to reduce prothrombin production (Shah and Suttie, 1975).

WATER-SOLUBLE VITAMINS

Biotin

Satisfactory growth of hamsters has been obtained with diets containing 0.82 μmol biotin/kg (Cohen et al., 1971) and 2.5 μmol biotin/kg (Rogers et al., 1974). A dietary concentration of 0.2 mg/kg (0.82 μmol/kg) seems to be a safe and adequate amount of biotin for hamsters. Under normal conditions golden hamsters do not require dietary biotin (Granados, 1968). Apparently, the biotin obtained through coprophagy is sufficient to meet the requirement.

Signs of Biotin Deficiency   A biotin deficiency is induced by feeding hamsters a diet containing both raw egg white and sulfaguanidine. Biotin-deficient animals developed dull rough coats, encrusted eyes, depigmented hair, and jerky movements. Daily injections of 16 nmol biotin (3.9 μg), equivalent to 0.66 mg/kg diet (2.7 μmol/kg), reversed deficiency signs within 4 to 6 weeks (Rauch and Nuting, 1958). Ten adult female hamsters fed a purified

diet containing 5.0 mg biotin/kg diet produced 118 normal live fetuses; but 11 animals fed a biotin-deficient diet had 20 live fetuses (Watanabe and Endo, 1989).

Choline

Hamsters fed a peanut meal diet deficient in choline developed poor appetite, reduced growth, and fatty livers (Handler and Bernheim, 1949). Investigators who fed hamsters diets containing more than 200 g casein/kg did not observe a requirement for choline (Hamilton and Hogan, 1944). Purified diets on which hamsters have achieved satisfactory growth have contained 14 μmol choline chloride/kg diet (Rogers et al., 1974) or 7.1 μmol choline bitartrate/kg diet (Cohen et al., 1971). Under most circumstances 1.8 g choline bitartrate/kg diet should provide a safe and adequate intake of choline.

Folates

Golden hamsters fed 2 mg folic acid/kg diet (Cohen et al., 1971) do not develop folate deficiency. Folate deficiency does develop when golden hamsters are fed a purified diet containing 60 percent sucrose or 1 percent sulfonamide with either cornstarch or sucrose as the carbohydrate. Hamsters resemble guinea pigs rather than rats in that a folate deficiency can be produced without the use of sulfonamide.

Signs of Folate Deficiency   In hamsters given a folate-deficient diet, liver folates are decreased to 10 percent of control and blood PCV and hemoglobin values are decreased 20 percent. In the deficient animal, increases are seen in urinary excretion of formiminoglutamic acid and aminoimidazolecarboxamide (Cohen et al., 1971).

Myo-inositol

Granados (1968) has stated that hamsters do not require *myo*-inositol for normal growth. Hamilton and Hogan (1944) also demonstrated that *myo*-inositol is not required for growth but is necessary for reproduction.

Niacin

Young hamsters fed diets containing 20 percent casein do not require dietary niacin for growth (Hamilton and Hogan, 1944; Granados, 1951). Niacin is necessary, however, for normal reproduction and adequate litter size (Hamilton and Hogan, 1944).

Signs of Niacin Deficiency   Hamsters fed a niacin-free, purified diet developed rough and denuded hair, suffered loss of weight and death. Hair quality and weight gain improved with daily administration of 100 μg niacin (Routh and Houchin, 1942). Niacin supplementation does not improve growth of hamsters fed 20 percent casein (Hamilton and Hogan, 1944; Granados, 1951). However, niacin probably will be required in diets with low concentrations of protein or diets in which tryptophan is first limiting. Animals fed purified diets containing 81 μmol niacin/kg achieved satisfactory growth rates over 11 weeks (Cohen et al., 1971).

Pantothenic Acid

Routh and Houchin (1942), Hamilton and Hogan (1944), Granados (1951), and Cohen et al. (1963) identified pantothenic acid as a nutrient required for normal growth. Nevertheless, the requirement for this vitamin has not been quantified in the hamster. Concentrations of dietary Ca-pantothenate from 21 μmol/kg diet (Hamilton and Hogan, 1944) to 84 μmol/kg diet (Rogers et al., 1974) have been used in purified diets. None of the investigators indicated whether the type used was Ca-*d*-pantothenate or Ca-*dl*-pantothenate. Thus, 21 μmol Ca-*d*-pantothenate/kg diet (10 mg/kg diet) seems to be a safe and adequate concentration of pantothenic acid activity.

Signs of Pantothenic Acid Deficiency In the absence of pantothenic acid, hamsters lost weight, developed a red encrustation around the mouth, and died in 20 days. Daily injections of 15 μg Ca-pantothenate supported maintenance, but larger doses were needed for growth (Routh and Houchin, 1942).

Vitamin B$_6$

Male weanling hamsters fed a pyridoxine-deficient diet for 2 to 3 weeks decreased their food and water intake and stopped growing. No quantitative requirement for vitamin B$_6$ can be set at this time.

Signs of Vitamin B$_6$ Deficiency In addition to weight loss, the hair of vitamin B$_6$-deficient hamsters was unkempt, and crusted lesions were occasionally observed on lips and mouth. Increased xanthurenic acid was found in urine. Atrophy of lymphoid tissue, particularly in the thymus, is an outstanding pathological change (Schwartzman and Strauss, 1949). In the absence of vitamin B$_6$, hamsters lose weight and develop an acrodynia-like condition around the mouth in less than 2 weeks. A daily dose of 3 μg pyridoxine cured the dermatitis and produced moderate growth (Rouch and Houchin, 1942).

Riboflavin

Riboflavin is required for normal growth and develop-

ment of hamsters, but no quantitative requirement has been established (Hamilton and Hogan, 1944; Granados, 1968). Deficiency signs did not occur in animals fed 20 mg/kg diet (Smith and Reynolds, 1961). Riboflavin depletion (measured by erythrocyte glutathione reductase) was produced in hamsters fed two concentrations of riboflavin—0.5 and 1.5 mg/kcal—in a liquid diet. Decreased growth was observed at the lower dose (Kim and Roe, 1985). Thus, a diet containing 15 mg/kg should be sufficient to support normal growth in hamsters.

Signs of Riboflavin Deficiency In the absence of dietary riboflavin, hamsters reduced their food and water intake, became inactive, showed stunted growth, and developed dull coats (Smith and Reynolds, 1961).

Thiamin

Thiamin is necessary for normal growth, but the specific requirement has not been established. Satisfactory growth has been obtained with purified diets containing 20 mg thiamin/kg diet (Arrington et al., 1966).

Signs of Thiamin Deficiency Hamsters fed 4 mg thiamin/kg diet developed a chronic deficiency (Salley et al., 1962). Hamsters fed a thiamin-deficient diet developed polyneuritis in 12 days. Oral administration of 3 μg thiamin/day reversed these signs in 2 days (Routh and Houchin, 1942).

Vitamin B$_{12}$

Early reports concluded that vitamin B$_{12}$ is not required for normal growth of golden hamsters (Scheid et al., 1950; Granados, 1951). Hamsters fed a diet high in soybean protein and cornstarch showed a mild vitamin B$_{12}$ deficiency identified by the presence of distinctive metabolites (methylmalonic acid and formiminoglutamic acid) in urine (Cohen et al., 1967). Metabolic changes in deficient animals were corrected by feeding them a diet containing 10 μg vitamin B$_{12}$/kg diet. The inclusion of inorganic cobalt (5 mg/kg diet) in the diet reversed deficiency changes and increased tissue storage of vitamin B$_{12}$ (Tseng et al., 1976).

POTENTIALLY BENEFICIAL DIETARY CONSTITUENTS

FIBER

Fiber-free diets containing high concentrations of purified sugars result in high mortality (Salley and Bryson, 1957). The substitution of cornstarch for glucose and sucrose or addition of 12 to 20 percent alfalfa to diets increased survival (Ershoff, 1956). Basal diets containing starch or lactose may not require fiber additions because these ingredients support favorable microflora in the colon (Snog-Kjaer et al., 1963; Hayes et al., 1989). Microorganisms present in the cecum and colon seem to be capable of degrading fiber sources (Banta et al., 1975; Vorontsov, 1979). Hamsters, like other rodents, practice coprophagy (von Frisch, 1990).

ASCORBIC ACID

Early work suggested that golden hamsters do not require a dietary source of ascorbic acid (Clausen and Clark, 1943). Male hamsters fed a purified diet supplemented with 4.0 mg ascorbic acid/g diet gained 1.07 g/day with a food intake of 3.5 g/day. Growth curves overlapped; curves for supplemented animals were slightly above those of controls, which were fed diets containing no ascorbic acid. None of the animals weighed more than 96 g after 120 days. Poiley (1972) reported average male weight as 122 g at 112 days (Table 5-4).

Forty female hamsters fed a diet scorbutic for guinea pigs were given daily supplements of ascorbic acid in water by pipette according to the weight/dose chart of Dann and Cowgill (1935). Controls received this diet and an equal amount of water. At 70 days average weights were 105.6 g with a 0.3 mg dose, 99.4 g with a 0.65 mg dose, and 91.0 g with 0.9 mg dose per 100-g animal per day. Controls fed water and a diet scorbutic for guinea pigs averaged 88.5 g. Animals fed water and a diet supplemented with lettuce averaged 92.3 g (Hovde, 1950). Poiley (1972) reported, on average, females weighed 103 g at 70 days.

REFERENCES

Adler, S. 1948. Origin of the golden hamster *Cricetus auratus* as a laboratory animal. Nature 162:256–257.

Adler, S., and O. Theodor. 1931. Investigations on Mediterranean kala azar. Proc. R. Soc. Biol. 108:453–463.

Anderson, M. C., and S. T. Shettleworth. 1977. Behavior adaptation to fixed-interval and fixed-time food delivery in golden hamster. J. Exp. Anal. Behav. 25:33–49.

Anderson, R. R., and K. N. Shina. 1972. Number of mammary glands and litter size in the golden hamster. J. Mammal. 53:382–384.

Anderson, S., and J. K. Jones, Jr., eds. 1984. Zoogeographic Map. P. 4 in Orders and Families of Recent Mammals of the World. New York: Wiley.

Arrington, L. R., J. K. Platt, and R. L. Shirley. 1966. Protein requirements of growing hamsters. Lab. Anim. Sci. 16:492–496.

Arrington, L. R., C. B. Ammerman, and D. E. Franke. 1979. Protein requirement of hamsters fed a natural diet. Lab. Anim. Sci. 29:469–471.

Avery, T. L. 1968. Observations on the propagation of Chinese hamsters. Lab. Anim. Care 18:151–159.

Balk, M. W., and G. M. Slater. 1987. Care and management. Pp. 63–64 in Laboratory Hamsters, G. L. Van Hoosier, Jr., and C. W. McPherson, eds. Orlando, Fla.: Academic Press.

Banta, C. A., R. G. Warner, and J. B. Robertson. 1975. Protein nutrition of the golden hamster. J. Nutr. 105:38–45.

Beems, R. B., L. van Beek, A. A. J. J. L. Rutten, and A. J. Speek. 1987. Subchronic (106-day) toxicology and nutrition studies with vitamin A and β-carotene in Syrian hamsters. Nutr. Rep. Int. 35:765–770.

Bieri, J. G. 1972. Kinetics of tissue α-tocopherol depletion and replication. Ann. N.Y. Acad. Sci. 203:181–191.

Bieri, J. G., and R. P. Evarts. 1974. Vitamin activity of γ-tocopherol in the rat, chick, and hamster. J. Nutr. 104:850–857.

Birt, D. F., and R. D. Conrad. 1981. Weight gain, reproduction and survival of Syrian hamsters fed five natural ingredient diets. Lab. Anim. Sci. 31:149–155.

Birt, D. F., P. Y. Baker, and D. S. Hruza. 1982. Nutritional evaluations of three dietary levels of lactalbumin throughout the life span of two generations of Syrian hamsters. J. Nutr. 112:2151–2160.

Birt, D. F., A. D. Julius, and C. E. Runice. 1986. Tolerance of low and high dietary selenium throughout the life span of Syrian hamsters. Ann. Nutr. Metab. 30:233–240.

Biven, W. S., G. A. Olsen, and K. A. Murray. 1987. Morphophysiology. Pp. 12–19 in Laboratory Hamsters, G. L. Van Hoosier, Jr., and C. W. McPherson, eds. Orlando, Fla.: Academic Press.

Borer, K. I. 1985. Regulation of energy balance in the golden hamster. Pp. 363–408 in The Hamster—Reproduction and Behavior, H. I. Siegel, ed. New York: Plenum Press.

Borer, R. T., N. Rowland, A. Mirow, R. C. Borer, Jr., and R. P. Kelch. 1979. Physiological and behavioral responses to starvation in the golden hamster. Am. J. Physiol. 236:E105–E112.

Boyer, C. C. 1968. Embryology. Pp. 73–90 in The Golden Hamster: Its Biology and Use in Medical Research, R. A. Hoffman, P. F. Robinson, and H. Magalhaes, eds. Ames: Iowa State University Press.

Bruce, H. M., and E. Hindle. 1934. The golden hamster, *Criceus (Mesocricetus auratus)* Waterhouse: Notes on its breeding and growth. Proc. Zool. Soc. London 104:361–366

Calland, C. J., S. R. Wightman, and S. B. Neal. 1986. Establishment of a Chinese hamster breeding colony. Lab. Anim. Sci. 36:183–185.

Cantrell, C. A., and D. A. Padovan. 1987. Other hamsters—Biology, care, and use in research. Pp. 369–387 in Laboratory Hamsters, G. L. Van Hoosier, Jr., and C. W. McPherson, eds. Orlando, Fla.: Academic Press.

Carleton, M. D., and G. G. Musser. 1984. Muroid rodents. Pp. 305, 310–314 in Orders and Families of Recent Mammals of the World, S. Anderson and J. K. Jones, Jr., eds. New York: Wiley.

Carpenter, S. J. 1982. Enhanced teratogenicity of orally administered lead in hamsters fed diets deficient in calcium or iron. Toxicology 24:259–271.

Chandler, D. B., J. C. Barton, D. D. Briggs III, T. W. Butler, J. I. Kennedy, W. E. Grizzle, and J. D. Fulmer. 1988. Effect of iron deficiency on bleomycin-induced lung fibrosis in the hamster. Am. Rev. Respir. Dis. 137:85–89.

Christensen, F., and H. Dam. 1952. A new symptom of fat deficiency in hamsters: Profuse secretion of cerumen. Acta Physiol. Scand. 27:204–205.

Clausen, D. F., and W. G. Clark. 1943. Vitamin C requirement of the Syrian hamster. Nature 152:300–301.

Cohen, N. L., L. Arnrich, and R. Okey. 1963. Pantothenic acid deficiency in cholesterol-fed hamsters. J. Nutr. 80:142–144.

Cohen, N. L., P. Reyes, J. T. Typpo, and G. M. Briggs. 1967. Vitamin B_{12} deficiency in the golden hamster. J. Nutr. 91:482–488.

Cohen, N. L., P. S. Reyes, and G. M. Briggs. 1971. Folic acid deficiency in the golden hamster. Lab. Anim. Sci. 21:350–355.

Cunnane, S. C., M. S. Manku, and D. F. Horrobin. 1985. Effect of ethanol on liver triglycerides and fatty acid composition in the Golden Syrian hamster. Ann. Nutr. Metab. 29:246–252.

Cunnane, S. C., Y. S. Huang, M. S. Manken, and D. F. Horrobin. 1986. Influence of different dietary fatty acids sources on erythocyte lipids and plasma and liver essential fatty acids in hamsters fed ethanol. Ann. Nutr. Metab. 30:81–86.

Cunnane, S. C., K. R. McAdoo, and F. D. Horrobin. 1987. Longterm ethanol consumption in the hamster: Effects on tissue lipids, fatty acids and erythrocyte hemolysis. Ann. Nutr. Metab. 31:265–271.

Daly, M. 1975. Behavioral development in three hamster species. Dev. Psychobiol. 9:315–323.

Dam, H., and F. Christensen. 1961. Alimentary production of gallstones in hamsters. 9. Influence of different carbohydrate sources on gallstone formation, diarrhea and growth. Zeit. Ernahrung. 2:91–103.

Dann, M., and G. R. Cowgill. 1935. Vitamin C requirement for the guinea pig. J. Nutr 9:507–519.

Diani, A., and G. Gerritsen. 1987. Chinese hamster—Use in research. Pp. 329–347 in Laboratory Hamsters, G. L. Van Hoosier, Jr., and C. W. McPherson. Orlando, Fla.: Academic Press.

Doull, J. A., and E. Megrall. 1939. Inoculation of human leprosy into the Syrian hamster. Int. J. Leprosy 7:509–512.

Ehle, F. R., and R. G. Warner. 1978. Nutritional implications of the hamster forestomach. J. Nutr. 108:1047–1053.

Ershoff, B. H. 1956. Beneficial effects of alfalfa, aureomycin and cornstarch on the growth and survival of hamsters fed highly purified rations. J. Nutr. 59:579–585.

Feldman, D. B., E. E. McConnell and J. J. Knapka. 1982. Growth, kidney disease and longevity of Syrian hamsters (Mesocricetus auratus) fed varying levels of protein. Lab. Anim. Sci. 32:613–618.

Fitts, D. A., and C. St. Dennis. 1981. Ethanol and dextrose preferences in hamsters. J. Stud. Alcohol 42:901–907.

Follis, R. H., Jr. 1959. Experimental colloid goiter in the hamster. Proc. Soc. Exp. Biol. Med. 100:203–206.

Follis, R. H., Jr. 1962. Iodinated protein patterns in thyroids from normal and iodine deficient hamsters. Proc. Soc. Exp. Biol. Med. 110:57–60.

Gamperl, R., G. Vistorin, and W. Rosenkranz. 1978. Comparison of chromosome banding patterns in five members of Cricetinae with comments on possible relationships. Caryologia 31:343–353.

Granados, H. 1951. Nutritional studies on growth and reproduction of the golden hamster (Mesocricetus auratus auratus). Chapter V. Qualitative requirements of the chemically identified vitamins for growth. Acta Physiol. Scand. 24(Suppl. 87):55–60.

Granados, H. 1968. Nutrition. Pp. 157–170 in The Golden Hamster: Its Biology and Use in Medical Research, R. A. Hoffman, P. F. Robinson, and H. Magalhaes, eds. Ames: Iowa State University Press.

Greaves, J. H., and P. Ayres. 1973. Warfarin resistance and vitamin K requirement in the rat. Lab. Anim. 7:141–148.

Gustafson, G., E. M. Stelling, E. Abramson, and E. Brunius. 1955. The cariogenic effect of different carbohydrates in dry and moist diets. Odont. Tid. 63:506–523.

Hamar, M., and M. Schutowa. 1966. New data on the geographic variability and the development of species of Mesocricetus. (Gr.) Zeitschrift Saugetierk. 31:237–251.

Hamilton, J. W., and A. G. Hogan. 1944. Nutritional requirements of the Syrian hamster. J. Nutr. 27:213–224.

Handler, P., and F. Bernheim. 1949. Choline deficiency in the hamster. Proc. Soc. Exp. Biol. Med. 72:569–571.

Harada, T., S. Yamashiro, P. D. Meade, P. K. Basrur, K. Maita, and Y. Shirasu. 1982. Stomach ulcers in vitamin A deficient Syrian golden hamsters. Jpn. J. Vet. Sci. 44:267–274.

Harkness, J. E., and J. E. Wagner. 1983. The Biology and Medicine of Rabbits and Rodents. Philadelphia: Lea & Febiger.

Hayes, K. C., A. Pronczuk, and J. S. Liang. 1993. Differences in the plasma transport and tissue concentrations of tocopherols and tocotrienols: Observations in humans and hamsters. Proc. Soc. Exp. Biol. Med. 202:353–359.

Hayes, K. C., Z. F. Stephan, A. Pronczuk, S. Lindsey, and C. Verdon. 1989. Lactose protects against estrogen-induced pigment gallstones in hamsters fed nutritionally adequate purified diets. J. Nutr. 119:1726–1736.

Hoffman, R. A., P. F. Robinson, and H. Magalhaes. 1968. The Golden Hamster—Its Biology and Use in Medical Research. Ames: Iowa State University Press.

Hoffmann, K. 1973. The influence of photoperiod and melantonin on testis size, body weight, and pelage colour in the Djungarian hamster (Phodopus sungorus). J. Comp. Physiol. 85:267–282.

Hoffmann, K. 1978. Effects of short photoperiods on puberty, growth, melatonin and moult in the Djungarian hamster (Phodopus sungorus). J. Reprod. Fertil. 54:29–35.

Holman, R. T. 1968. Essential fatty acid deficiency. Proc. Chem. Fats Other Lipids. 9:318–325.

Horowitz, I., and H. A. Waisman. 1966. Some biochemical changes in the hamsters fed excess phenylalanine diets. Proc. Soc. Exp. Biol. Med. 122:750–755.

Houchin, O. B. 1942. Vitamin E and muscle degeneration. Fed. Proc. 1:117–118 (abstr.).

Hovde, C. H. 1950. The vitamin C requirement for normal growth, pregnancy, and lactation of the golden hamster. M.S. thesis. Bucknell University, Lewisburg, Pennsylvania.

Hsieh, E. T. 1919. A new laboratory animal (Cricetulus griseus). Natl. Med. J. China 5:20–24.

Iakovenko, V. V. J. J. 1974. The estrous cycle of the Djungarian hamster. Arch. Anat. Physiol. Embryol. 66:32–36. Estral'nyl Tsikl Dzungarskogo Khomiachka [J. J. Emmerieck, trans.]. Arkhiv Anatomii, Gistologii Embriologii. 66:32–36.

Jones, J. H. 1945. Experimental rickets in the hamster. J. Nutr. 30:143–146.

Kane, G. G., and C. M. McCay. 1947. Calcium requirement of old and young hamsters and rats. J. Gerontol. 2:244–248.

Keeler, R. F., and S. Young. 1979. Role of vitamin E in the etiology of spontaneous hemorrhagic necrosis of the central nervous system of fetal hamsters. Teratology 20:127–132.

Kim, C. I., and D. A. Roe. 1985. Development of riboflavin deficiency in alcohol-fed hamsters. Drug-Nutr. Interact. 3:99–107.

Knapka, J. J., and F. J. Judge. 1974. The effects of various levels of dietary fat and apple supplements on growth of golden hamsters (Mesocricetus auratus). Lab. Anim. Sci. 24:318–325.

Krueger, H., and R. Rieschel, Jr. 1950. Valves of the hamster caecum. Fed. Proc. 9:72–73 (abstr.).

Kunstyr, I. 1974. Some quantitative and qualitative aspects of the stomach microflora of the conventional rat and hamster. Zblt. Vet. Med. A 21:553–561.

Lavappa, K. S., and G. Yerganian. 1970. Spermatogonial and meiotic chromosomes of the Armenian hamster, Cricetus migratorius. Exp. Cell Res. 61:159–172.

Lyman, C. P., and R. C. O'Brien. 1977. A laboratory study of the Turkish hamster Mesocricetus brandti. Brevora 442:1–27.

Lyman, C. P., R. C. O'Brien, G. C. Greene, and E. D. Papafrangos. 1981. Hibernation and longevity in the Turkish hamster Mesocricetus brandti. Science 212:668–670.

Lyman, C. P., R. C. O'Brien, and W. H. Bossert. 1983. Differences in tendency to hibernate among groups of Turkish hamsters (Mesocricetus brandti). J. Thermal Biol. 8:255–257.

Magalhaes, H. 1968. Gross anatomy. Pp. 91–109 in The Golden Hamster—Its Biology and Use in Medical Research, R. A. Hoffman, P. F. Robinson, and H. Magalhaes. Ames: Iowa State University Press.

Mason, K. E., and S. I. Mauer. 1975. Reversible testis injury in the vitamin E-deficient hamster. J. Nutr. 105:484–490.

Matsumoto, T. 1955. Nutritive value of urea as a substitute for feed protein. I. Utilization of urea by the golden hamster. Tohoku J. Agr. Res. 6:127–131.

Micheli, M. O., and C. W. Malsbury. 1982. Availability of a food hoard facilitates maternal behaviour in virgin female hamsters. Physiol. Behav. 28:855–856.

Mohr, U., and H. Ernst. 1987. The European hamster—Biology, care, and use in research. Pp. 351–366 in Laboratory Hamsters, G. L. Van Hoosier, Jr., and C. W. McPherson, eds. Orlando, Fla.: Academic Press.

Moore, W., Jr. 1965. Observations on the breeding and care of the Chinese hamster, Cricetulus griseus. Lab. Anim. Care 15:94–101.

Murphy, M. R. 1977. Intraspecific sexual preferences of female hamsters. J. Comp. Physiol. Psychol. 91:1337–1346.

Murphy, M. R. 1985. History of the capture and domestication of the Syrian golden hamster. Pp. 7–14 in The Hamster—Reproduction and Behavior, H. I. Siegel, ed. New York: Plenum Press.

Musser, G. G., and M. D. Carleton. 1993. Family Muridae. Pp. 536–539 in Mammal Species of the World, 2nd Ed., D. E. Wilson and D. M. Reeder, eds. Washington, D.C.: Smithsonian Institution Press.

Parkening, T. A. 1982. Reproductive senescence in the Chinese hamster (Cricetus griseus). J. Gerontol. 37:283–287.

Pogosianz, H. E., and O. L. Sokova. 1967. Maintaining and breeding of the Djungarian hamster under laboratory conditions. Zeit. Versuchtier 9:292–297.

Poiley, S. M. 1950. Breeding and care of the Syrian hamster, Cricetus auratus. Pp. 118–152 in The Care and Breeding of Laboratory Animals, E. J. Farris, ed. New York: Wiley.

Poiley, S. M. 1972. Growth tables for 66 strains and stocks of laboratory animals. Lab. Anim. Sci. 22:759–779.

Pond, C. M., D. Sadler, and C. A. Mattacks. 1987. Sex differences in the distribution of adipose tissue in the Djungarian hamsters Phodopus sungorus. Nutr. Res. 7:1325–1328.

Popescu, N. C., and J. A. DiPaolo. 1980. Chromosomal interrelationship of hamster species of the genus Mesocricetus. Cytogent. Cell Genet. 28:10–23.

Rauch, H., and W. B. Nuting. 1958. Biotin deficiency in the hamster. Experientia 14:382–383.

Reznik, G., H. Reznik-Schüller, and U. Mohr. 1978. Clinical anatomy of the European hamster (Cricetus cricetus L.), P. C. Water and P. Dodson, eds. Washington, D.C.: U.S. Government Printing Office.

Reznik-Schüller, H., G. Reznik, and U. Mohr. 1974. The European hamster (Cricetus cricetus L.) as an experimental animal: Breeding methods and observations of their behavior in the laboratory. Zeit. Versuchstier. 16:48–58.

Robens, J. F. 1968. Influence of maternal weight on pregnancy, number of corpora lutea, and implantation sites in the golden hamster (Mesocricetus auratus). Lab. Anim. Care 18:651–653.

Rogers, A. E., G. H. Anderson, G. M. Lenhardt, G. Wolf, and P. M. Newberne. 1974. A semisynthetic diet for long-term maintenance of hamsters to study the effects of dietary vitamin A. Lab. Anim. Sci. 24:495–499.

Routh, J. I., and O. B. Houchin. 1942. Some nutritional requirements of the hamster. Fed. Proc. 1:191–192 (abstr.).

Rowland, N. E., and M. J. Fregly. 1988. Sodium appetite: Species and strain differences and role of renin-angiotensin-aldosterone system. Appetite 11:143–178.

Ruf, T., M. Klingenspor, H. Preis, and G. Heldmaier. 1991. Daily torpor in the Djungarian hamster (Phodopus sungorus) interactions with food intake, activity, and social behavior. J. Comp. Physiol. B 160:609–615.

Rutten, A. A. J. J. L., and A. P. de Groot. 1992. Comparison of cereal-based with purified diet by short-term feeding studies in rats, mice, and hamsters, with emphasis on toxicity characteristics. Food Chem. Toxicol. 30:601–610.

Sakaguchi, E., J. Itoh, H. Shinohara, and T. Matsumoto. 1981. Effects of removal of the forestomach and caecum on the utilization of dietary urea in golden hamsters (Mesocricetus auratus) given two different diets. Br. J. Nutr. 46:503–512.

Salley, J. J., and W. F. Bryson. 1957. Vitamin A deficiency in the hamster. J. Dent. Res. 36:935–944.

Salley, J. J., J. R. Eshlemann, and J. H. Morgan. 1962. Effect of chronic thyroid deficiency on oral carcinogens. J. Dent. Res. 41:1405–1412.

Scheid, H. E., B. H. McBride, and B. S. Schweigert. 1950. The vitamin B_{12} requirements of the Syrian hamster. Proc. Soc. Exp. Biol. Med. 75:236–239.

Schwartzman, G., and L. Strauss. 1949. Vitamin B_6 deficiency in the Syrian hamster. J. Nutr. 38:131–153.

Selle, R. M. 1945. Hamster sexually mature at twenty-eight days of age. Science 102:485–486.

Shah, D. V., and J. W. Suttie. 1975. Vitamin K requirement and Warfarin tolerance in hamsters. Proc. Soc. Exp. Biol. Med. 150:126–128.

Siegel, H. I., ed. 1985. The Hamster—Reproduction and Behavior. New York: Plenum Press.

Slater, G. M. 1972. The care and feeding of the Syrian hamster. Prog. Exp. Tumor Res. 16:42–49.

Smith, C. 1957. The introduction and breeding of the Chinese striped hamster (Cricetus griseus) in Great Britain. J. Anim. Tech. Assoc. 7:59–60.

Smith, R. E., and I. M. Reynolds. 1961. Leptospirosis in hamsters on diets containing various levels of riboflavin. Am. J. Vet. Res. 22:800–806.

Snog-Kjaer, A., I. Prange, F. Christensen, and H. Dam. 1963. Alimentary production of gallstones in hamsters. 12. Studies with rice starch diets with and without antibiotics. Zeit. Ernahrung. 4:14–25.

Steinlechner, S., G. Heldmaier, and H. Becker. 1983. The seasonal cycle of body weight in the Djungarian hamster: Photoperiodic control and the influence of starvation and melatonin. Oecologia (Berlin) 60:401–405.

Stralfors, A. 1961. Inhibition of dental caries in hamsters. V. The effect of dibasic and monobasic phosphate. Odont. Rev. 12:236–256.

Takahashi, S., and H. Tamate. 1976. Light and electron microscopic observation of the forestomach mucosa in the golden hamster. Tohoku J. Agr. Res. 27:26–39.

Thompson, R. 1971. The water consumption and drinking habits of a few species and strains of laboratory animals. J. Inst. Anim. Tech. 22:29–36.

Todd, N. B., G. W. Nixon, D. A. Mulvaney, and M. E. Connelly. 1972. Karyotypes of Mesocricetus brandti and hybridization within the genus. J. Hered. 63:73–77.

Tseng, R. Y., L. N. L. Cohen, P. S. Reyes, and G. M. Briggs. 1976. Metabolic changes in golden hamsters fed vitamin B_{12}-deficient diets. J. Nutr. 106:77–85.

Van Hoosier, Jr., G. L., and C. W. McPherson, eds. 1987. Laboratory Hamsters. Orlando: Academic Press.

von Frisch, O. 1990. Hamsters. Hauppauge, N.Y.: Barron's Educational Series.

Vorontsov, N. N. 1979. Evolution of the Alimentary System in Myomorph Rodents. Washington, D.C.: Smithsonian Institution and National Science Foundation; New Delhi: Indian National Scientific Document Centre.

Wade, G. N., and T. J. Bartness. 1984. Effects of photoperiod and gonadectomy on food intake, body weight, and body composition in Siberian hamsters. Am. J. Physiol. 246:R26–R30.

Watanabe, T., and A. Endo. 1989. Species and strain differences in teratogenic effects of biotin deficiency in rodents. J. Nutr. 119:255–261.

West, W. T., and K. E. Mason. 1958. Histopathology of muscular dytrophy in the vitamin E-deficient hamster. Am. J. Anat. 102:323–363.

Wilson, D. E., and D. M. Reeder. 1993. Mammal Species of the World: A Taxanomic and Geographic Reference, 2nd Ed. Washington, D.C.: Smithsonian Institution Press.

Yerganian, G. 1958. The striped-back hamster. J. Natl. Cancer Inst. 20:705–727.

Yerganian, G. 1977. History and cytogenetics of hamsters. Prog. Exp. Tumor Res. 16:2–41.

6 Nutrient Requirements of the Gerbil

The Mongolian gerbil, *Meriones unguiculatus*, used in laboratories in the United States, originates from animals captured in the Amur River basin in eastern Mongolia in 1935 (Rich, 1968; Thiessen and Yahr, 1977). The Mongolian gerbil is one of 13 similar species of gerbils and jirds of the genus *Meriones* distributed in North Africa, the Middle East, and central and eastern Asia (Corbet, 1980). The name gerbil is also applied to other genera in the subfamily Gerbillinae. The gerbils as a group are typically arid-adapted inhabitants of deserts and dry steppes (Corbet, 1980). As arid-adapted species they produce concentrated urine and have low water turnover rates (Burns, 1956; Holleman and Dieterich, 1973). In the remainder of this report the name gerbil will be reserved for the Mongolian gerbil, unless otherwise indicated.

REPRODUCTION AND DEVELOPMENT

Some aspects of the reproduction and development of gerbils are summarized in Table 6-1. Adult gerbils weigh about 70 to 135 g. Pups at birth weigh about 2.5 g and are normally weaned at 14 to 18 g. An acceptable weight gain for the postweaning period from 3.5 to 7 weeks is 1 g/day. After reaching sexual maturity at 65 to 85 days, females are polyestrus, have a gestation period of 24 to 26 days, and exhibit postpartum mating (Marston and Chang, 1965; Loew, 1968; Rich, 1968; Gulotta, 1971; McManus, 1971; Schwentker, 1971; Norris and Adams, 1972; Thiessen and Yahr, 1977).

The stomach of the gerbil is simple and the cecum and colon are not especially well developed, suggestive of a species that in nature consumes mostly low-fiber foods such as seeds (Gulotta, 1971; Vorontsov, 1979). Gerbils generally have had acceptable growth and reproduction when fed pelleted natural-ingredient diets formulated for other rodent species such as rats, mice, and guinea pigs.

Sometimes supplementary cereals and/or seeds have been used but these are not necessary (Marston and Chang, 1965; Arrington, 1968; Loew, 1968; Rich, 1968; McManus, 1971, 1972; McManus and Zurich, 1972; Norris and Adams, 1972).

ENERGY AND WATER

Growing gerbils consume about 5 to 6 g dry diet/day or 8 to 10 g diet/100 g BW. Dietary energy intake averaged 36 to 40 kcal gross energy/100 g BW/day (150 to 170 kJ/100 g BW/day) (Harriman, 1969; McManus and Zurich, 1972; Mele, 1972). Digestibility of energy was 93 to 94 percent when ambient temperature was 0° to 15° C, and both intake and digestibility decreased at temperatures of 20° to 35° C (Mele, 1972).

Although gerbils are arid-adapted and able to subsist on relatively low water intakes, Rich (1968) observed high

TABLE 6-1 Reproductive and Developmental Indices for the Mongolian Gerbil

Variable	Unit	Amount
Minimum breeding age	Week	10–12
Estrous cycle	Day	4–6
Gestation period	Day	24–26
Litter size	Pups	4.5 (range 1–12)
Birth weight	g	2.5
Incisors erupt	Day	10–16
Eyes open	Day	16–20
Weaning age	Day	21–24
Weaning weight	g	14–18
Litters per lifetime	Number	≥6
Life span	Year	>4

SOURCES: Gulotta (1971), Thiessen and Yahr (1977), and references cited in text.

mortality when gerbils were restricted for long periods to dry-type diets without water or succulent feeds such as carrots or lettuce. Thus free access to water and/or succulent feeds should be provided when dry diets are used (Marston and Chang, 1965; Rich, 1968; McManus, 1971; Mele, 1972; Norris and Adams, 1972). Gerbils will voluntarily consume 4 to 10 mL water/100 g BW/day (Winkelman and Getz, 1962; Harriman, 1969; McManus, 1972). Total daily water intake (including free water in food and metabolic water) has been estimated as 8 to 13 percent of body weight (McManus, 1972; Holleman and Dieterich, 1973).

LIPIDS

Purified diets fed to gerbils have contained 2 to 20 percent fat (Zeman, 1967; Arrington, 1968; Harriman, 1969; Arrington et al., 1973; Hegsted et al., 1973; Kroes et al., 1973; Hegsted et al., 1974), but the growth response to variation in dietary fat per se has not been quantitated. No minimum requirements for fat and essential fatty acids have yet been determined, although gerbils have been maintained for prolonged periods on purified diets containing as little as 1 to 2 percent of metabolizable energy from 18:2 [Pronczuk et al., 1994 (in press)].

Although the gerbil can convert linoleic acid to arachidonic acid, arachidonic acid is minimally present in plasma cholesterol esters and comparatively low in the body fat of the gerbil. The body fat of gerbils is higher in oleic and palmitic acid than is the body fat of rats (Gordon and Mead, 1964).

The gerbil responds to high-fat, high-cholesterol diets with increased HDL- and LDL-cholesterol concentrations, especially when diets contain casein, and may prove to be a useful model for the study of cholesterol metabolism (Nicolosi et al., 1976; Forsythe, 1986; DiFrancesco and Mercer, 1990). High dietary cholesterol leads to excess deposits in several body organs but not in arteries. However, older breeder animals fed natural-ingredient diets show spontaneous arteriosclerosis (Gordon and Cekleniak, 1961; Wexler et al., 1971; D'Elia et al., 1972).

PROTEIN

Weight gains of 1 g/day were obtained when weanling gerbils (18 g) were fed purified diets containing 16 percent or more protein, but weight gains were lower (0.6 to 0.8 g/day) when the gerbils received diets containing 12 to 14 percent protein (Arrington et al., 1973). Young gerbils (38 g) fed purified diets with 13 percent protein as casein gained only 0.69 g/day as compared to gains of 0.81 to 0.88 g/day when fed 17 to 25 percent protein (Hall and Zeman, 1968). Based on these studies, the protein requirement of growing

gerbils seems to be about 16 percent when dietary fat is 2 to 5 percent.

Little specific information is available on amino acid requirements of gerbils. However, purified diets based on amino acids have been fed to gerbils with mixed success (Otken and Garza, 1983). Gerbils fed an amino acid-based purified diet had greatly improved growth when taurine was added to the diet at a concentration of 4.5 g/kg (36 mmol/kg) (Otken et al., 1985). Taurine added at concentrations of 7 g/kg diet (60 mmol/kg diet) resulted in lower growth rates.

MINERALS

MACROMINERALS

Calcium and Phosphorus

The amounts of minerals in natural-ingredient rodent diets commonly fed to gerbils (e.g., Table 2-3) are apparently sufficient to meet the needs of gerbils. In the absence of specific data on the requirements of gerbils, the recommended dietary concentrations of calcium (5.0 g Ca/kg diet) and phosphorus (3.0 g P/kg diet) are the same as for the rat (Table 2-2).

Magnesium

A low incidence (≤ 2 percent) of convulsive seizures, especially in response to handling, environmental change, or other stimulation, has been noted in many gerbil colonies (Marston and Chang, 1965; Zeman, 1967; Thiessen et al., 1968; Harriman, 1974; Loskota et al., 1974; McCarty, 1975). Gerbils fed a low-magnesium, purified diet had an elevated susceptibility to seizure in a novel environment; the seizures were eliminated when magnesium was added to the diet at 1.39 g/kg (Harriman, 1974). Gerbils fed purified diets low in calcium, sodium, or vitamin B_6 did not have seizures. Gerbils develop some degree of alopecia when fed purified diets containing ≤ 1.0 g Mg/kg, with the severity related to the extent of magnesium deprivation. Alopecia became noticeable after 14 days when dietary magnesium was less than 0.12 g/kg, and a mortality rate of 70 to 83 percent occurred within 40 days when magnesium was 0.06 to 0.12 g/kg diet. A dietary concentration of 0.25 g Mg/kg prevented weight loss and death (A. E. Harriman, 1976, Oklahoma State University, personal communication). These results indicate a dietary magnesium requirement of ≥ 1.0 g/kg diet. A dietary concentration of 1.5 g/kg is recommended. This is higher than the requirement of the rat.

Sodium Chloride

A purified diet that did not include added sodium chloride produced alopecia within 30 days, but no weight loss. Recovery was dramatic when NaCl was provided (Cullen and Harriman, 1973; A. E. Harriman, 1976, Oklahoma State University, personal communication). Gerbils are able to tolerate relatively high sodium intakes because of their ability to produce concentrated urine. Gerbils that receive a 0.75 M sodium chloride solution as the only liquid maintained body weight, but food intake declined progressively as sodium chloride solutions increased from 0.5 to 1.5 M (McManus, 1972). However, Rowland and Fregley (1988) found that gerbils are reluctant to ingest NaCl either spontaneously or after treatment with several of the natriogexigenic stimuli that are effective in rats. As there is no reason to expect that gerbils require more sodium or chloride than rats, the recommended minimal dietary concentrations for both sodium and chloride are 0.5 g/kg (see Table 2-2). By analogy to the rat, the estimated potassium requirement for gerbils is 3.6 g/kg.

TRACE MINERALS

No studies could be located that specifically addressed the iron, copper, zinc, or manganese requirements of gerbils. Patt et al. (1990) reported that adult gerbils fed a low-iron diet (concentration not stated) for 8 weeks developed low brain and serum iron concentrations compared to controls. The observed reductions in tissue iron were apparently functionally significant in that they were associated with a decreased risk of brain reperfusion injury.

Until data specific to the gerbil are available, the recommended dietary concentrations for iron, copper, zinc, and manganese are the same as for the rat (Table 2-2). The recommended concentration of iron is 35 mg/kg diet for growing and adult animals and 75 mg/kg diet for pregnant and lactating dams. The recommended copper concentration is 5 mg/kg diet for growth and maturity and 8 mg/kg for reproduction. Based on the use of soybean protein-based diets by some investigators (DiFrancesco et al., 1990a,b), the recommended zinc concentration is 25 mg/kg diet for all stages of life. The recommended concentration for manganese is 10 mg/kg diet.

By analogy to the rat, it is assumed that the gerbil requires about 150 μg I/kg diet, 150 to 400 μg Se/kg diet, and 150 μg Mo/kg diet (Table 2-2). Other potentially beneficial mineral constituents are discussed in Chapter 2, and in the absence of information on gerbils, it is assumed that the same conclusions apply to this species.

VITAMINS

FAT-SOLUBLE VITAMINS

The concentrations of fat-soluble vitamins provided in natural-ingredient diets developed for rats and mice (e.g., Table 2-3) appear to be adequate for gerbils, although the fat-soluble vitamin requirements of gerbils have not been studied.

WATER-SOLUBLE VITAMINS

Relatively little is known about the requirements of gerbils for water-soluble vitamins. Hall and Zeman (1968) reported growth retardation and urinary riboflavin excretion when gerbils were fed diets containing riboflavin at 0.46 to 0.70 mg/kg diet as compared to 3.5 to 7.7 mg/kg. This is consistent with the estimated requirement of 3.0 mg riboflavin/kg diet for growth in rats. For the rat, the recommended riboflavin concentration for reproduction is somewhat higher (4.0 mg/kg diet); it is not known if this is true for the gerbil.

The gerbil appears to have a definite requirement for choline that can not be replaced by a high intake of methionine (Otken and Garza, 1983). The best growth was obtained with diets containing 2.3 g choline chloride/kg (16.5 mmol/kg) as well as 11.7 g methionine/kg and 5 g cystine/kg (Otken, 1984; Otken and Garza, 1983).

Male gerbils fed diets containing about 2 percent fat do not have a dietary requirement for *myo*-inositol because of intestinal synthesis of that vitamin. Females require more than 20 mg/kg diet. This requirement increases to 70 mg/kg when diets contain 20 percent saturated fat. The male requires *myo*-inositol when fed this same amount of saturated fat, but the amount has not been determined. Addition of sufficient *myo*-inositol to these diets prevented weight loss or decreased weight gain, hyperkeratosis of the skin and accumulation of fat in the intestinal tissue, and increased *myo*-inositol content of intestinal tissue (Hegsted et al., 1973; Kroes et al., 1973; Hegsted et al., 1974). Cholesterol added to the diet apparently increased the need for dietary *myo*-inositol.

Until further information becomes available, it is recommended that the concentrations of other water-soluble vitamins in gerbil diets meet or exceed the concentrations recommended for the rat (Table 2-2). In the previous edition of this report, concentrations of vitamins and minerals that had been used in purified diets for gerbils were summarized; but as some of these concentrations may have been suboptimal, they do not provide a useful model and are omitted from this edition.

REFERENCES

Arrington, L. R. 1968. Nutrition of Mongolian gerbils and golden hamsters—An evaluation of two commercially available rodent rations. Lab. Anim. Dig. 4:7–9.

Arrington, L. R., C. B. Ammerman, and D. E. Franke. 1973. Protein requirement of growing gerbils. Lab. Anim. Sci. 23:851–854.

Burns, T. W. 1956. Endocrine factors in the water metabolism of the desert mammal, *G. gerbillus*. Endocrinology 58:243–254.

Corbet, G. B. 1980. The mammals of the Palearctic region: A taxonomic review. Ithaca, N.Y.: Cornell University Press.

Cullen, J. W., and A. E. Harriman. 1973. Selection of NaCl solutions by sodium-deprived Mongolian gerbils in Richter-type drinking tests. J. Psychol. 83:315–321.

D'Elia, J., G. S. Bazzano, and G. Bazzano. 1972. The effect of cholesterol supplementation on glutamate-induced hypocholesterolemia in the Mongolian gerbil. Lipids 7:394–397.

DiFrancesco, L., and N. H. Mercer. 1990. Plasma cholinesterase and lipid levels as coronary heart disease risk factors in Mongolian gerbils fed casein or soy protein. Nutr. Res. 10:173–182.

DiFrancesco, L., O. B. Allen, and N. H. Mercer. 1990a. Long-term feeding of casein or soy protein with or without cholesterol in Mongolian gerbils. II. Plasma lipid and liver cholesterol response. Acta Cardiol. 45:273–290.

DiFrancesco, L., D. H. Percy, and N. H. Mercer. 1990b. Long-term feeding of casein or soy protein with or without cholesterol in Mongolian gerbils. I. Morphologic effects. Acta Cardiol. 45:257–271.

Forsythe, W. A., III. 1986. Comparison of dietary casein or soy protein effects on plasma lipids and hormone concentrations in the gerbil (*Meriones unguiculatus*). J. Nutr. 116:1165–1171.

Gordon, S., and W. P. Cekleniak. 1961. Serum lipoprotein pattern of the hypercholesteremic gerbil. Am. J. Physiol. 201:27–28.

Gordon, S., and J. F. Mead. 1964. Conversion of linoleic-1-C_{14} acid to arachadonic acid in the gerbil. Proc. Soc. Exp. Biol. Med. 116:730–733.

Gulotta, E. F. 1971. *Meriones unguiculatus*. Mammal. Species 3:1–5.

Hall, S. M., and Zeman, F. J. 1968. The riboflavin requirement of the growing Mongolian gerbil. Life Sci. 7:99–106.

Harriman, A. E. 1969. Food and water requirements of Mongolian gerbils as determined through self-selection of diet. Am. Midl. Nat. 82:149–156.

Harriman, A. E. 1974. Seizing by magnesium-deprived Mongolian gerbils given open field tests. J. Gen. Psychol. 90:221–229.

Hegsted, D. M., K. C. Hayes, A. Gallagher, and H. Hanford. 1973. Inositol deficiency: An intestinal lipodystrophy in the gerbil. J. Nutr. 103:302–307.

Hegsted, D. M., A. Gallagher, and H. Hanford. 1974. Inositol requirement of the gerbil. J. Nutr. 104:588–592.

Holleman, D. F., and R. A. Dieterich. 1973. Body water content and turnover in several species of rodents as evaluated by the tritiated water method. J. Mammal. 54:456–465.

Kroes, J. F., D. M. Hegsted, and K. C. Hayes. 1973. Inositol deficiency in gerbils: Dietary effects on the intestinal lipodystrophy. J. Nutr. 103:1448–1453.

Loew, F. M. 1968. Differential growth rates in male Mongolian gerbils (*Meriones unguiculatus*). Can. Vet. J. 9:237–238.

Loskota, W. J., P. Lomax, and S. T. Rich. 1974. The gerbil as a model for the study of epilepsies. Epilepsia 15:109–119.

Marston, J. H., and M. C. Chang. 1965. The breeding, management and reproductive physiology of the Mongolian gerbil (*Meriones unguiculatus*). Lab. Anim. Care 15:34–48.

McCarty, R. 1975. Magnesium deprivation and seizures in Mongolian gerbils. J. Gen. Psychol. 92:3–4.

McManus, J. J. 1971. Early postnatal growth and the development of temperature regulation in the Mongolian gerbil, *Meriones unguiculatus*. Comp. Biochem. Physiol. A 43:959–967.

McManus, J. J. 1972. Water relations and food consumption of the Mongolian gerbil, *Meriones unguiculatus*. J. Mammal. 52:782–792.

McManus, J. J., and Zurich, W. M. 1972. Growth, pelage development and maturational molts of the Mongolian gerbil, *Meriones unguiculatus*. Am. Midl. Nat. 87:264–271.

Mele, J. H. 1972. Temperature regulation and bioenergetics of the Mongolian gerbil. Am. Midl. Nat. 87:272–282.

Nicolosi, R. J., M. G. Herrera, M. el Lozy, and K. C. Hayes. 1976. Effect of dietary fat on hepatic metabolism of ^{14}C-oleic acid and very low density lipoprotein-triglyceride in the gerbil. J. Nutr. 106:1279–1285.

Norris, M. L., and C. E. Adams. 1972. The growth of the Mongolian gerbil, *Meriones unguiculatus*, from birth to maturity. J. Zool. London 166:277–282.

Otken, C. C. 1984. Liver lipid levels in gerbils fed diets containing varying proportions of methionine, cystine, and choline. Nutr. Rep. Int. 29:1–10.

Otken, C. C., and Y. Garza. 1983. Some quantitative dietary comparisons for Mongolian gerbils: The need for choline. Nutr. Rep. Int. 28:1393–1402.

Otken, C. C., Dougherty, S. V., and Servin, M. E. 1985. A possible need for taurine in the diet of the gerbil. Nutr. Rep. Int. 31:955–962.

Patt, A., I. R. Horesh, E. M. Berger, A. H. Harken, and J. E. Repine. 1990. Iron depletion or chelation reduces ischemia/reperfusion-induced edema in gerbil brains. J. Pediat. Surg. 25:224–228.

Pronczuk, A., P. Khosla, and K. C. Hayes. 1994. Dietary myristic, palmitic, and linoleic acids modulate cholesterolemia in gerbils. FASEB (in press).

Rich, S. T. 1968. The Mongolian gerbil (*Meriones unguiculatus*) in research. Lab. Anim. Care 18:235–243.

Rowland, N. E., and M. J. Fregley. 1988. Sodium appetite: Species and strain differences and role of renin-angiotensin-aldosterone system. Appetite 11:143–148.

Schwentker, V. 1971. The Gerbil: An Annotated Bibliography. Brent Lake, N.Y.: Tumblebrook Farms.

Thiessen, D., and P. Yahr. 1977. The Gerbil in Behavioral Investigations. Austin, Tex.: University of Texas Press.

Thiessen, D. D., G. Lindzey, and H. C. Friend. 1968. Spontaneous seizures in the Mongolian gerbil (*Meriones unguiculatus*). Psychon. Sci. 11:227–228.

Vorontsov, N. N. 1979. Evolution of the alimentary system in myomorph rodents [translated from Russian]. New Delhi: Indian National Scientific Documentation Centre.

Wexler, B. C., J. T. Judd, R. F. Lutmer, and J. Saroff. 1971. Spontaneous arteriosclerosis in male and female gerbils (*Meriones unguiculatus*). Atherosclerosis 14:107–119.

Winkelman, J. R., and L. L. Getz. 1962. Water balance in the Mongolian gerbil. J. Mammal. 43:150–154.

Zeman, F. J. 1967. A semipurified diet for the Mongolian gerbil (*Meriones unguiculatus*). J. Nutr. 91:415–420.

7 Nutrient Requirements of the Vole

It has been demonstrated that voles are useful as a small-animal model for testing the quality of forages and other agricultural crops (Elliot, 1963; Keys and Van Soest, 1970; Shenk et al., 1971; Shenk et al., 1974, 1975; Shenk, 1976). More recently, laboratory and field research on voles has focused on the effect of nutritional and other environmental factors on reproduction and population growth (e.g., Cole and Batzli, 1979; Nelson et al., 1983; Batzli, 1985, 1986; Hasbrouck et al., 1986; Spears and Clarke, 1987; Hall et al., 1991). In nature, voles undergo significant annual and cyclical changes in population density that are not fully understood (Taitt and Krebs, 1985). Despite the ecological importance of voles and their apparent responsiveness to nutritional variables, little is known about the nutrient requirements of these abundant rodents.

BIOLOGICAL CHARACTERISTICS

Voles (*Microtus* sp.) represent a successful radiation of small herbivores that are particularly abundant in grassy areas in Asia, Europe, and North America. The genus *Microtus* encompasses about 60 species, if *Pitymys* is accepted as a subgenus of *Microtus* (Corbet and Hill, 1980; Nowak and Paradiso, 1983; Anderson, 1985). The most common species for laboratory study in North America are the prairie vole (*M. ochrogaster*), the meadow vole (*M. pennsylvanicus*), and the pine vole [*M. (Pitymys) pinetorum*].

Taxonomically, voles are members of the rodent subfamily Microtinae (sometimes called Arvicolinae) that includes lemmings, muskrats, and nutria (Anderson, 1985). Other microtine genera also are referred to as voles, such as red-backed voles (*Clethrionymys* sp.), mountain voles (*Alticola* sp.), and water voles (*Arvicola* sp.). It is not known if the nutritional requirements of these genera differ from *Microtus*, but because of differences in natural diet, body size, and physiological adaptations to the environment (Batzli, 1985; Woodall, 1989) they will not be included in this discussion.

In nature, voles rely on grasses both for shelter and as a primary food source (Getz, 1985). Although all *Microtus* consume primarily the vegetative parts of plants, the types of plants eaten vary among species, habitats, and seasons (Batzli, 1985). It seems that voles select food plant species on the basis of availability, composition (particularly nitrogen and fiber fractions), and deterrent secondary compounds such as phenolics and tannins (Batzli, 1985; Lindroth et al., 1986; Marquis and Batzli, 1989; Bucyanayandi and Bergeron, 1990). Unlike many small rodents, voles remain active in winter months—tunneling under snow, if necessary, and feeding on senescent grasses, rhizomes, seeds, and other plant material (Batzli, 1985; Getz, 1985). Ability to survive in cold conditions on foods of low digestible energy content appears to be a key adaptive feature of voles (Wunder, 1985; Hammond and Wunder, 1991).

Rapid reproductive rates are also characteristic of voles when food is abundant. Because female voles typically enter into estrus and are inseminated shortly after giving birth, concurrent pregnancy and lactation are common, leading to the production of litters at about 3-week intervals (Kudo and Oki, 1984; Keller, 1985). Developmental and reproductive indices of some of the common laboratory voles, as well as scientific binomials of the species discussed in this chapter, are listed in Table 7-1. Most of the data in Table 7-1 are from laboratory colonies maintained on natural-ingredient diets and presumably represent animals at a high plane of nutrition. Postnatal growth rates were measured in the first weeks postpartum; daily growth rates after weaning can be expected to be similar or somewhat greater. For example, Shenk (1976) considered a weanling growth rate of 0.9 to 1.1 g/day to be maximal and indicative of dietary adequacy in meadow voles; preweaning growth rates in this species average about 0.7 g/day (Table 7-1).

TABLE 7-1   Reproductive and Developmental Indices of Voles

Species Name	Common Name	Adult Weight, g	Gestation, days	Litter Size	Birth Weight, g	Postnatal Growth, g/day	Approx. Weaning Age, Days
M. arvalis	Common vole	35	20	4.8	2.1	—	—
M. californianus	California vole	50	21	4.7	2.8	1.1	14
M. montanus	Montane vole	50	21	6.0	2.5	1.0	15–17
M. montebelli	Japanese field vole	41	21	4.7	2.7	0.9	21
M. ochrogaster	Prairie vole	40	20–22	3.9	2.8	1.0	14–21
M. oeconomus	Tundra vole	45	20.5	4.0	3.0	0.8	18
M. pennsylvanicus	Meadow vole	40	20–21	4.4	2.3	0.7	14
M. pinetorum	Pine vole	26	24	2.2	2.0	—	17

SOURCES: Lee and Horvath (1969), Richmond and Conaway (1969), Lindroth et al. (1984), Kudo and Oki (1984), Nadeau (1985).

HUSBANDRY AND FORM OF DIET

Vole colonies are usually founded by the capture of wild animals. Successful breeding colonies have been established for at least 10 species of voles in North America (Mallory and Dieterich, 1985) and several other species in Europe and Japan (Kudo and Oki, 1984). Different colonies of the same species may be genetically distinct because of substantial variation among the founding wild populations. Extensive morphological variation over the ranges of most species has resulted in the description of a plethora of subspecies. At the extreme, 27 subspecies have been described for *Microtus pennsylvanicus* (Hoffman and Koeppel, 1985). Populations of voles may vary in characteristics such as body mass and litter size, although the extent to which this variation is a result of genetic or environmental effects is not clear (Keller, 1985). Care must be exercised in generalizing from one colony to another; the values listed in Table 7-1 may not apply to all colonies.

Voles have been maintained in a variety of cages, but plastic mouse cages with solid bottoms and added bedding material are especially suitable (Richmond and Conaway, 1969; Mallory and Dieterich, 1985). Details of lighting, ambient temperature, social groupings, and other aspects of the husbandry of voles have been reviewed by Lee and Horvath (1969), Richmond and Conaway (1969), Dieterich and Preston (1977), and Mallory and Dieterich (1985).

NATURAL-INGREDIENT AND PURIFIED DIETS

After an initial adaptation period, voles adapt well to captivity and can be fed on various natural-ingredient diets. Pelleted natural-ingredient diets developed for rabbits, mice, rats, and guinea pigs are apparently the most commonly used diets, either with or without supplementation with succulent foods such as lettuce or apples. Of the 10 species listed by Mallory and Dieterich (1985), 7 have been maintained on rabbit diets and 7 on mouse breeder diets.

Sole or heavy use of succulent items (such as barley sprouts, carrots, lettuce, and apples) or of unsupplemented seeds and grains (such as sunflower seed, grass seed, and oats) is not recommended because of the likelihood of unintended mineral or vitamin deficiencies (e.g., Batzli, 1986).

Purified diets based on vitamin-free casein, a purified cellulose source, a mixture of starch and sugars, and vitamin and mineral premixes and fed in the form of wafers supported adequate weight gains in short-term (6 to 10 days) experiments with weanling meadow voles as long as the cellulose source was at least 25 percent of the diet (Shenk et al., 1971). This may be too short a period to adequately assess response, however (Lindroth et al., 1984). Both prairie voles and tundra voles exhibited normal growth after weaning when fed either a purified diet (based on 20 percent casein and 40 percent cellulose) or a commercial natural-ingredient diet (formulated for rabbits); but the purified diet subsequently led to greater fat deposition, reduced production of litters, and reduced litter size (Lindroth et al., 1984). Sugawara and Oki (1988) noted that purified diets that contained less than 20 percent casein impaired female fertility in the common vole. It has yet to be demonstrated that a purified diet can maintain long-term reproduction in a breeding colony of voles.

DIET DIGESTIBILITY AND INTAKE

Voles exhibit a number of anatomical features that are associated with herbivory and fermentation of plant fiber. The cheek teeth are high-crowned and contain a complex array of cusps; the stomach is separated into two compartments, one of which (sometimes termed the esophageal pouch) is lined with stratified squamous epithelium and harbors anaerobic microorganisms; the cecum is enlarged and separated in pockets by projecting isthmuses; the colon is both elongate and arranged into spirals that facilitate particle segregation (Vorontsov, 1979; Kudo and Oki, 1984; Stevens, 1988). Although volatile fatty acids are produced in the esophageal pouch when high-fiber diets are consumed (Kudo and Oki, 1984), most fiber fermentation seems to occur in the colon and cecum. According to Bjornhag (1987), voles and allied microtines utilize a colonic separation mechanism to retain fermentative bacteria, enabling the bacteria to proliferate before being washed out of the colon. Coprophagy is also an important part of the digestive strategy of voles; in a study of meadow and pine voles, prevention of coprophagy resulted in a decrease in energy digestibility (Cranford and Johnson, 1989).

When fed natural-ingredient diets, voles have energy digestibilities of about 55 to 75 percent, with the lower digestibilities occurring on high-fiber diets (Cherry and Verner, 1975; Batzli, 1986; Hammond and Wunder, 1991). Energy digestibilities of various forages are even more varied and may be less than 50 percent with some grasses (Batzli, 1986; Batzli and Cole, 1979; Johanningsmeier and Goodnight, 1969). The rapid transit of digesta ($\approx$12-hour turnover time) precludes extensive fiber digestion. For example, Hammond and Wunder (1991) found neutral detergent fiber (NDF) and acid detergent fiber (ADF) digestibilities of 17 to 18 percent and 6 to 9 percent, respectively, when meadow voles were fed a diet containing 16 percent NDF and 8 percent ADF. However, when fiber fractions were increased (to 39 percent NDF and 23 percent ADF), fiber digestibilities also increased, suggesting some adaptation of digestive function. Voles are able to compensate for an increase in dietary fiber concentrations by an increase in mass of the gastrointestinal tract and especially the cecum (Gross and Wang, 1985; Hammond and Wunder, 1991).

Food intake is affected by both energy requirements and digestible energy content of the diet. For example, the daily dry-matter intake of adult prairie voles was 13 percent of body weight when they were fed low-fiber (about 8 percent ADF) diets and maintained at an ambient temperature of 23° C, but when fiber was increased to 23 percent ADF and temperature reduced to 5° C, dry-matter intake rose to 32 percent of body weight (Hammond and Wunder, 1991). The corresponding digestible energy intakes were about 175 kcal/$BW_{kg}^{0.75}$/day (732 kJ/$BW_{kg}^{0.75}$/day) and 370 kcal/$BW_{kg}^{0.75}$/day (1,548 kJ/$BW_{kg}^{0.75}$/day). Dry-matter and energy intakes of female voles more than double during lactation (Migula, 1969; Innes and Millar, 1981).

PROTEIN AND AMINO ACIDS

A limited amount of research has been conducted on the growth responses of weanling voles to dietary protein. Shenk et al. (1970) fed weanling meadow voles purified diets containing 3 to 25 percent casein and various proportions of carbohydrates, oil, and cellulose. Voles consuming diets of 9 percent casein (8.3 percent crude protein) or

less had subnormal growth rates, whereas voles consuming diets of 12 percent or more casein (11 percent or more crude protein) and intermediate energy densities had apparently normal growth rates ($\geq$0.9 g/day). Sugawara and Oki (1988) observed decreased growth in weanling common voles fed 5 percent casein in a purified diet, as compared to diets containing 10, 15, and 20 percent casein. Spears and Clarke (1987) found no difference between growth rates of field voles fed closed-formula natural-ingredient diets containing 8, 16, and 24 percent protein, but the growth rates of all animals were apparently depressed ($\approx$0.3 g/day); animals fed a 4 percent protein diet gained less than 0.2 g/day. Although these studies indicate that low protein concentration decreases growth rate, they do not permit a quantitative assessment of the minimum protein requirement of voles, in part because the amino acid patterns may not be ideal.

Shenk (1976) asserted that maximal growth of weanling meadow voles could be achieved with purified diets containing mixtures of amino acids providing 13 percent of dry matter as protein and 0.9, 3.2, and 5.9 percent of protein as tryptophan, methionine, and lysine, respectively. Unfortunately these experiments were never published, but unpublished tables provided by Shenk (personal communication, 1989) indicate that the concentrations of amino acids tested, as a percent of total protein, were 0.2, 0.5, 0.9, and 1.3 percent tryptophan; 0.5, 1.2, 3.2, and 5.2 percent methionine; and 0.9, 1.9, 5.9, and 9.9 percent lysine. Given that animals receiving 0.5 percent tryptophan and 1.2 percent methionine had adequate growth rates (0.91 g/day), and that detailed methods and statistics were not reported, this study must be considered preliminary. Animals receiving 0.9, 3.2, and 5.9 percent of protein as tryptophan, methionine, and lysine, respectively, but only 7 percent of dry matter as protein, had somewhat lower rates of growth (0.79 g/day).

On the basis of available data, it seems that 11 to 13 percent of high-quality protein is sufficient for maximal weanling growth; however, such diets have not supported maximal reproduction. Sugawara and Oki (1988) noted higher fertility for common voles fed a purified diet containing 20 percent casein than for voles fed diets containing 10 and 15 percent casein. Further study is needed to determine whether higher protein concentrations are required by voles for reproduction than for growth.

MINERALS

The few studies of mineral metabolism in voles have been stimulated by apparent mineral imbalances in natural foods. Batzli (1986) demonstrated that reproductive performance of California voles was impaired when the voles were fed solely seeds of ryegrass (*Lolium* sp.) containing

0.017 to 0.018 percent calcium and 0.01 to 0.02 percent sodium. Supplementation with sodium chloride improved the frequency and size of litters, but survival and growth of suckling pups were low. Supplementation with calcium chloride did not affect litter frequency or size but did improve postnatal growth and survivorship. When both salts were supplemented, reproductive performance (including postnatal growth and survival) was normal (Batzli, 1986). Thus deficiencies of both sodium and calcium impair reproduction but in different ways. As the amounts of supplements consumed were not determined, it is not possible to estimate requirements.

Many of the plants consumed by meadow voles are low in sodium and high in potassium concentrations, especially in spring (Hastings et al., 1991). It has been shown that both low dietary sodium (0.001 percent) and high potassium ($\geq$3.0 percent) concentrations induce hypertrophy of the zona glomerulosa of the adrenal gland of this species (Christian, 1989; Hastings et al., 1991) but actual intakes by wild voles have not been measured. In experimental trials adult ($\geq$20 g), singly housed meadow voles consumed diets containing up to 2.8 percent potassium for 4 weeks without adverse results (Mickelson and Christian, 1991).

In the absence of direct estimates of the mineral requirements of voles, and in view of the fact that voles have been successfully maintained when fed diets formulated for rats, mice, and rabbits, it is suggested that diets for voles should contain mineral concentrations similar to those that have proven adequate for these species.

VITAMINS

Nothing is known about the vitamin requirements of voles other than that both the meadow vole and prairie vole have substantial hepatic activity of l-gulonolactone oxidase (Jenness et al., 1980) and hence do not appear to require a dietary source of ascorbic acid. The fact that colonies of various species of voles have been maintained successfully on natural-ingredient diets formulated for rabbits, rats, and mice suggests that supplementation concentrations of vitamins used for these species may be adequate.

REFERENCES

Anderson, S. 1985. Taxonomy and systematics. Pp. 52-83 in Biology of New World *Microtus*, Special Publication No. 8, R. H. Tamarin, ed. Provo, Utah: American Society of Mammologists.

Batzli, G. O. 1985. Nutrition. Pp. 779-811 in Biology of New World *Microtus*, Special Publication No. 8, R. H. Tamarin, ed. Provo, Utah: American Society of Mammologists.

Batzli, G. O. 1986. Nutritional ecology of the California vole: Effects of food quality on reproduction. Ecology 67:406–412.

Batzli, G. O., and F. R. Cole. 1979. Nutritional ecology of microtine rodents: Digestibility of forage. J. Mammal. 60:740–750.

Bjornhag, G. 1987. Comparative aspects of digestion in the hindgut of mammals. The colonic separation mechanism (CSM). Dtsch. Tierärztl. Wochenschr. 94:33–36.

Bucyanayandi, J., and J. Bergeron. 1990. Effects of food quality on feeding patterns of meadow voles (*Microtus pennsylvanicus*) along a community gradient. J. Mammal. 71:390–396.

Cherry, R. H., and L. Verner. 1975. Seasonal acclimation to temperature in the prairie vole, *Microtus ochrogaster*. Am. Midl. Nat. 94:354–360.

Christian, D. P. 1989. Effects of dietary sodium and potassium on mineral balance in captive meadow voles (*Microtus pennsylvanicus*). Can. J. Zool. 67:168–177.

Cole, F. R., and G. O. Batzli. 1979. Nutrition and population dynamics of the prairie vole, *Microtus ochrogaster*, in central Illinois. J. Anim. Ecol. 48:455–470.

Corbet, G. B., and J. E. Hill. 1980. A world list of mammalian species. Ithaca, N.Y.: Cornell University Press.

Cranford, J. A., and E. O. Johnson. 1989. Effects of coprophagy and diet quality on two microtine rodents (*Microtus pennsylvanicus* and *Microtus pinetorium*). J. Mammal. 70:494–502.

Dieterich, R. A., and D. J. Preston. 1977. The meadow vole (*Microtus pennsylvanicus*) as a laboratory animal. Lab. Anim. Care 27:494–499.

Elliott, F. C. 1963. The meadow vole (*Microtus pennsylvanicus*) as a bioassay test organism for individual forage plants. Michigan Agric. Exp. St. Q. Bull. 46:58–72.

Getz, L. L. 1985. Habitats. Pp. 286-309 in Biology of New World *Microtus*, Special Publication No. 8, R. H. Tamarin, ed. Provo, Utah: American Society of Mammalogists.

Gross, J. E., and Z. Wang. 1985. Effects of food quality and energy needs: Changes in gut morphology and capacity of *Microtus ochrogaster*. J. Mammal. 66:661–667.

Hall, A. T., P. E. Woods, and G. W. Barrett. 1991. Population dynamics of the meadow vole (*Microtus pennsylvanicus*) in nutrient-enriched old-field communities. J. Mammal. 72:332–342.

Hammond, K. A., and B. A. Wunder. 1991. The role of diet quality and energy need in the nutritional ecology of a small herbivore, *Microtus ochrogaster*. Physiol. Zool. 64:541–567.

Hasbrouck, J. J., F. A. Servello, and R. L. Kirkpatrick. 1986. Influence of photoperiod and nutrition on pine vole reproduction. Am. Midl. Nat. 116:246–255.

Hastings, J. J., D. P. Christian, T. E. Manning, and C. C. Harth. 1991. Sodium and potassium effects on adrenal-gland indices of mineral balance in meadow voles. J. Mammal. 72:641–651.

Hoffman, R. S., and J. W. Koeppl. 1985. Zoogeography. Pp. 84-115 in Biology of New World *Microtus*, Special Publication No. 8, R. H. Tamarin, ed. Provo, Utah: American Society of Mammalogists.

Innes, D. G. L., and J. S. Millar. 1981. Body weight, litter size, and energetics of reproduction in *Clethrionomys gapperi* and *Microtus pennsylvanicus*. Can. J. Zool. 59:785–789.

Jenness, R., E. C. Birney, and K. L. Ayaz. 1980. Variation of l-gulonolactone oxidase activity in placental mammals. Comp. Biochem. Physiol. B 67:195–204.

Johanningsmeier, A. G., and C. J. Goodnight. 1969. Digestibility of nitrogen, cellulose, lignin, dry matter and energy in *Microtus pennsylvanicus* on *Agrostis stolonifera* and *Poa ratensis*. Agron. Abstr. 61:59.

Keller, B. L. 1985. Reproductive patterns. Pp. 725-778 in Biology of New World *Microtus*, Special Publication No. 8, R. H. Tamarin, ed. Provo, Utah: American Society of Mammalogists.

Keys, J. E., and P. J. Van Soest. 1970. Digestibility of forages by the meadow vole (*Microtus pennsylvanicus*). J. Dairy Sci. 53:1502–1508.

Kudo, H., and Y. Oki. 1984. *Microtus* species as new herbivorous laboratory animals: Reproduction, bacterial flora and fermentation in the digestive tracts, and nutritional physiology. Vet. Res. Commun. 8:77–91.

Lee, C., and D. J. Horvath. 1969. Management of the meadow vole (*Microtus pennsylvanicus*). Lab. Anim. Care 19:88–91.

Lindroth, R. L., G. O. Batzli, and S. I. Avildsen. 1986. *Lespedeza* phenolics and *Penstemon* alkaloids: Effects on digestion efficiencies and growth of voles. J. Chem. Ecol. 12:713–728.

Lindroth, R. L., G. O. Batzli, and G. R. Guntenspergen. 1984. Artificial diets for use in nutritional studies with microtine rodents. J. Mammal. 65:139–143.

Mallory, F. F., and R. A. Dieterich. 1985. Laboratory management and pathology. Pp. 647-684 in Biology of New World *Microtus*, Special Publication No. 8, R. H. Tamarin, ed. Provo, Utah: American Society of Mammalogists.

Marquis, R. J., and G. O. Batzli 1989. Influence of chemical factors on palatability of forage to voles. J. Mammal. 70:503–511.

Mickelson, P. A., and D. P. Christian. 1991. Avoidance of high-potassium diets by captive meadow voles. J. Mammal. 72:177–182.

Migula, P. 1969. Energetics of pregnancy and lactation in European common vole. Acta Theriol. 14:167–179.

Nadeau, J. H. 1985. Ontogeny. Pp. 254-285 in Biology of New World *Microtus*, Special Publication No. 8, R. H. Tamarin, ed. Provo, Utah: American Society of Mammalogists.

Nelson, R. J., J. Kark, and I. Zucker. 1983. Influence of photoperiod, nutrition and water availability on reproduction of male California voles (*Microtus californicus*). J. Reprod. Fertil. 69:473–477.

Nowak, R. M., and J. L. Paradiso. 1983. Walker's Mammals of the World, 4th Ed. Baltimore: The Johns Hopkins University Press.

Richmond, M., and C. H. Conaway. 1969. Management, breeding, and reproductive performance of the vole, *Microtus ochrogaster*, in a laboratory colony. Lab. Anim. Care 19:80–87.

Shenk, J. S. 1976. The meadow vole as an experimental animal. Lab. Anim. Sci. 26:664–669.

Shenk, J. S., R. F. Barnes, J. D. Donker, and G. C. Marten. 1975. Weanling meadow vole and dairy cow responses to alfalfa hay. Agron. J. 67:569–571.

Shenk, J. S., F. C. Elliot, and J. W. Thomas. 1970. Meadow vole nutrition studies with semisynthetic diets. J. Nutr. 100:1437–1446.

Shenk, J. S., F. C. Elliot, and J. W. Thomas. 1971. Meadow vole nutrition studies with alfalfa diets. J. Nutr. 101:1367–1372.

Shenk, J. S., M. L. Risius, and R. F. Barnes. 1974. Weanling meadow vole responses to crownvetch forage. Agron. J. 67:569–571.

Spears, N., and J. R. Clarke. 1987. Effect of nutrition, temperature and photoperiod on the rate of sexual maturation of the field vole (*Microtus agrestis*). J. Reprod. Fertil. 80:175–181.

Stevens, C. E. 1988. Comparative physiology of the vertebrate digestive system. New York: Cambridge University Press.

Sugawara, M., and Y. Oki. 1988. The influence of casein levels in semisynthetic diets on the growth and reproduction of common voles (*Microtus arvalis* Pallas). Jpn. J. Zootech. Sci. 59:929–935.

Taitt, M. J., and C. J. Krebs. 1985. Population dynamics and cycles. Pp. 567-620 in Biology of New World *Microtus*, Special Publication No. 8, R. H. Tamarin, ed. Provo, Utah: American Society of Mammalogists.

Vorontsov, N. N. 1979. Evolution of the alimentary system in myomorph rodents [translated from Russian]. New Delhi, India: Indian National Scientific Documentation Centre.

Woodall, P. F. 1989. The effects of increased dietary cellulose on the anatomy, physiology and behaviour of captive water voles, *Arvicola terrestris*. Comp. Biochem. Physiol. A 94:615–621.

Wunder, B. A. 1985. Energetics and thermoregulation. Pp. 812-844 in Biology of New World *Microtus*, Special Publication No. 8, R. H. Tamarin, ed. Provo, Utah: American Society of Mammologists.

Yokota, H. 1988. Digestibility of crude fiber and nitrogen in forages by common voles (*Microtus arvalis*). Jpn. J. Zootech. Sci. 59:565–567.

Appendix

APPENDIX TABLE 1 Fatty Acid Composition (%) of Some Common Fats Used in Rodent Diets

Fatty Acids[a]	Canola Oil[b]	Cocoa Butter[c]	Coconut (hydrogenated-96°) Oil[d]	Corn Oil[e]	Cottonseed Oil	Fish (Menhaden) Oil[f]	Tallow	Linseed Oil[g]	Olive Oil[h]	Peanut Oil	Sunflower Oil[i]	Soybean Oil[j]
8:0			6.8									
10:0			5.8									
12:0			46.7			0.2	0.1					
14:0		0.1	18.5	0.1	0.8	7.3	3.4		0.1	0.1		0.1
14:1			0.2				0.7					
15:0							0.6					
16:0	3.9	25.4	9.5	11.0	23.4	19.0	24.8	5.3	12.0	10.9	6.7	10.5
16:1	0.2	0.2		0.1	0.6	9.1	3.4		1.2	0.1	0.1	
17:0						0.9	1.3		0.1	0.1		0.1
17:1							0.8		0.1	0.1		
18:0	2.0	33.2	3.1	2.2	2.4	4.2	18.6	4.1	2.3	2.5	4.8	3.9
18:1(n-9)	61.8	32.6	7.1	25.5	17.9	13.2	42.2	20.2	72.9	45.8	18.7	22.6
18:2(n-6)	20.0	2.8	1.9	59.5	53.6	1.3	2.8	12.7	9.6	33.0	67.3	54.2
18:3(n-3)	9.3	0.1	0.1	1.0	0.5	1.3	0.8	53.3	1.0	0.5	1.1	7.7
20:0	0.4			0.4	0.3	0.4	0.3		0.1	1.3	0.2	0.3
20:1(n-11)	1.5			0.3		2.0			0.3	1.0	0.2	
20:4(n-6)						0.2						
20:5(n-3)						11.0						
22:0	0.4					0.2			0.1	2.9	0.6	0.3
22:1(n-11)	0.5					0.6						
22:6(n-3)						9.1						
Other	0.0	5.6	0.3	0.0	0.5	20.0	0.2	4.4	0.3	1.7	0.3	0.2
n-6:n-3	2.2	28.0	19.0	59.5	107.2	0.1	3.5	0.2	9.6	66.0	61.2	7.0

NOTE: Unless otherwise indicated, fatty acid composition was supplied by E. Wayne Emmons, AOCS Chromatography Chairman, 1991 (personal communication).

[a] 14:1 and 17:1 contain double bond in n-9 or n-11 position, exact position unknown.

[b] Abundance of n-3 and monounsaturated fat.

[c] Abundance of long-chain saturated and monounsaturated fatty acid; fatty acid composition from U.S. Department of Agriculture Handbook No. 8-4 (1979).

[d] A highly saturated fat; contains predominantly medium-chain triglycerides and very little *trans* fatty acid.

[e] Good source of essential fatty acid, n-6; suggested for use in the AIN-76 rodent reference diet (American Institute of Nutrition, 1977).

[f] Contains large amounts of very long chain n-3 fatty acids; usually needs to be supplemented with an n-6 containing oil for use in a rodent diet (fatty acid composition from Ackman, R. G., 1982).

[g] Contains large quantity n-3 fatty acid; fatty acid composition from U.S. Department of Agriculture Handbook No. 8-4 (1979).

[h] Contains large quantity of monounsaturated fat.

[i] Greatest source of n-6 fatty acid of these fat sources.

[j] Contains significant quantity of n-6 and n-3 fatty acid; partially hydrogenated soybean oil will also contain some *trans* fatty acids.

APPENDIX TABLE 2   Amino Acid Composition (mg/g nitrogen) of Purified Proteins Used in Laboratory Animal Diets

Amino Acid	Acid Casein	ANRC Casein	Lactal-bumin	Whey Protein Concen-trate	Soy-bean Pro-tein Isolate
Alanine	188	188	369	331	269
Arginine	231	231	188	175	475
Asparagine	431	469	769	681	725
Cystine	25	25	200	169	81
Glutamic acid	1,306	1,563	925	1,094	1,194
Glycine	113	125	144	125	263
Histidine	181	200	131	131	163
Isoleucine	288	313	319	338	306
Leucine	569	594	875	744	513
Lysine	481	488	669	588	394
Methionine	181	200	131	156	81
Phenylalanine	319	331	244	219	325
Proline	650	706	319	331	319
Serine	363	381	313	338	325
Threonine	269	300	369	413	238
Tryptophan	75	75	144	125	81
Tyrosine	344	369	250	206	238
Valine	356	388	313	319	313

NOTE: Data was obtained from New Zealand Milk Products, Inc., and Protein Technologies International. All calculations assume protein contained 16 percent nitrogen.

APPENDIX TABLE 3   Molecular Weights of Vitamins

Compound	Molecular Weight
Vitamin A	
Retinol	286.46
3,4-Dehydroretinol (vitamin A_2)	284.44
Retinaldehyde (Retinal)	284.44
Retinoic acid	300.44
Retinyl acetate	328.50
Retinyl palmitate	524.88
Retinyl stearate	552.93
Retinyl oleate	550.91
β-Carotene	536.89
Vitamin D	
Ergocalciferol (vitamin D_2)	396.66
Cholecalciferol (vitamin D_3)	384.65

(Table continues in next column.)

APPENDIX TABLE 3   (Continued)

Compound	Molecular Weight
25-Hydroxycholecalciferol	400.65
1,25-Dihydroxycholecalciferol	416.65
Vitamin E	
α-Tocopherol	430.72
α-Tocopheryl acetate	472.76
β-Tocopherol	416.69
γ-Tocopherol	416.69
δ-Tocopherol	402.66
α-Tocotrienol	424.67
Vitamin K	
Phylloquinone (vitamin K_1)	450.71
Menaquinone-4	444.66
Menaquinone-7	649.02
Menaquinone-8	717.14
Meanquinone-9	785.26
Menadione	172.18
Menadione sodium bisulfite	276.25
Menadione dimethylpyridinol bisulfite	378.41
Vitamin C	
Ascorbic acid	176.13
Dehydroascorbic acid	174.11
Biotin	
Biotin	244.32
Choline	
Choline (free base)	104.17
Choline bitartrate	253.25
Choline chloride	139.63
Folate	
Folic acid	441.41
Niacin	
Nicotinic acid	123.11
Nicotinamide	122.13
Pantothenic acid	
Pantothenic acid	219.24
Sodium pantothenate	241.22
Calcium pantothenate	476.54
Calcium pantothenate monohydrate	494.56
Vitamin B_6	
Pyridoxine hydrochloride	205.64
Pyridoxal	167.17
Pyridoxal hydrochloride	203.63
Pyridoxamine dihydrochloride	241.12
Pyridoxal phosphate	247.15
Riboflavin	
Riboflavin	376.37
Sodium riboflavin phosphate (FMN)	478.33
Flavin adenine dinucleotide (FAD)	785.56
Thiamin	
Thiamin hydrochloride	337.27
Thiamin diphosphate chloride	460.77
Thiamin mononitrate	327.36
Thiamin triphosphate	504.29
Vitamin B_{12}	
Cyanocobalamin	1,355.39
Hydroxocobalamin	1,346.38
5′-deoxyadenosylcobalamin	1,579.61

NOTE: Data are based on atomic weights from Pure and Applied Chemistry (1991; 63:978–979).

APPENDIX TABLE 4 Conversion Factors

Element	Traditional Unit	Converted Equivalent
Vitamin A	1 international unit	0.3 μg retinol
	1 international unit	0.34 μg retinyl acetate
	1 international unit	0.55 μg retinyl palmitate
	1 international unit	0.6 μg β-carotene
	1 retinol equivalent	1 μg retinol
	1 retinol equivalent	6 μg β-carotene
	1 retinol equivalent	12 μg other provitamin A carotenoids
	1 retinol equivalent	3.33 IU vitamin A activity from retinol
	1 retinol equivalent	10 IU vitamin A activity from β-carotene
Vitamin D	1 international unit	25 ng cholecalciferol (vitamin D_3)
Vitamin E	1 international unit	1 mg all-*rac*-α-tocopheryl acetate
	1 international unit	0.74 mg *RRR*-α-tocopheryl acetate
	1 international unit	0.91 mg all-*rac*-α-tocopherol
	1 international unit	0.67 mg *RRR*-α-tocopherol
Thiamin	1 international unit	3 μg thiamin hydrochloride

NOTE: These terms are now obsolete; the preferred expression is the molar concentration.

SOURCE: John Edgar Smith, Penn State University, personal communication, 1994.

Authors

Norlin J. Benevenga (*Chair*) is professor of meat and animal science and nutritional sciences at the University of Wisconsin at Madison. He received his Ph.D. in nutrition in 1965 from the University of California, Davis. His research interests include nutrition and metabolism of amino acids, especially methionine and lysine and the nutrition and metabolism of the newborn. Benevenga also served as a member of the Board on Agriculture's Committee on Animal Nutrition and the previous subcommittee on laboratory animal nutrition.

Christopher Calvert is currently a professor of animal science at the University of California, Davis, where he has served since 1980. He received his Ph.D. in animal science from Purdue University. His research interests include protein and energy metabolism, growth, lactation, and systems analysis.

Curtis D. Eckhert is professor and vice-chair of the Department of Environmental Health Sciences at the University of California, Los Angeles. He received his Ph.D. in nutritional biochemistry from Cornell University. Eckhert was head of the Division of Nutritional Sciences at the University of California, Los Angeles, from 1989 to 1993. Research interests include the physiology and biochemistry of microvascular selenium.

George C. Fahey is professor of animal sciences and nutritional sciences at the University of Illinois, Urbana-Champaign. He received the Ph.D. in animal nutrition from West Virginia University in 1976. Research interests include comparative carbohydrate nutrition, ruminal and lower gut fermentation processes, and plant cell wall biochemistry as it applies to nutrition.

Janet L. Greger is professor of nutrition at the University of Wisconsin at Madison. From 1975 to 1989, Greger was a principal grants investigator for the National Institutes of Health, U.S. Department of Agriculture, and the W. K. Kellogg Foundation. She received her Ph.D. in nutrition from Cornell University. Research interests include mineral metabolism from a nutritional and toxicological point of view; minerals studied include aluminum, calcium, zinc, tin, and copper.

Carl L. Keen is a professor of nutrition and internal medicine at the University of California, Davis, from where he also received his Ph.D. in nutrition. He is currently chair of the Department of Nutrition. Research interests include mineral metabolism and developmental nutrition.

Joseph J. Knapka, since 1967, has served as a nutritionist at the National Institutes of Health's Division of Research Services. Currently he is special assistant to the acting director of the Veterinary Resources Program. He holds a Ph.D. in animal science from the University of Tennessee. Research interests include physiology and feeding management of laboratory animals.

Hulda Magalhaes is professor emerita of zoology at Bucknell University, where she began as an assistant professor of physiology in 1946. She received her Ph.D. in marine zoology from Duke University. Research interests include nutrition and metabolism of human and hamster subjects.

Olav T. Oftedal is a research nutritionist in the zoological research department at the National Zoological Park, Smithsonian Institution, in Washington, D.C. He received his Ph.D. in nutrition from Cornell University. Research interests include the nutrient requirements of zoo animals and wildlife, the lactation strategies and early development of mammals, and the interactions of nutrition and ecology in reptiles, seals, primates, and other species.

154

Philip G. Reeves currently serves as supervisory research chemist with the U.S. Department of Agriculture's Nutrient Function Management Unit of the Grand Forks Human Nutrition Research Center in Grand Forks, North Dakota. He received his Ph.D. in nutrition from the University of Illinois. Research interests include zinc and reproduction, trace element metabolism and interaction, and laboratory animal nutrition.

Helen Anderson Shaw chairs the Department of Food, Nutrition and Food Service Management at the University of North Carolina at Greensboro. Previously a professor of nutrition at the University of Missouri, Shaw received her Ph.D. in nutrition from the University of Wisconsin. Research interests include amino acid nutrition in humans and small animals.

John Edgar Smith is an associate professor of human nutrition at Pennsylvania State University's College of Health and Human Development. He received his Ph.D. in nutrition from the University of Nebraska-Lincoln. Research interests include fat-soluble vitamin transport in blood and protein metabolism.

Robert D. Steele is assistant professor of nutrition at the University of Wisconsin at Madison, from where he also received his Ph.D. in nutrition and biochemistry. Research interests include investigating the central role of the liver in amino acid and protein metabolism in mammals, especially in conditions of altered function such as liver disease and cirrhosis.

Index

157

Other Titles
in the Series

Related Publications